FUNDAMENTALS OF RISK MANAGEMENT FOR PROCESS INDUSTRY ENGINEERS

FUNDAMENTALS OF RISK MANAGEMENT FOR PROCESS INDUSTRY ENGINEERS

MAUREEN HASSALL
The University of Queensland, St Lucia, Brisbane, QLD, Australia

PAUL LANT
The University of Queensland, St Lucia, Brisbane, QLD, Australia

ELSEVIER

Elsevier
Radarweg 29, PO Box 211, 1000 AE Amsterdam, Netherlands
The Boulevard, Langford Lane, Kidlington, Oxford OX5 1GB, United Kingdom
50 Hampshire Street, 5th Floor, Cambridge, MA 02139, United States

ISBN: 978-0-12-820320-0

For Information on all Elsevier publications
visit our website at https://www.elsevier.com/books-and-journals

Publisher: Candice G. Janco
Acquisitions Editor: Anita Koch
Editorial Project Manager: Andrea Dulberger
Production Project Manager: Kumar Anbazhagan
Cover Designer: Mark Rogers

Typeset by MPS Limited, Chennai, India

Contents

Acknowledgements ix

1. Introduction to risk in the process industries 1

1.1 Introduction 2
1.2 What is risk? 2
1.3 Some guiding principles 3
1.4 Case studies: some real consequences of inadequate risk management 5
1.4.1 The Fonterra whey protein crisis 5
1.4.2 Buncefield petrol overflow and vapour cloud explosion 6
1.4.3 Deepwater Horizon drilling rig explosion 7
1.4.4 Synopsis 8
1.5 Why is risk management so important? 9
1.6 What types of risks should engineers consider? 11
1.7 Engineering decision scenarios 14
1.7.1 Scenario 1 14
1.7.2 Scenario 2 15
1.7.3 Scenario 3 15
1.7.4 Scenario 4 15
1.8 Summary 16
Review questions 16
References 17

2. Fundamentals of risk management 19

2.1 Introduction 20
2.2 The risk language 21
2.3 The risk management process 22
2.4 Who is responsible for risk management? 28
2.5 A brief history of operational risk management in industry 28
2.6 Two approaches to modern risk management 33
2.7 Case studies illustrating two approaches to risk management 35
2.7.1 Case study 1: process plant operation 35
2.7.2 Case study 2: the availability of safety PPE 36
2.8 Summary 36
Review questions 37
References 37

3. Humans and risk 41

3.1 Introduction 42
3.2 The changing role of the human operator in industry 43
3.2.1 Era 1 – localised, direct and physical control 43
3.2.2 Era 2: regional, centralised and more cognitive control 45
3.2.3 Era 3: global, collaborative, cognitive 47
3.2.4 What have we learnt from the changing nature of industrial work? 50
3.3 The role of humans – risk perceivers, analysers and controllers 50
3.3.1 Risk perceivers 51
3.3.2 Risk analysers 52
3.3.3 Risk controllers 52
3.4 Risk communication 53
3.5 The human decision-making process 56
3.5.1 Situation assessment 56
3.5.2 Response strategies 58
3.6 Improving human performance 61
3.6.1 An example of designing for improved human performance – overfilling vessels 64
3.7 Summary 65
Review questions 65
References 67

4. Professional practice 71

4.1 Introduction 72

4.2 What is professional practice? 72
4.3 What is a professional engineer? 73
4.3.1 Ethics 74
4.3.2 Competence 75
4.3.3 Performance 78
4.4 Obligations, accountabilities and responsibilities 78
4.4.1 Legal and regulatory obligations 78
4.4.2 Accountabilities 80
4.4.3 Responsibilities 81
4.5 Case study: dieselgate – VW emissions scandal 82
4.6 Summary 82
Review questions 83
References 83

5. Identify, assess and treat risks 85

5.1 Introduction 86
5.2 Establishing the context 88
5.3 Risk assessment 90
5.3.1 Risk identification 90
5.3.2 Risk analysis 95
5.3.3 Risk evaluation 99
5.4 Risk treatment and management 103
5.4.1 Overview of risk treatment 103
5.4.2 Bowtie analysis 105
5.4.3 Unwanted event characterisation 108
5.4.4 Unwanted event identification 108
5.4.5 Determine the scope of the analysis 109
5.4.6 Identify the threats that can cause the unwanted event 111
5.4.7 Identify the possible consequences that could result from the unwanted event 112
5.4.8 Control analysis 114
5.4.9 Management of controls 118
5.5 Summary 122
Review questions 122
References 123

6. Monitor and review risks 125

6.1 Introduction 126
6.2 Why perform event investigations? 128
6.3 Purpose and theory behind investigations 130
6.4 Incident investigation techniques and application considerations 132
6.4.1 User beware 134
6.4.2 Timeline 134
6.4.3 5 Whys analysis 135
6.4.4 Fishbone 138
6.4.5 Human factors analysis and classification system 142
6.4.6 Bowtie analysis 145
6.4.7 Accident analysis mapping 148
6.4.8 Strategies Analysis for Enhancing Resilience 150
6.5 Integration of learning back into the business 154
6.6 Summary 154
Review questions 155
References 157

7. Health and safety risks 161

7.1 Introduction 162
7.2 What are health and safety risks? 163
7.3 Managing safety risks 166
7.3.1 Inherently safer design 173
7.3.2 Defence-in-depth 174
7.3.3 Resilience engineering 177
7.4 Safety culture 178
7.5 Process safety effectiveness 180
7.6 Safety management systems 181
7.6.1 Overview of ISO45001 181
7.6.2 Overview of process safety management systems 183
7.7 Case study – process safety consideration for filling bulk storage tanks 183
7.8 Summary 187
Review questions 188
References 188

8. Environmental and social risk 193

8.1 Introduction 194
8.2 What is environmental and social risk? 196
8.3 The project lifecycle 198
8.4 Impact assessment 200
8.5 Social licence to operate 202
8.6 Tools to identify, assess, treat and monitor environmental and social risks 204
8.6.1 Preliminary stakeholder analysis and management 204
8.6.2 Stakeholder risk assessment 207
8.6.3 Social impact management plan 208

Acknowledgements

This book is the culmination of almost a decade of work. It started as a new undergraduate course at The University of Queensland and progressively evolved into this book. Many people played a role in this journey. We should first start by acknowledging the key role played by Bob Hannah in challenging us to develop a new course (subject) that addressed contemporary risk management issues for process industry engineers. Bob helped to establish the vision that has become this book.

We were blessed to be able to work with a team of experienced and passionate professional engineers to conceive, design and teach the new material that ultimately became this book. Clive Killick was a key member of our teaching team when we started and his material was key in framing the material on project risk in Chapter 9. Andrew Murphy and Matthew Ames contributed significant resource sector experience and contemporary practice in the field of environmental and social risk. Many of their ideas form the basis of Chapter 8. Kelly Smith and Chris Lilburne have contributed significantly through their involvement as teaching assistants over several years. Kelly and Chris had many years of contemporary experience in oil refining and helped us to develop teaching resources that were not only engaging but also of direct practical relevance for graduating chemical engineers. Jannie Grove and Amy Hodson also provided many useful insights and experiences when working as teaching assistants. We thank all of you for giving this book contemporary industry relevance and for providing lots of practical examples and insights. But mostly, we thank you all for your endless support, enthusiasm and encouragement.

We would also like to thank the many guest lecturers and many cohorts of UQ Chemical Engineering students in CHEE4002 (Risk in Process Industries) and Minerals Processing Engineering students in MINE4200 (Risk in the Minerals Industries) who have helped us to further refine and develop this material. To the guest lecturers, we thank you for your time, enthusiasm and support. To the students, we also thank you for engaging with the material and providing your opinions and feedback on it. We also apologise that we didn't always get everything right! It is a real joy for us to see you learn and to develop high-level competencies and understanding about contemporary risk management practices.

Maureen Hassall would like to thank friends and family who inquired about and encouraged me to keep writing this book. Thank you for inspiring me to turn a wish into a reality.

Paul Lant would like to thank Maureen Hassall for driving this to completion. I would also like to thank my many new Catalan friends at ICRA in Girona. Thank you for hosting my sabbatical in 2019 and for giving me such a stimulating environment to work on this book. Jan and I loved our time in Girona. Finally, I would like to acknowledge my two beautiful four-legged office mates who kept me company for much of the writing of this book but didn't quite make it to the end – Stella and Poppy.

CHAPTER

1

Introduction to risk in the process industries

OUTLINE

1.1 Introduction 2

1.2 What is risk? 2

1.3 Some guiding principles 3

1.4 Case studies: some real consequences of inadequate risk management 5

1.4.1 The Fonterra whey protein crisis 5

1.4.2 Buncefield petrol overflow and vapour cloud explosion 6

1.4.3 Deepwater Horizon drilling rig explosion 7

1.4.4 Synopsis 8

1.5 Why is risk management so important? 9

1.6 What types of risks should engineers consider? 11

1.7 Engineering decision scenarios 14

1.7.1 Scenario 1 14

1.7.2 Scenario 2 15

1.7.3 Scenario 3 15

1.7.4 Scenario 4 15

1.8 Summary 16

Review questions 16

References 17

Fundamentals of Risk Management for Process Industry Engineers
DOI: https://doi.org/10.1016/B978-0-12-820320-0.00011-8

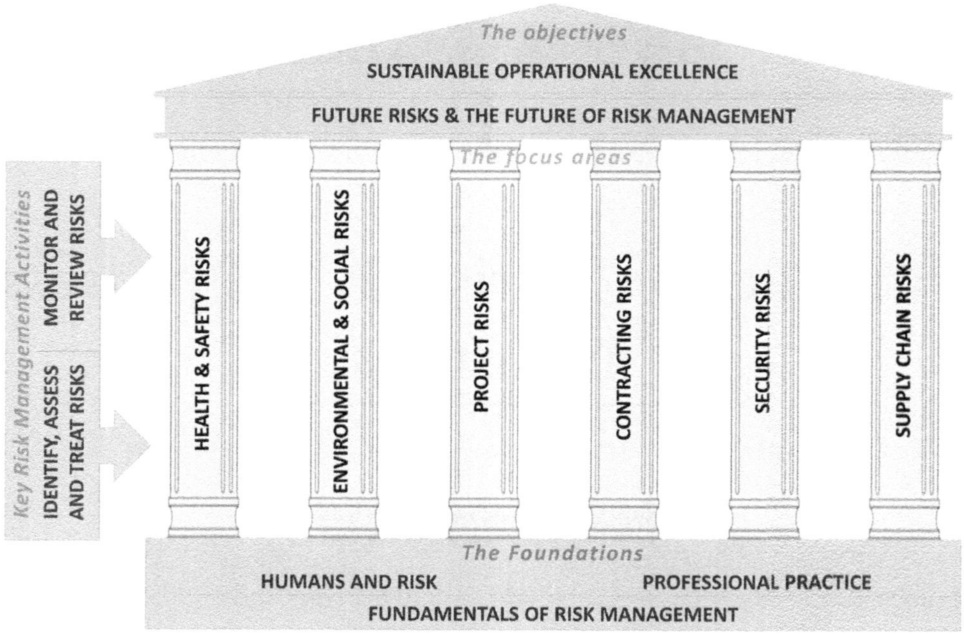

'Risk is a situation or event where something of human value (including humans themselves) has been put at stake and where the outcome is uncertain' (Rosa, 1998).

1.1 Introduction

What is risk? What does it 'look' like in the process industries? Why is it critically important for engineers to be able to skilfully manage it?

In this book, we will argue that risk management is one of the most critical tasks performed by engineers, and that it is critical for engineers to learn how to manage risk effectively. This is our motivation for writing the book.

Whilst practicing engineers manage risk on a daily basis in the workplace, there are few structured learning resources available, and until recently, very few universities even included a dedicated risk management course in their engineering curricula. With this book, we hope to make a small beneficial contribution to the education of engineers, helping to prepare them for the real challenges of managing risks in their workplace.

1.2 What is risk?

You have undoubtedly been exposed to the concept of risk. You are, of course, managing risk every day as you make your way through life. But what does 'risk' really mean?

There are many definitions of risk. The international standard for risk management – ISO 31000:2018 – defines risk as the 'effect of uncertainty on objectives' or the potential for positive and negative deviations that create uncertainty that matters because it can impact on the achievement of objectives. In this book, we use a more detailed definition of risk which is:

> 'Uncertainty that matters because it can affect the attainment of objectives'.

'Risk is a situation or event where something of human value (including humans themselves) has been put at stake and where the outcome is uncertain' (Rosa, 1998, p. 28). Risk is created by variability, incomplete knowledge, known and unknown threats and opportunities.

Some key characteristics of risk that underpin this book are:

1. Risk exists when there is uncertainty about the things that will happen in the future.
2. Risk is NOT bad. Indeed, it is only by taking, and managing, risk that companies can provide the products and services that society demands, be innovative, and provide commercial returns to their shareholders.
3. Managing risk is really about collecting intelligence, forming perceptions, making decisions and taking actions.

The very nature of the proposed definition of 'risk' has probably already got you asking several pertinent questions. What is uncertainty that matters? What uncertainties? Matters to whom? How do I know if it matters? How do I know if I have done enough to manage the risk? These are all good questions. We hope that this book will help you to answer them.

1.3 Some guiding principles

We live in a time where the sustainability of the processing industries is being threatened by the inability to address current and future threats and opportunities that impact operational competitiveness. These threats and opportunities stem from a business environment where sustaining leading-edge performance is becoming more challenging because of increasing risks and uncertainties which are being driven by the following trends:

1. Faster rate of change which is making the future more difficult to predict and gives business less time to respond (Aon, 2019; Elahi, 2010; Withers et al., 2015). Examples of the increasing rate of change can be found in the adoption rates for new technology as shown in Fig. 1.1.
2. Increasing complexity of business processes, technologies and supply chains which increases the chance of hidden faults and unexpected outcomes (Elahi, 2010; World Economic Forum, 2015) which is driven, at least in part, by the sociotechnical disruptions emerging from the fourth Industrial Revolution (Industry 4.0).

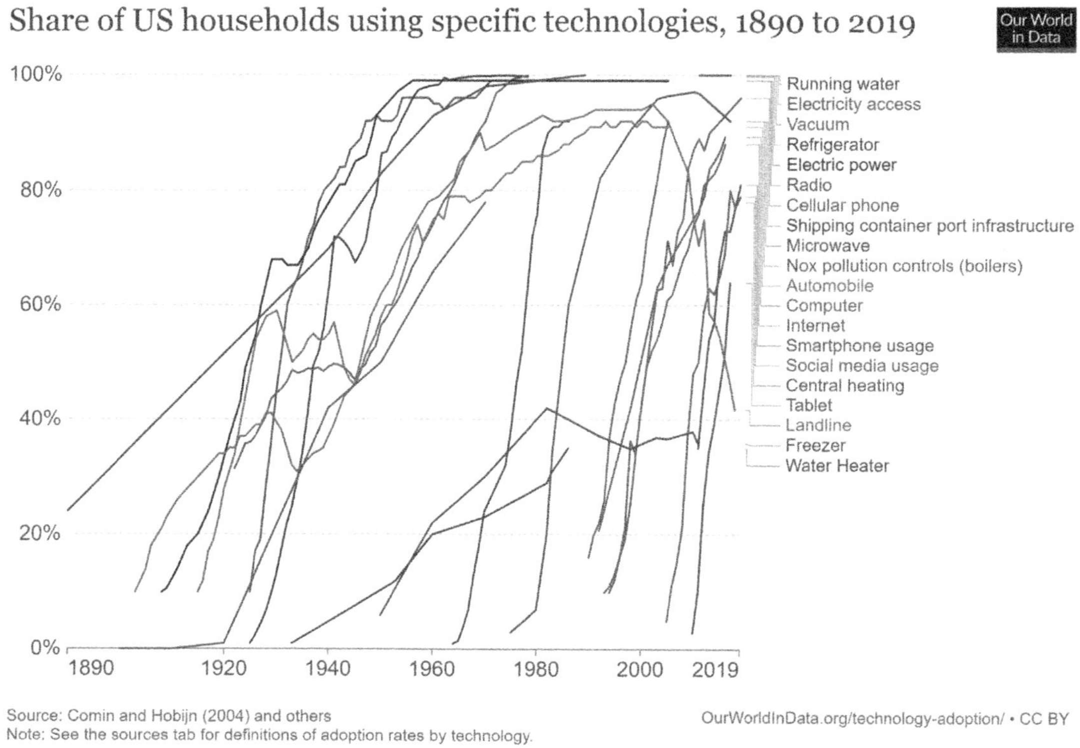

FIGURE 1.1 Technology adoption (Ritchie and Roser, 2017). *Source: Adapted from Ritchie, H., & Roser M. (2017). Technology adoption. Published online at OurWorldInData.org. https://ourworldindata.org/technology-adoption [Online Resource].*

3. Globalisation which is increasing the interactions and interdependencies between businesses and means that risks and impacts from one region or business can quickly spread to others (Elahi, 2010; Withers et al., 2015; World Economic Forum, 2015).
4. Changing regulatory requirements (Allianz, 2015; Aon, 2014) which includes pressure to transition from fossil fuels to renewables, hydrogen and nuclear energy (McKinsey & Company, 2022) and to deliver net zero carbon emissions (Naughtin et al., 2022; EY, 2020).
5. Increasing cost competitiveness associated with imported alternatives and the corresponding loss of local suppliers and customers (Allianz, 2015; Aon, 2014; Mooney, 2014).
6. Increasing stakeholder expectations (Allianz, 2015; Ernst & Young, 2015) as illustrated with the sample of media articles alluded to in Fig. 1.2.
7. Increasing frequency and severity of climate change related impacts on the weather, water, food supplies, and suitable locations for domestic and commerical infrastructure (Manyika et al., 2020; Naughtin et al., 2022).

Without effective risk management, adverse unwanted events will occur. This is evidenced in the incident investigation findings from the following case studies.

FIGURE 1.2 Sample of protests from around the world. *Source: Image 1: Antinuclear protests—photo courtesy of Daeyong Wee sourced from https://pixabay.com/photos/glory-hanbit-nuclear-power-plant-620584/; Image 2: Australian march on climate change – photo courtesy of John Englart sourced from https://flic.kr/p/9qPyyb; Image 3: Australian anti-coal protests – photo courtesy of John Englart sourced from https://flic.kr/p/EWZPCM; Image 4: American anti-fracking protests – photo courtesy of CREDO.FRACKING sourced from https://flic.kr/p/diftNs.*

1.4 Case studies: some real consequences of inadequate risk management

1.4.1 The Fonterra whey protein crisis

In August 2013, the Ministry for Primary Industries in New Zealand publicly announced that a whey protein product (WPC80) produced by Fonterra might be contaminated with *Clostridium botulinum*, which can cause botulism. Whey protein is used in the production of a variety of products, including infant formula. Fonterra announced product recalls of the WPC80.

Fonterra is one of the world's largest dairy companies. The announcement and product recall generated widespread international concern from consumers, and several countries (including China) ceased importing dairy products from New Zealand.

Tests conducted a few weeks later in the United States concluded that the contaminant was actually a nonpathogenic bacterium, *Clostridium sporogenes*. The incident had been a 'false alarm', although the consequences were very significant, including serious financial losses across the NZ dairy sector, loss of brand reputation, and significant supply chain disruption.

The New Zealand Government inquiry into the Fonterra whey protein incident(Government Inquiry into the Whey Protein Concentrate Contamination Incident., 2014) highlighted the importance of risk identification and conducting risk assessments. In particular:

- Underlying many of the recommended improvements was the idea that the dairy industry must anticipate future risks as well as counter existing threats, and
- A key immediate cause was '*improvised reprocessing of WPC80, without a risk assessment and in breach of its risk management programme*' (Government Inquiry into the Whey Protein Concentrate Contamination Incident., 2014).

1.4.2 Buncefield petrol overflow and vapour cloud explosion

In December 2005, as a result of a faulty level gauge and an inoperable independent high-level switch, petrol overflowed from a storage tank that was being filled at the Buncefield oil storage depot (UK) (Buncefield Major Incident Investigation Board, 2008). The resulting vapour cloud explosion was catastrophic and resulted in an enormous fire that engulfed 20 storage tanks and lasted for 5 days (Fig. 1.3). There were 40 casualties and significant property damage in the surrounding community (Buncefield Major Incident Investigation Board, 2008). Moreover, failures in the site bunding meant that large volumes of liquid (including water, fire-fighting foam and fuel) flowed from the site into the surrounding environment. Five companies were subsequently charged with offences arising out of the investigation, and the collective fine was more than 4 million pounds.

In their report on the incident, the Competent Authority for the Control of Major Accident Hazards (COMAH, 2011) noted that:

FIGURE 1.3 Buncefield tank farm fire. *Source: Image 1: Buncefield fire from air on 11th December 2005 - photo courtesy of Stuart Axe sourced from https://www.flickr.com/photos/stuart166axe/3250420727; Image 2: Buncefield fire from 1.2km away on 11th December 2005 - photo courtesy of Lvivan sourced from https://upload.wikimedia.org/wikipedia/commons/3/3a/Buncefield2.jpg.*

FIGURE 1.4 Deepwater Horizon fire, oil spill and cleanup by burning spilt oil. *Source: From Image 1: Deepwater Horizon on fire taken on April 21, 2010 - photo courtesy of Ideum - ideas + media sourced from https://www.flickr.com/photos/ideum/4711481781, Image 2: Oil spil footprint to July 16,2010 - graphic courtesy of SkyTruth sourced from https://www.flickr.com/photos/skytruth/4835555232; Image 3: Controlled burning of oil in Gulf - photo courtesy of Deepwater Horizon Response sourced from https://www.flickr.com/photos/deepwaterhorizonresponse/4624487176.*

> There should be a clear understanding of major accident risks and the safety critical equipment and systems designed to control them. This understanding should exist within organisations from the senior management down to the shop floor, and it needs to exist between all organisations involved in supplying, installing, maintaining and operating these controls. (COMAH, 2011, p. 5).

1.4.3 Deepwater Horizon drilling rig explosion

In April 2010 in the Gulf of Mexico, the Deepwater Horizon drilling rig experienced an uncontrolled release of gas (a blowout) while trying to cap the Macondo well. The gas ignited, and the resulting explosion and fire caused catastrophic damage (Fig. 1.4). The rig sunk 2 days later, and the unsealed well continued to spill oil into the Gulf of Mexico for 3 months. Eleven men were killed. The oil spill was the largest recorded in U.S. history, and the environmental impact was massive. The rig was owned by Transocean, and was leased by BP. It is estimated that the disaster will cost BP in excess of $65bn (Lee et al., 2018).

There have been several investigations into the incident, including a President's National Commission (Graham et al., 2011), and many articles written. Of these, the write-up by Andrew Hopkins highlights a couple of points that are pertinent to process engineers. First, he highlights how engineers were predominantly focussed on commercial risks and lost sight of safety risks. He writes:

> The risks that were supposed to be considered included safety, environmental, scheduling, production and cost. However, the risk register that was compiled for the Macondo well made no mention of safety risks.... In short, the development of a risk register for the Macondo well was not a safety risk management process (Hopkins, 2012, p. 16).

Hopkins also highlights how operations personnel and leadership were predominantly focussed on personal safety and not process safety.

1.4.4 Synopsis

As is often the case with serious process accidents, and as these case studies demonstrate, incident scenarios are complex, and incidents occur as a consequence of inadequate risk assessments, several failed controls, and gaps in the management systems required to support adequate risk assessment.

The insights gleaned from incident case studies such as these are also reflected in contemporary research into industry operational performance. The research reveals the following key findings, which also form the key principles underpinning this book:

1. **Most risks faced by companies emerge from combinations of, and changes in, technology, human and environmental factors**. So approaches to risk management need to reflect that systems have become more complex (nonlinear) and more integrated.
2. **Humans are crucial to effective risk management**. Humans can be the risk analyser, risk controller and/or risk perceiver. Adopting a human-centred approach to risk management is needed to deliver real improvements in risk identification, understanding, analysis, control, communication and governance. However, most risk management approaches used in industry focus of technology, procedural compliance or financial outcomes. Expanding risk management to include and leverage the human factor should deliver more effective risk identification, control and oversight.
3. **Most industrial incidents and production disruptions are the result of foreseeable, repeat or reoccurring events**. Moreover, investigations of these incidents usually do not find new causes for the events. So risk management needs to expand to include an increased focus on the selection, optimisation and management of human and technological risk controls to prevent adverse outcomes and to ensure beneficial outcomes.
4. **It is necessary to build good risk management capability** in order to achieve sustainable step change improvements in risk management and operational performance outcomes.

FIGURE 1.5 Risk Maturity and Financial Performance Link between risk maturity and financial performance. *Source: Adapted from* (Ernst & Young, 2013; Federation of European Risk Management Associations et al., 2012).

5. **Effective risk management drives competitive advantage** as evidenced by Ernst & Young research which found that companies with more mature risk management practices generated higher growth as shown in Fig. 1.5.

Thus better management of risk is a priority area for industry. Successfully addressing the risk challenges will require a fundamental understanding of human performance, operational risks, enterprise risks and impacts in order to determine how best to optimise them.

This book has been written to introduce engineers (and other processing industry professionals) to the fundamentals of risk management that you need to successfully identify, and address, the range of contemporary risks that challenge the sustainable competitiveness of companies operating within the process industries. In particular, it describes the risks that matter, and how you should manage them.

1.5 Why is risk management so important?

ISO 31000:2018 'Risk Management – Principles and Guidelines' describes some pertinent reasons why risk management is important. These include:

- Risk management creates and protects value
- Risk management is part of decision making
- Risk management explicitly addresses uncertainty
- Risk management takes human and cultural factors into account
- Risk management facilitates continual improvement of the organisation

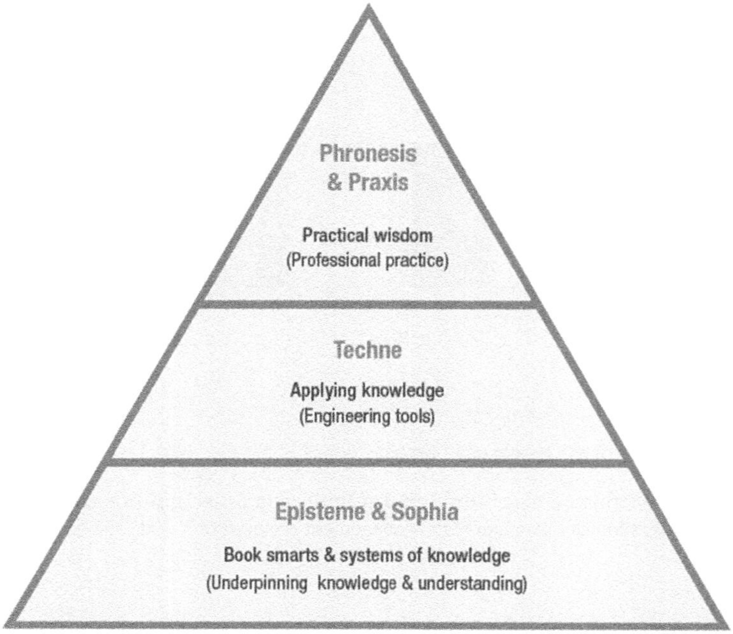

FIGURE 1.6 Three-tiered system of knowledge.

Risk management is also a critically important aspect of professional engineering practice. Engineering is about doing things! It is about understanding the foundational knowledge, learning how to apply tools and techniques and being able to do so in wise and professional ways in order to identify and solve real-world challenges. Thus good risk management requires the development and use of three tiers of knowledge described by Aristotle (the Greek philosopher) and Oliver and Dennison (2013) as shown in Fig. 1.6.

This book aims to help you to build your *episteme* and *sophia* – your understanding of knowledge and systems of knowledge usually described in risk management books. It also focusses on informing you about risk tools and techniques that can be used to solve problems. However, ultimately, good risk management is about building *phronesis* and demonstrating *praxis* – it is about building practical wisdom and demonstrating professional practice.

Using *phronesis* and *praxis* to identify and manage risks is about making decisions with incomplete knowledge and taking actions to manage uncertainties. It is fundamentally about you, your attitudes, your abilities and your professionalism. It is about your ability to form perceptions, make decisions and take actions as shown in Fig. 1.7.

Fig. 1.7 highlights that risk is the uncertainty that matters and it is managed by human decisions and actions as shown in the centre of the diagram and with the inner green ring. Human decisions and actions impact on, and are impacted by, the risk management tools and techniques selected, influences and advice provided by others, and professional ethics and standards used to detect, assess, treat, and review risks that specify the quality of work produced as shown in the middle blue ring. The decisions, actions, and work produced to manage risks and uncertainty, impacts and are impacted by an organisation's human performance, technical performance, financial performance and its ability to comply with legal

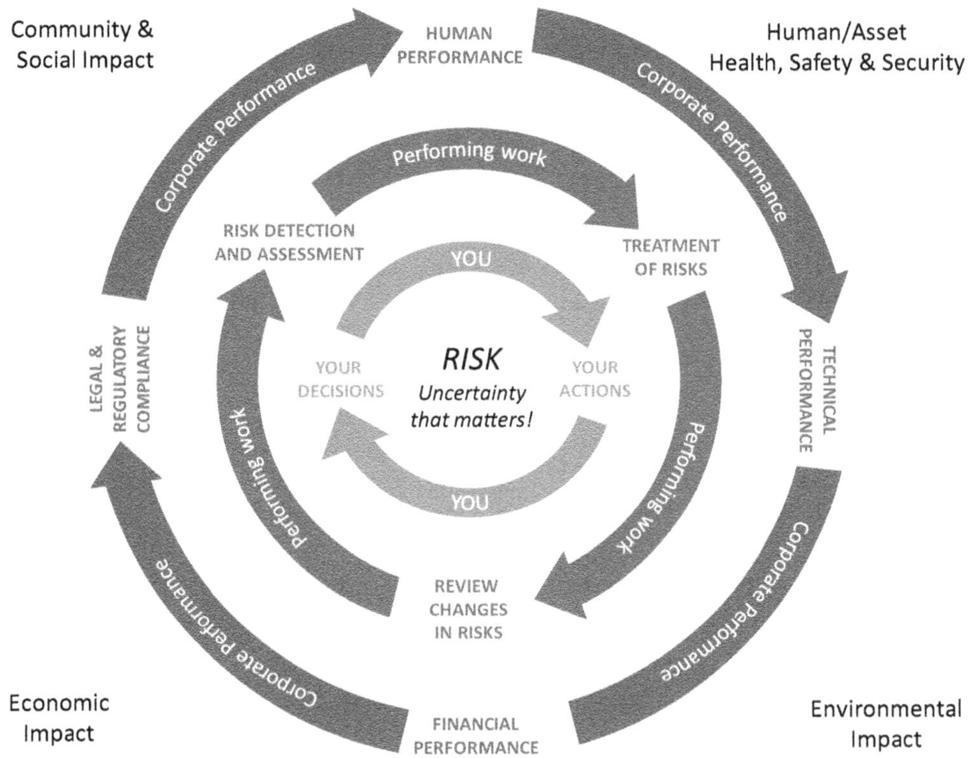

FIGURE 1.7 The influences and impacts on/of industrial risks.

and regulatory requirements as shown in the outer red ring. Last, the internal workings of an organisation, as shown in all three rings, impacts and are impacted by external factors including community and social factors, human health and safety and asset security compliance requirements, economic conditions and environmental conditions.

Risk management is a core competency of all engineers. Engineers are employed to facilitate continual improvement of the organisation. To do so, professional engineers are responsible for the identification, assessment and management of risks associated with all aspects of a business. This will include the planning and execution of projects, the operation of processing plants, and management of other aspects of the business including its environment and community interactions to obtain and sustain operation excellence.

1.6 What types of risks should engineers consider?

Many different types of risks can impact a business. Some of these risks have been identified and ranked by many organisations including the World Economic Forum, and several global leaders in insurance and professional services (e.g. Allianz, BDO, and Ernst and Young). One might wonder why insurance companies and professional services businesses (consulting firms) may be interested in process industry risks. Why would these organisations care, and why are they interested in how industry manages their risks? A summary of the results from some of the surveys that relate to the processing industries is shown in Table 1.1. Table 1.1 also shows an assessment of overall business risks (for a broad range of business sectors). This is a confronting and diverse range of risks, and it is quite likely that you may never have considered them before. You may never have even heard of some of these terms before! You are certainly not alone if some of these terms are new to you. It is not usual for us to discuss issues such as supply chain disruption, regulation uncertainty, licence to operate, price variability, digital effectiveness and labour concerns in our engineering courses. Yet, these are some of the highest ranked risks identified by industry leaders and practitioners. Surely it is important, then, for current and future

TABLE 1.1 Top 10 risk concerns for industry.

	Specific industry risks			Overall business risks	
Rank	**Manufacturing risks 2017 (BDO, 2017a)**	**Mining and metals risks 2019–2020 (Ernst & Young, 2018)**	**Oil and gas risks 2017 (BDO, 2017b)**	**Top business risks for 2019 (Allianz, 2019)**	**Top business risks for 2021 (Allianz, 2021)**
1	Supplier, vendor disruption	Licence to operate	Volatile prices	Business interruption	Business interruption
2	Regulations	Digital effectiveness	Regulatory and legislative change and cost	Cyber incidents	Pandemic outbreak
3	Labour concerns	Maximising portfolio returns	Disruptions due to natural disasters	Natural catastrophes	Cyber incidents
4	General economic conditions	Cyber	Environmental and health requirements	Changes in legislation	Market developments
5	Competition and pricing pressure	Rising costs	Global economic conditions	Market developments	Changes in legislation
6	Environmental regulations and liability	Energy mix	Changes in demand	Fire, explosion	Natural catastrophes
7	Cybersecurity	Future of workforce	Supply risks	New technologies	Fire, explosion
8	Threats to international operations	Disruption	Inaccurate reserve estimates	Climate change	Macroeconomic developments
9	Failure to execute growth strategy	Fraud	Operational risks	Loss of reputation or brand value	Climate change
10	Foreign currency exposure	New world commodities	Limited access to capital	Shortage of skilled workforce	Political risks and violence

employees to be knowledgeable about these risks, and to be knowledgeable about contemporary methods for managing them?

It is very instructive to compare the overall business risks listed for 2019 and 2021. Whilst most of them are similar, it is significant to note that 'pandemic outbreak' was the second highest ranked business risk in 2021. This was not included in the earlier lists. This clearly illustrates how risks are transient, that risk profiles can quickly change, and that organisations need to be continuously scanning for pending risks and need to have dynamic approaches to risk management.

The scope of the risks listed in Table 1.1 is so broad, that it is simply impossible to cover them all in one book. However, the foundational theory and practical tools that you will learn, are generically applicable to all forms of risk, and at the end of this book you should be adept at identifying, assessing and treating risks, of all kind, in a systematic manner.

It is the responsibility of people at all levels in organisations to manage risks. This is certainly true for processing and engineering companies. However, the risk management responsibilities for an individual will vary depending on their role, and level, in the company. Simplistically, it is the role of the board and senior executives to manage the broader business risks (e.g. exposure to price variability, regulation change, exchange rate fluctuations) and to have oversight and accountability for operational risks such as health and safety, environmental risk and project risk. It is then the role of the operations staff to manage these operational risks. This will often be done with the assistance of specialists in certain fields (e.g. environmental risk). We have thus attempted to restrict the scope in this book to operational risks, as these are the risks that you will most likely be faced with in

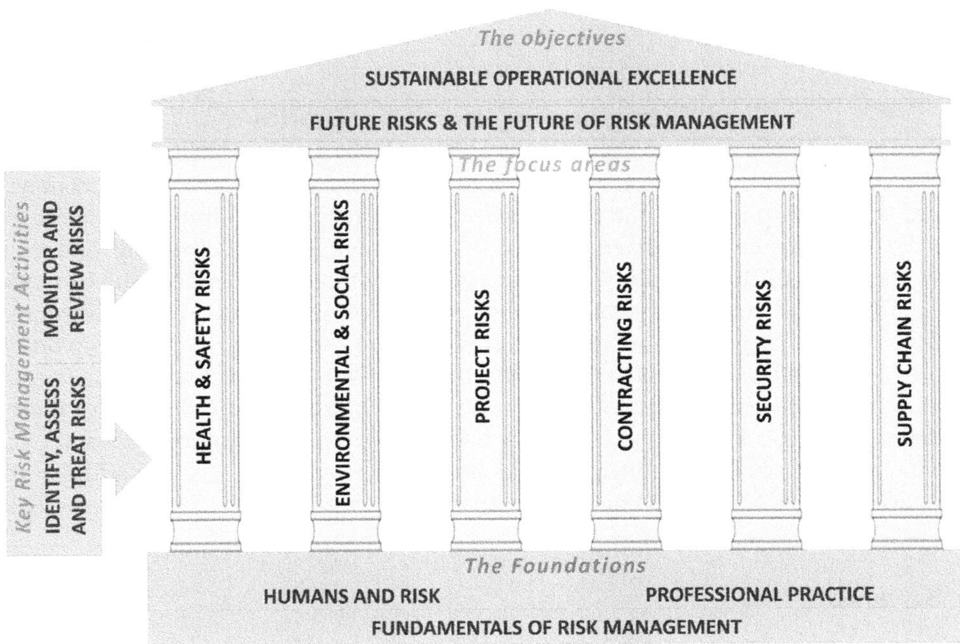

FIGURE 1.8 Scope of operational risk management.

the workplace, and the risks that you will most likely be responsible for managing in the near future. The scope of these risks is shown pictorially in the pillar diagram in Fig. 1.8. We use this figure to organise the book, and you will see it a lot throughout coming chapters. We have explicitly excluded financial risk, as this is a complete course of study all on its own. It is critically important, but beyond the scope of this book.

We have broken the foundations of the book into three core components: Fundamentals of risk management, Professional practice and Humans and risk. These three topics cover the foundational information that underpins how we perceive and manage risk. To the left of the figure, are located the two key activities of risk management, namely *Identify, assess and treat risks* and *Monitor and review risks*. These two activities are illustrated to 'cut across' the major risk factors that are shown as pillars in the centre of the figure. The pillars represent the range of the operational risks that need to be managed in order for an organisation to achieve its objectives. A risk-based approach is needed to optimise the trade-offs between these risks in such a way as to reduce the likelihood of negative outcomes and increase the likelihood of positive outcomes. Sitting on top of the pillars are the overarching activities that are required to sustain operating excellence. These activities include integrating the reporting of risk management work so it can be communicated in a manner that provides transparent insights into the things that matter by informing decision makers on their risk exposures and the effectiveness of risk mitigation efforts. Equally important to sustaining operational excellence, is the ability to develop foresight into potential future risks and future approaches to risk management. Using foresighting techniques will help future-proof process industries against future risks and changes. At the top or apex of the diagram is the objective of risk management work and of organisations themselves, namely to achieve *sustainable operational excellence*.

1.7 Engineering decision scenarios

So what does all of this mean for you as a junior engineer? The preceding sections have introduced the language of risk and attempted to justify why it is so important for engineers to have the competence to manage risk in an ever more rapidly changing and complex world. In this section, through some introductory case studies, we endeavour to make risk more meaningful and tangible. As an engineer, you will be confronted with decision making on a daily basis that may focus on one or several different aspects of risk. That is, you will be required to decide on a course of action when there is no 'one correct solution'. These decisions will be in the face of uncertainty. They will also require you to rely on a lot more than just technical expertise. As you will hopefully find, the decision making in each of the following scenarios will implicitly require you to identify and assess risks. As you address each of these scenarios, consider the approach that you are adopting in order to manage the risks.

1.7.1 Scenario 1

You are in a design team which is developing a conceptual design for a proposed succinic acid plant. As part of your scope of work, you are required to design a hydrochloric acid storage tank. The HCl is a feed for a fermentation process where the acid is dosed to control the pH. You have to specify the size of the tank. What factors would you consider in determining the size of the storage tank?

1.7.2 Scenario 2

You are part of a design team that is designing a scrubbing system which is part of a high-pressure venting system from a liquid storage tank. The liquid (MIC) is a highly volatile compound. It is produced onsite as an intermediate, and is subsequently reacted to produce a pesticide. The scrubbing system is a safety system that is required in the unlikely event that there will be a release of vapour from the storage tank. The design team has identified two possible options for the scrubbing system, option A and option B. Option B is significantly more expensive because it also includes a flare after the scrubber to burn off any of the gas that might not be removed in the scrubber. The design team has assured you that, in the unlikely event there will be a release of gas through the vent, the scrubber can absorb the gas to the specified level. How would you decide which option to select?

1.7.3 Scenario 3

You are a junior engineer working at a large factory. Your factory has a wastewater stream which is discharged to the sewer (this is called trade waste). Your company pays a fee to discharge the trade waste. They paid more than $1 million last year in trade waste charges. The trade waste specifications are regulated by the trade waste agreement that your factory has with the local council, and the trade waste is monitored by trade waste officers. It is your job to manage the operations of the wastewater treatment plant. You have noticed that the trade waste officers collect their samples on regular days each month, and on those days the effluent is within the specifications, largely because the operators are modifying the operations on those particular days to improve the water quality (e.g. by blending with higher-quality effluent that has been stored in a spare storage tank). You are also aware that the factory tends to run batches that result in lower-quality effluent on the days when the trade waste officers are not in attendance. The factory has very few licence breaches. What would you do?

1.7.4 Scenario 4

You work for a large oil and gas company. The company is in the early stages of a new coal bed methane project (sometimes also referred to as coal seam gas). You have been working in the project team that has been evaluating disposal options for the water that is produced from the gas wells. As part of your role, you have been requested to attend a public meeting in the local community. Apparently there is some unrest, as the local farmers are concerned about the impact of the gas processing on their farm, and several have reported gas being emitted from groundwater on their properties. What would you do?

1.8 Summary

Risk is *'uncertainty that matters because it can affect the attainment of objectives'*. In the process industries, these objectives are often associated with safety, health, technical, project and contractor management, supply chain, environmental and social impact, political and financial performance. Risk is created by variability, incomplete knowledge and known and unknown threats and opportunities.

Managing risks in order to deliver optimum performance is what engineers do! So as an engineer, you will be constantly faced with making decisions, in environments that are 'under-specified' – that is, there is not one correct solution to the problem.

Risk management is the approach that engineers can use to create and protect value, explicitly address uncertainty, take human and cultural factors into account, and to facilitate continual improvement of the organisation. This book has been written to help engineers understand the scope and approaches for managing contemporary risks in the process industries. The following sections of the book have been organised to follow the structure of the pillar diagram, namely: Section 1.2: The Foundations (Chapters 2–4); Section 1.3: Key Risk Management Activities (Chapters 5–6); Section 1.4: The Operational Risks (Chapters 7–12); and Section 1.5: Sustaining Operational Excellence (Chapter 13).

The process engineering scenarios described at the end of this chapter have hopefully illustrated how engineering decision making can be challenging. It should now be apparent that we require a diverse range of skills and knowledge in order to effectively manage risk, that the consequences can be extremely significant, and that there is a need to employ a systematic approach to identifying, assessing and treating the risks. Chapter 2 will introduce this methodology.

Review questions

1. In process engineering, risk refers to unwanted events that could impact on what?
2. What is the definition of risk?
3. A study by Ernst and Young (2013) examined how company financial performance varied with the level of risk maturity of the company. What did it reveal?
4. Which of the following statements are incorrect?
   a. In engineering projects, risk must be considered rationally, and stakeholder perception should be kept outside of decision making.
   b. It is only appropriate to progress a project when all of the identified risks have been eliminated.
   c. Companies who operate and/or construct process plants use external contractors to perform many roles. This is because they are the experts, and thus are less likely to be involved in safety incidents on the site.
   d. A significant number of actions taken in response to incidents in industry are technical. This is because the majority of incidents are caused by technical faults.

5. Before proceeding to Chapter 2, now is a good time to reflect on your 'starting knowledge', and to consider the knowledge, skills and competencies that you want to learn. Reflect on this question: *What do **you** need to know about risk management to work effectively, safely and professionally as an engineer?*

References

Allianz. (2015). *Allianz risk barometer: Top business risks 2015.* Allianz and Allianz Global Corporate & Specialty SE. Available from https://www.agcs.allianz.com/content/dam/onemarketing/agcs/agcs/reports/Allianz-Risk-Barometer-2015.pdf.

Allianz. (2019). *Allianz risk barometer: Top business risks for 2019.* Allianz. https://www.agcs.allianz.com/content/dam/onemarketing/agcs/agcs/reports/Allianz-Risk-Barometer-2019.pdf.

Allianz. (2021). *Allianz risk barometer: Identifying the major business risks for 2021.* Allianz. https://agcs.wufoo.com/forms/download-the-allianz-risk-barometer-2021-report/.

Aon. (2014). *Aon's 2014 Australasian risk survey.* Aon Corporation. Available from http://www.aon.com.au/australia/thought-leadership/risk-survey.jsp.

Aon. (2019). *Global risk management survey 2019.* Aon. Available from https://www.aon.com/getmedia/8d5ad510-1ae5-4d2b-a3d0-e241181da882/2019-Aon-Global-Risk-Management-Survey-Report.aspx.

BDO. (2017a). *2017 BDO manufacturing riskfactor report.* BDO USA. https://www.bdo.com/getattachment/07362692-3d80–4640-ac0a-d310bc60b946/attachment.aspx?2017-Manufacturing-Riskfactor-Report-Brochure_WEB.pdf.

BDO. (2017b). *2017 BDO oil & gas riskfactor report.* BDO USA. https://www.bdo.com/getattachment/a1bf67be-1beb-42b1–8f0c-f3db2446c6ed/attachment.aspx?2017-Oil-Gas-Riskfactor-Report-Brochure_WEB.pdf.

Buncefield Major Incident Investigation Board. (2008). The Buncefield Incident 11 December 2005: The final report of the Major Incident Investigation Board,. Surrey, UK. 1.

COMAH. (2011). *Buncefield: Why did it happen?* https://www.hse.gov.uk/comah/buncefield/buncefield-report.pdf.

Elahi, E. (2010). *How risk management can turn into competitive advantage.* College of Management Working Papers and Reports. http://scholarworks.umb.edu/cgi/viewcontent.cgi?article = 1006&context = management_wp.

Ernst & Young. (2013). *Turning risk into results: How leading companies use risk mangement to fuel better performance.* Ernst & Young Global Limited. Available from http://www.ey.com/Publication/vwLUAssets/Turning_risk_into_results/$FILE/Turning%20risk%20into%20results_AU1082_1%20Feb%202012.pdf.

Ernst & Young. (2015). *Business risks facing mining and metals 2015–2016: Moving from the back seat to the driver's seat.* Ernst & Young Global Limited.

Ernst & Young. (2018). *Top 10 business risks facing mining and metals in 2019–20.* EYGM Ltd. Available from https://assets.ey.com/content/dam/ey-sites/ey-com/en_gl/topics/mining-metals/mining-metals-pdfs/ey-top-10-business-risks-facing-mining-and-metals-in-2019-20_v2.pdf.

EY. (2020). *Are you reframing your future or is the future reframing you? Megatrends 2020 and beyond.* Retrieved from https://assets.ey.com/content/dam/ey-sites/ey-com/en_gl/topics/megatrends/ey-megatrends-2020-report.pdf.

Federation of European Risk Management Associations., AXA Corporate Solutions., & Ernst & Young. (2012). *FERMA Risk Management Benchmarking Survey 2012.* FERMA - Federation of European Risk Management Associations.

Graham, B., et al. (2011). Deep water: The Gulf oil disaster and the future of offshore drilling: Report to the President. USA, National Commission on the BP Deepwater Horizon Oil Spill and Offshore Drilling.

Government Inquiry into the Whey Protein Concentrate Contamination Incident. (2014). *The WPC80 incident: Causes and responses.* November 2014. ISBN: 978-0-473-30935-0.

Hopkins, A. (2012). *Disastrous decisions: The human and organisational causes of the Gulf of Mexico blowout.* CCH Australia Limited.

ISO 31000:2018. *Risk management – Guidelines.* Available from https://www.iso.org/standard/65694.html.

Lee, Y. G., Garza-Gomez, X., & Lee, R. M. (2018). Ultimate costs of the disaster: Seven years after the deepwater horizon oil spill. *Journal of Corporate Accounting & Finance*. Available from https://doi.org/10.1002/jcaf.22306.

Manyika, J., Smit, S., & Woetzel, J. (2020). *Climate risk and response: Physical hazards and socioeconomic impacts*. Retrieved from https://www.mckinsey.com/capabilities/sustainability/our-insights/climate-risk-and-response-physical-hazards-and-socioeconomic-impacts.

McKinsey & Company. (2022). *Global energy perspective 2022: executive summary*. Retrieved from https://www.mckinsey.com/~/media/McKinsey/Industries/Oil%20and%20Gas/Our%20Insights/Global%20Energy%20Perspective%202022/Global-Energy-Perspective-2022-Executive-Summary.pdf.

Mooney, S. (2014). Asia risk report: The top concerns for Asian risk managers - 2015 edition Ausatralia. Newsquest Specialist Media Ltd.

Naughtin, C., Hajkowicz, S., Schleiger, E., Bratanova, A., Cameron, A., Zamin, T., & Dutta, A. (2022). *Our Future World: Global megatrends impacting the way we live over coming decades*. Retrieved from Brisbane, Australia: Available from https://www.csiro.au/en/research/technology-space/data/Our-Future-World.

Oliver, P., & Dennison, W. (2013). *Dancing with dugongs: Having fun and developing a practical philosophy for environmental teaching and research*. USA: University of Maryland.

Ritchie, H., & Roser, M. (2017) – *Technology adoption*. Published online at OurWorldInData.org. Available from https://ourworldindata.org/technology-adoption [Online Resource].

Rosa, E. A. (1998). Metatheoretical foundations for post-normal risk. *Journal of Risk Research*, *1*(1), 15–44. Available from https://doi.org/10.1080/136698798377303.

Withers, G., Gupta, N., Curtis, L., & Larkins, N. (2015). *Securing Austalia's future: Australia's comparative advantage*. Australian Council of Learned Academies (ACOLA).

World Economic Forum. (2015). *Global risks 2015*. World Economic Forum.

CHAPTER

2

Fundamentals of risk management

OUTLINE

2.1 Introduction 20

2.2 The risk language 21

2.3 The risk management process 22

2.4 Who is responsible for risk management? 28

2.5 A brief history of operational risk management in industry 28

2.6 Two approaches to modern risk management 33

2.7 Case studies illustrating two approaches to risk management 35

2.7.1 Case study 1: process plant operation 35

2.7.2 Case study 2: the availability of safety PPE 36

2.8 Summary 36

Review questions 37

References 37

Fundamentals of Risk Management for Process Industry Engineers
DOI: https://doi.org/10.1016/B978-0-12-820320-0.00012-X

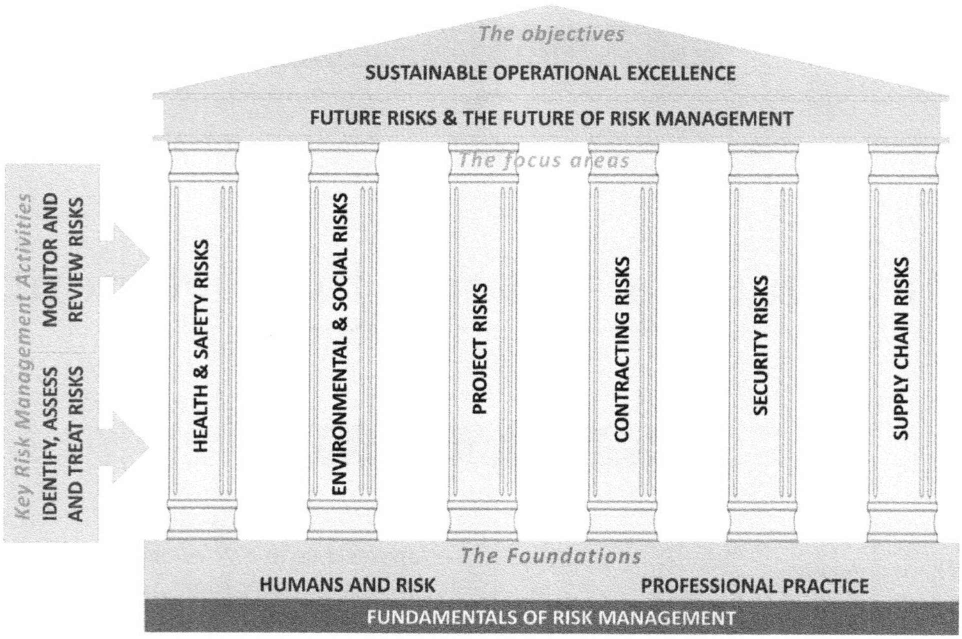

If the future is predetermined then the term risk makes no sense, risk denotes the uncertainties associated with the future that have the potential to impact on the things that humans value (Renn, 1992, 2008).

2.1 Introduction

Chapter 1 introduced us to risks that engineers in industry have to manage on a daily basis. We saw that risk is not necessarily bad. Companies that manage risks well tend to perform better, but that poor management of risks can lead to devastating outcomes. So how do we manage risks well? This chapter will introduce a structured process which can guide how we approach managing all kinds of risks—The Risk Management Process.

Risk management is about identifying, assessing and treating the uncertainties that matter because they can affect the achievement of objectives as stated in the standard ISO 31000 (ISO 31000, 2018). Risk management is inherently future focussed—it seeks to identify those future uncertainties that can affect the achievement of objectives. This chapter covers the following fundamentals of risk management for process industry contexts:

- Know the core risk management terms and definitions
- Understand the risk management process (ISO 31000)
- Review a brief history of risk management
- Learn about the two main risk management approaches used in process industries

At this early stage, we would like to emphasise that the risk management process is merely a framework within which we can systematically identify and address risks. It is not a system of rules, nor a check-list of activities. Professional judgement, thought and action must always be used. The people with accountabilities in the risk management process are crucial, and they (and you) must be suitably competent and knowledgeable.

2.2 The risk language

These are some of the key words and phrases that we use in the language of risk.

Risk is uncertainty that matters because it can positively or negatively affect the achievement of objectives.

Risk identification: Identifying sources of uncertainty that matter.

- ***Hazard:*** A potential source of harm (e.g. electricity, gas at pressure, hot fluids).
- ***Threat:*** Something that can release a hazard (e.g. corrosion).
- ***Opportunity:*** Something that can lead to the exceeding of objectives or produce better than expected performance outcomes.

Risk analysis: An estimation of the likelihood and consequence of something happening. Sometimes it also involves an estimation of exposure and vulnerability.

Likelihood: How often something might happen (e.g. 1/d, 1/y).

Consequence: The outcome or impact of an event (e.g. injury, death, damage/loss of assets, environmental destruction, social unrest, profit, growth, improved quality).

Exposure: The number of entities of value that might be present when risk materialises.

Risk evaluation: The process of deciding whether a risk is acceptable or tolerable, or whether there needs to be some treatment of the risk to make it acceptable or tolerable.

Risk treatment: A process of selecting the options for treating the risk. In the process industries, the risk treatment options may include:

- avoiding, eliminating or substituting the hazard
- isolating the hazard
- reducing exposure to the hazard
- eliminating the threats
- identifying the controls and control management systems needed to address the risks by preventing unwanted events or mitigating their consequences to an acceptable level.

Unwanted event: An unplanned release of a hazard (e.g. loss of containment of a hazardous material, loss of control of a hazardous energy source or loss of awareness of the situation).

Control: A device and/or human action that, of itself, will arrest or mitigate an unwanted event sequence and whose performance is specifiable, measurable and auditable.

Prevention control: An object and/or human action that, of itself, prevents or hinders an unwanted event.

Mitigating control: An object and/or human action that, of itself, reduces the severity of the consequences of an unwanted event sequence.

Control management systems: The organisational activities required to ensure that a control is implemented, maintained and operating effectively as required when required.

2.3 The risk management process

The International Standards AS/NZS ISO31000:2018 'Risk Management—Guidelines' and ANSI/ASSE Z690.2-2011 'Risk Management-Principles and Guidelines' describes the risk management process and defines the key language underpinning risk management (ISO 31000:2018). There are also many other standards and industry guidelines that provide guidance on risk management. Examples are shown in Table 2.1. These standards seek to set out vocabulary and criteria for achieving consistency and reliability of activities associated with the identification, analysis, evaluation, treatment and overall management of risk. It is recommended that people read and reference the standards and industry guidelines relevant to their work. The remainder of this section provides an overview of ISO31000.

In ISO31000:2018, the risk management process that is described is the one depicted in Fig. 2.1. Fig. 2.1 is simple and a useful reference to begin to understand some of the fundamentals of risk. It highlights that managing risk first requires an assessment of the risks

TABLE 2.1 Examples of risk-based standards and industry guidelines.

Source	Title	Description
International standards	ISO 31000	Risk management – Guidelines
	ANSI/ASSE Z690.2-2011	Risk management – Principles and guidelines
	BS/EN/IEC/ISO31010 and ANSI/ASSE Z690.3-2011	Risk management – Risk assessment techniques
	ISO22000	Food safety management standard
	ISO27000 series	Security risk management standards
	ISO Guide 73 and ANSI/ASSE Z690.1-2011	Details the vocabulary and definitions for generic risk management terms
Industry guidelines	COSO Enterprise risk management – integrated framework	The Committee of Sponsoring Organizations of the Treadway Commission (COSO) guide that defines a framework and essential elements for managing enterprise risk
	Guidance for industry – Q9 quality risk management	Guidance on managing risks that can impact product quality produced by U.S. Food and Drug Administration (FDA)
	Hazard identification and risk assessment	Guidance note published by The National Offshore Petroleum Safety and Environmental Management Authority (NOPSEMA)
	Guidelines for the governance of systemic risks	Guidance on governance required to manage risks in dynamic systems published by International Risk Governance Council (IRGC)
	Risk management: Leading practice sustainable development programme for the mining industry	Risk management handbook for mining published by the Australian Government

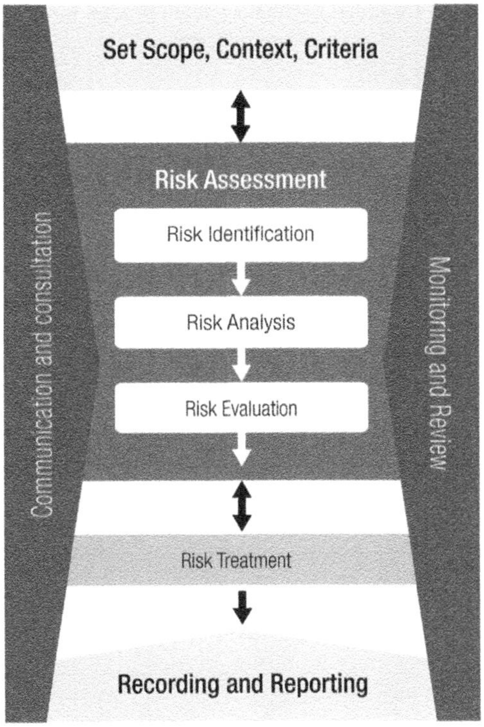

FIGURE 2.1 The risk management process (AS/NZS ISO31000:2018).

then a determination of how to treat the risks. However, in the process industries, one of the things we have learned from incidents that there needs to be equal emphasis put into risk assessment activities and risk treatment activities as shown in Fig. 2.2.

As these Figs. 2.1 and 2.2 highlight, establishing the scope of risk management activities is a very important first step. This step involves determining what things will be considered within the scope of the risk management activities, and what things will be considered out of scope. Using a framework such as PLEATESO (People, L = Locations, E = Equipment and plant, A = Activities, T = Timeframes, E = Environment, S = Scenarios and O is for other assumptions) can help ensure that a more complete range of factors will be considered. See Table 2.2 for example of a template for a scope table. More information on completing scope tables can be found Chapter 5.

Being transparent about what is in and out of scope, and what other assumptions have been made, is important as it ensures those participating in, implementing, reviewing and updating any of the elements of the risk management process will have a common understanding of the basis on which the risk assessment and risk treatment work was done.

Once the context has been established, the next step in the risk management process is risk assessment—that is, identify, analyse and evaluate the risks. Risk identification is performed to identify the sources of uncertainty that matter. This includes identifying the threats that could harm an entity's performance and/or the opportunities which could produce better than expected performance outcomes. After the risks are identified, the

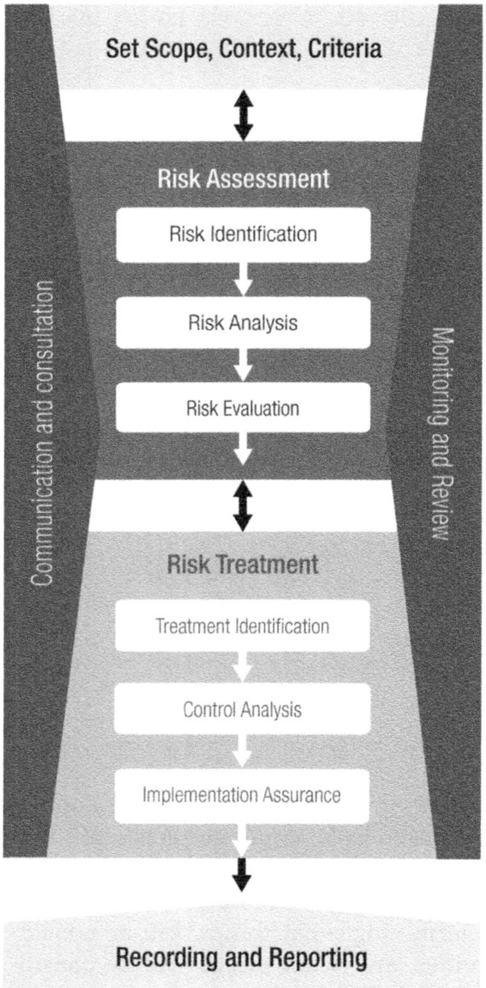

FIGURE 2.2 Extended risk management process for process industries.

next step is risk analysis. Risk analysis involves determining the range of consequences or impacts that might result if the risk materialises, and the likelihood that these different consequences will occur. The outcome of the risk analysis step is a risk rating often derived from the estimated likelihood and consequences either quantitatively, semi-quantitatively or qualitatively (e.g. using a risk matrix). The risk rating is used to evaluate the risk. The risk evaluation is a process for determining if the risk is acceptable, tolerable or unacceptable. It guides us in deciding if some treatment of the risk is required to make it acceptable or tolerable. Often risk analysis and evaluation guidance is provided within the risk matrix as shown in Fig. 2.3 with detailed descriptors for impact shown in Fig. 2.4 and for likelihood in Fig. 2.5. Risk can also be evaluated quantitatively and then displayed on ALARP graphs like that shown in Fig. 2.6. The concept of ALARP (as low as reasonably practicable) can be used

TABLE 2.2 Example scope table.

Description	Included	Excluded
People: People involved in management of the risk and people potentially impacted if risks are not managed		
Locations: Locations or areas where the risk exists or that could be impacted by the risk event if it occurs		
Equipment: Equipment and plant exposed to the risk and associated with managing the risk		
Activities: Activities include operations, maintenance, startup, shutdown, normal and abnormal tasks and functions that are exposed to, could impact or be impacted by the risk		
Timeframes: To capture time considerations (e.g. the times of the day or days of year) that risk needs to be managed plus other considerations such as shift changes and how far into the future is being considered		
Environment: Known weather and climate-related conditions (heatwave, tsunami, bush fire, earthquake, landslides, hurricanes, etc.)		
Scenarios: Known risk scenarios that need to be considered		
Other assumptions: Captures other assumptions such as information on the hazardous chemicals and energies and environmental issues being considered		

to guide our assessment of whether a risk is tolerable. Simply, ALARP means that all practicable controls to reduce the risk have been implemented, and that further risk reduction is impracticable (i.e. the cost is disproportionate to the risk reduction gained). More details on risk assessment processes are presented in Chapter 5.

The next step is to identify the **risk treatment** options to address the event scenarios highlighted in the risk assessment phase. In the process industries, options for risk treatment include those specified in inherently safe design then defence in depth philosophies. Inherently safe design advises that risks should be treated first by seeking to eliminate, isolate or minimise hazards, then by seeking to remove threats and then by selecting and optimising risk controls. Controls are selected that address remaining threats to prevent or reduce the likelihood of the unwanted events from occurring, and/or to enhance or increase the likelihood of wanted events occurring. Controls are also selected to mitigate adverse consequences and to promote desirable outcomes should the event scenario occur. To ensure the selected risk controls are implemented, monitored and maintained, control assurance management plans also should be developed. Control Assurance Management involves evaluating the support activities and management systems elements that are required to ensure the controls are implemented, monitored and maintained in a manner that ensures they would work as required, when required, and are effective at addressing the risk to an acceptable level.

For example, the controls required to stop a tank from overfilling should include level control systems and the controls to mitigate consequences of an overflowing tank from causing irreparable harm (e.g. a bund). The Control Assurance Management of these

	LIKELIHOOD				
IMPACT	**Rare**	**Unlikely**	**Possible**	**Likely**	**Almost Certain**
Catastrophic	Very High	Very High	Very High	Very High	Very High
Major	Medium	Significant	Significant	Very High	Very High
Moderate	Low	Medium	Significant	Significant	Very High
Minor	Low	Medium	Medium	Medium	Significant
Insignificant	Low	Low	Low	Medium	Medium

Risk rating

High	Unacceptable – Operations do not to continue until risk is reduced. Risk treatment considerations should include inherently safe design and defense in depth analyses. Assign to Executive Team Member
Significant	Tolerable if actioned as a high priority and industry recognised risk controls measures are identified, implemented, monitored and reported on with respect to their ability to reduce risk. Assign senior manager responsible to action, monitor and review.
Medium	Tolerable if actioned with industry recognised risk controls measures are identified, implemented, monitored and reported on with respect to their ability to reduce risk. Assign manager responsible to action, monitor and review.
Low	Generally Acceptable - Manage with regular monitoring and review.

FIGURE 2.3 Example of a risk ranking matrix.

controls should include operational checks and routine maintenance to ensure that these controls are implemented and fully functional.

The risk management framework (shown in Figs. 2.1 and 2.2) also highlights two very important activities in the darker blue boxes on each side of the diagram, namely 'communication and consultation' and 'monitoring and review'. Communication and consultation should take place with all stakeholders to get their input and to help them understand the risk and risk management requirements, and their accountabilities in delivering them. The monitoring and review processes focus on ensuring quality and effectiveness of risk management activities for current risks and potential future changes and challenges. The approaches, tools and techniques to use for each of these steps will become more evident in later chapters.

IMPACT				
Impact	**Injury and Disease (includes workers and community)**	**Business Cost (assets losses, fines, etc)**	**Environment**	**Social/Reputation**
Catastrophic	Multiple fatality and/or significant irreversible exposure to a health risk that affects greater than 10 people	> $25 m	Permanent effects on the environment (e.g. air, water, soil, plants, animals). Potential loss of licence to operate. Prosecution of company and directors possible.	Very serious widespread/ irrepairable damage to significant relationships. Extremely adverse media internationally. Reputation impacted with majority of key stakeholders. Will not gain approval to operate.
Major	Single fatality, permanent disabling injury or exposure to a health risk that is permanently disabling and/or reduces life expectancy	$5m - $25 m	Major effects on environment (e.g. air, water, soil, plants, animals). Substantial cleanup costs. Personal & business prosecution possible	On-going serious social issues with significant relationships. Reputation impacted by adverse media and with significant number of key stakeholders. May lose licence to operate or not gain approval.
Moderate	Serious injury/exposure that require hospitalisation with moderate permanent effects (e.g. internal/head injuries, toxic inhalations).	$0.5m - $5m	Pollution with moderate effects on environment (e.g. air, water, soil, plants, animals). Required to report to authorities. Extensive cleanup may be required	Ongoing social issues. Attention from media. Heightened concern by local community. Criticism by local NGOs. Significant difficulties in gaining approvals. Reputation impacted with some key stakeholders.
Minor	Significant injury. Medical treatment required from which recovery is likely (e.g. burns, broken bones, severe rashes/reactions, cuts).	$50k - $500k	Pollution with minor effects on enivronment. Required to inform authorities. May involve a cleanup. Visible discharge observed offsite.	Minor medium term impacts on local population / items of significance. Mostly repairable. Minor, adverse local public or media attention. Reputation is impacted with a small number of people.
Insignificant	Minor injury not requiring medical treatment (e.g. cuts, mild skin irritations, no measurable physical effects).	< $50 k	Low Pollution = No observable effect to environment. No requirement to inform authorities. No observable discharges offsite.	Low-level social or cultural impacts. Public concern restricted to local complaints and individual concerns. No discernable impact on reputation.

FIGURE 2.4 Examples of impact categories used in a risk ranking matrix.

LIKELIHOOD				
Rare	**Unlikely**	**Possible**	**Likely**	**Almost Certain**
Event may occur under exceptional circumstances. (Estimated frequency between 1 every 10,000 years and 1 every 1000 years)	**Event could occur at some time.** (Estimated frequency between 1 every 1000 years and 1 every 100 years)	**Event should occur at some time** (May happen in plant life based on industry events. Est frequency between 1 every 100 years and 1 every 10 years)	**Event will probably occur in most circumstances** (Probably occur once based on history of near misses. Est freq = 1 every 10 years or 1 per year)	**Event expected to occur in most circumstances** (e.g. does occur, definitely history of occurrence or estimated frequency = 1-10 times per year)

FIGURE 2.5 Examples of likelihood categories used in a risk ranking matrix.

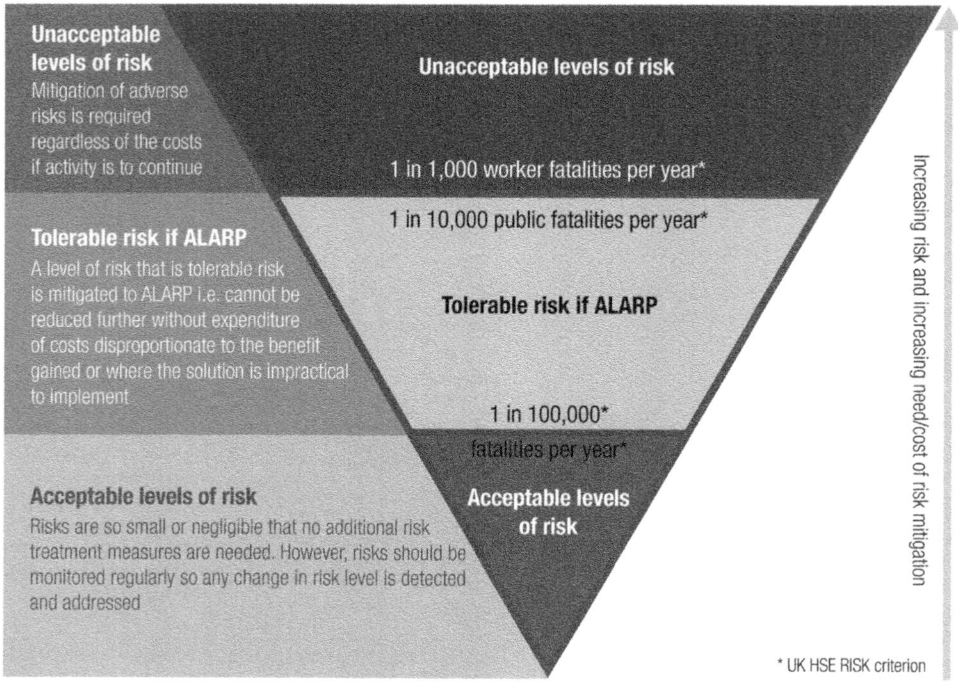

FIGURE 2.6 Risk levels mapped against ALARP standards.

2.4 Who is responsible for risk management?

Risk management involves everyone at all levels of the organisation. Decisions at all levels of an enterprise determine how risks are identified, assessed and treated. The outcome of these decisions results in value being eroded, preserved or additional value created. Typically, the board and senior executives are responsible for strategic risks and senior managers and staff are responsible for operational risks as shown in Table 2.3. In addition, people at the board and executive levels of an organisation are responsible for defining risk, setting the entities' risk management appetite, establishing the risk management frameworks, and assigning roles and responsibilities to execute the risk management system. This is explored in detail in Chapter 3. Further, the importance of taking personal responsibility for managing risks is formally recognised in a number of laws, regulations and professional competency requirements. The professional responsibilities for managing risk in process industries are explored in Chapter 4.

2.5 A brief history of operational risk management in industry

This section provides a brief history of industrial work and examines the evolving role of the human operator. This is critical for understanding the contemporary context within

TABLE 2.3 Strategic versus operational risk.

	Strategic risk	Operational risk
Definition	Risks that affect or are created by the business strategy	Risks that affect the organisations ability to deliver the strategic plan
Timeframe	Longer term (years)	Short to medium term (immediate to months)
External or internal drivers	External	Internal and external
Key decision makers	Board and senior executives	Staff managing and delivering the operations (incl. the engineers)
Impact scale	Very significant	Minor to very significant
Example	Government intervention in a project in response to major community outrage and loss of social licence to operate	Project environmental risks such as noise, air and water emissions

which operational staff work, and for appreciating the key factors that will inform human-centred approaches to managing risk.

There are numerous ways that the history of risk management can be presented. In this book we look at a summary of the modern history of risk management as it relates to the process industries. Risk management techniques first emerged as a means to addressing poor technical performance and poor safety performance. The changes in management of technical and safety performance over time is discussed by Hale and Hovden (1998), Borys et al. (2009), and Hollnagel (2011). This history has also been represented in diagrams as shown in Fig. 2.7.

Another perspective on the history of unwanted event management is shown in Fig. 2.8. As Fig. 2.8 shows, our understanding of what causes unwanted events in industry has changed, which has led to the development of different types of risk management techniques.

After the industrial revolution, the unreliable nature of plant and equipment often resulted in unwanted events such as equipment malfunctions, plant breakdowns, and catastrophic failures. During this period, risk assessment techniques were developed to help practitioners identify and address potential technology failures. Some of these techniques included:

- Failure mode effects analysis (FMEA): This analysis involves identifying the system components that could fail, the causes and consequences of the failure, and if required the actions needed to address risks associated with the failure.
- Fault tree analysis (FTA): This type of risk assessment involves using deductive thinking to decompose an unwanted event or undesired system state into its possible causal subevents using Boolean logic to quantify the risks associated with a given event or system state.
- Event tree analysis (ETA): ETA involves developing a tree of the possible outcomes of an event. It is sometimes referred to as a Consequence Diagram. ETA analysis often involves the quantification of the different outcomes in terms of the probability of occurrence.

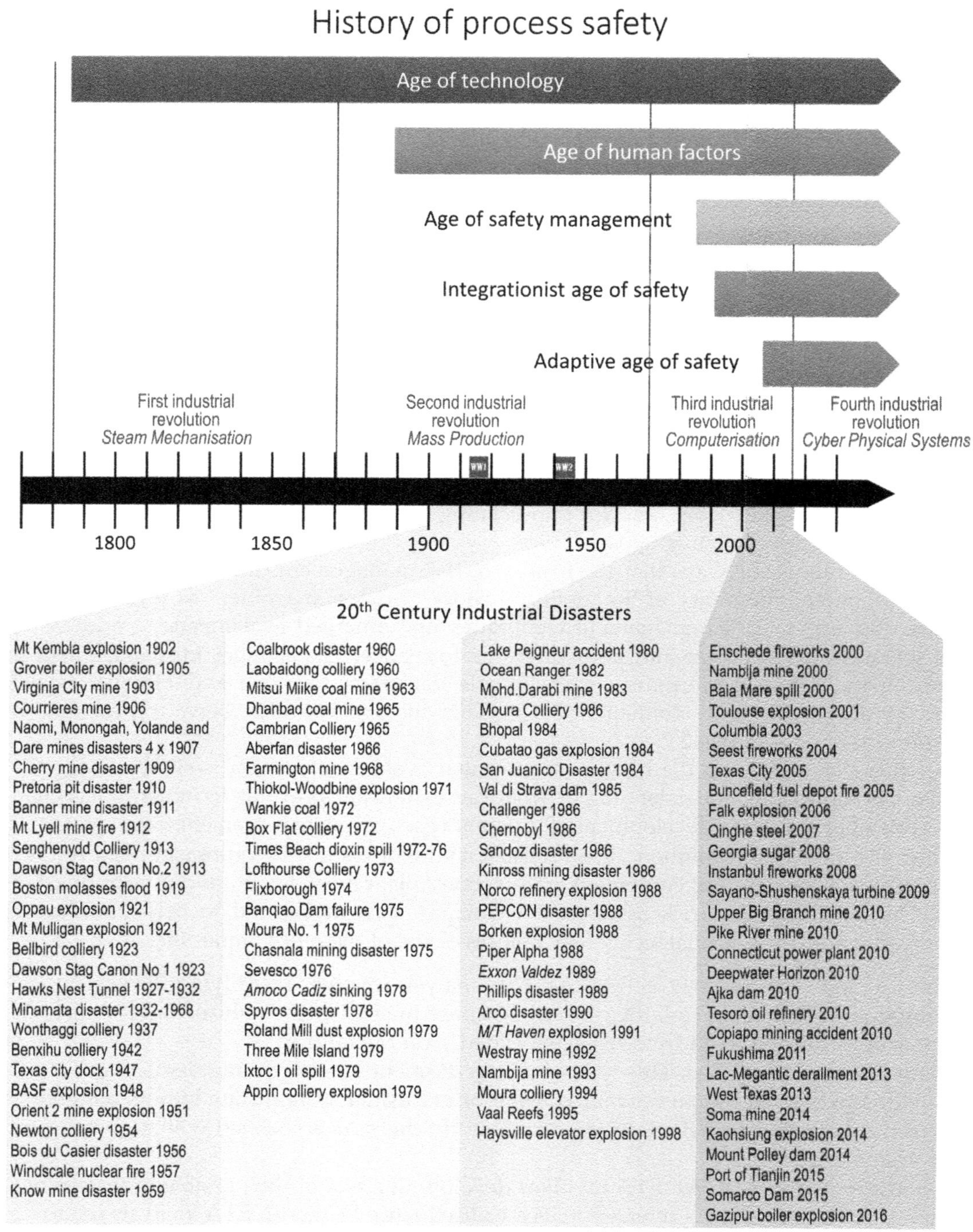

FIGURE 2.7 History of poor technology and safety performance in hazardous industries.

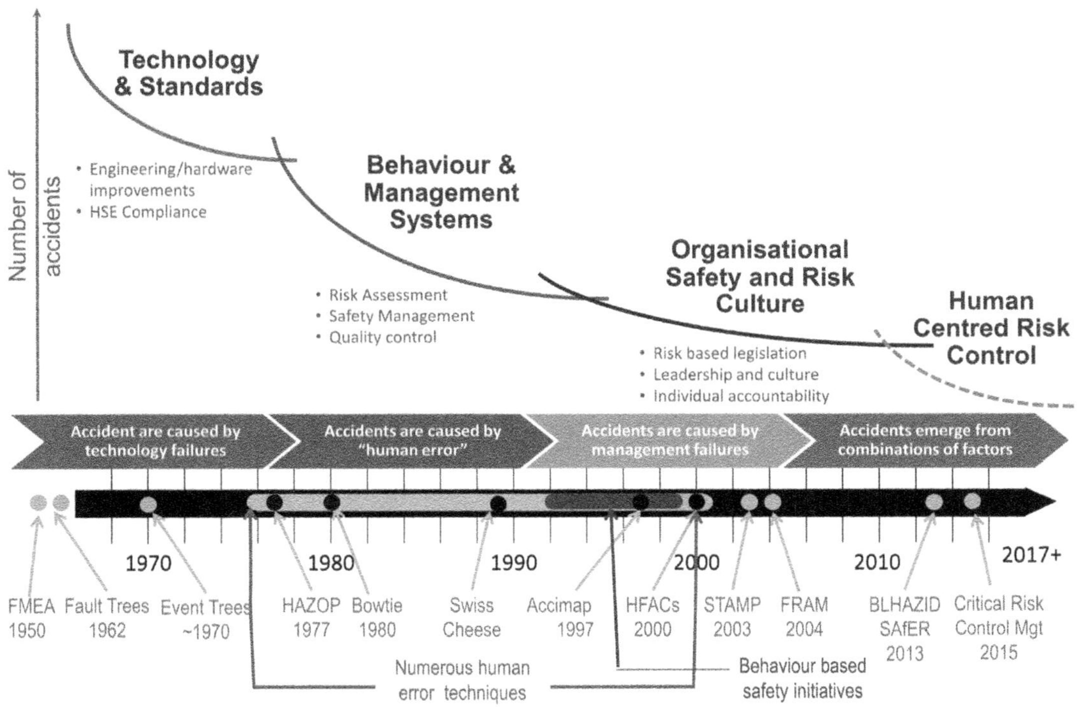

FIGURE 2.8 History of industrial performance (Hassall, 2014; Hassall, 2015). *Source: Adapted from Hassall, M. E. (2014). MINE4200 Humans and Risk-Cognitive Human Factors Lecture Slides March 31, 2014. Brisbane Australia, The University of Queensland. Hassall, M. E. (2015). Improving human control of hazards in industry. 19th Triennial Congress of the IEA 9-14 August 2015, Melbourne, Australia.*

- Hazard and operability studies (HAZOP): This type of analysis involves selecting a subsection of a process (a node) and systematically identifying and assessing the risks that deviations in process parameters might present.
- Bowtie analysis (BTA): This type of risk assessment involves visually representing an unwanted event, its causes and consequences, and the prevention and mitigation controls. It has the physical appearance of a bowtie. The bowtie diagram helps decision makers determine whether they have adequate controls to address the risk.

More detailed information on each of these techniques is provided in several sections later in the book.

Results from these risk assessment techniques combined with other initiatives led to improved technology and standards, which in turn led to a reduction in unwanted events (Fig. 2.8). However, the focus on improving reliability didn't eliminate unwanted events which continued to occur at an acceptably high level.

Analysis on the unwanted events that continued to occur highlighted that **humans** were playing an important role in initiating and/or escalating these unwanted events. For example, the Exxon Valdez oil spill happened when the third mate sailed the ship outside the normal shipping lane and collided with Bligh Reef causing a massing oil spill in Prince William Sound

(National Transport Safety Board, 1990). A more recent example is the Texas City refinery explosion which occurred when operators overfilled a column releasing sufficient hydrocarbon to form a vapour cloud that found an ignition source and caused a violent explosion (U.S. Chemical Safety & Hazard Investigation Board, 2007). Attributing incident causes to 'human error' led to the development of numerous – over 50–risk assessment techniques aimed at helping practitioners identify and address potential sources of 'human error' (Stanton et al., 2005). Examples of 'human error' risk assessment approaches include HEART (Williams, 1986), SHERPA (Embrey, 1986), SPEAR (AIChE Centre for Chemical Process Safety, 1994), Human Error HAZOP (Whalley-Lloyd, 1998), CREAM (Hollnagel, 1998), THEA (Pocock et al., 1999), TRACer (Shorrock & Kirwan, 2002) and human factors checklists (Bellamy et al., 2008). Towards the end of this era, behaviour-based safety initiatives also began to emerge. Behaviour-based safety programs focus on reinforcing workers to behave safely and to see safety as their responsibility, and not just a management issue (Tuncel et al., 2006).

The focus on human error and behaviour-based safety led to the realisation that most human behaviour was shaped by organisational and system factors (as discussed in Chapter 3). An example of organisational contributions to accidents is highlighted in the Challenger space shuttle disaster, which occurred because of the decision to launch in unsafe conditions when the actual temperature was below the safe temperature for o-rings (Committee on Science & Technology, 1986).

This insight led to the development of risk assessment approaches to help practitioners identify and address organisational factors that could cause unwanted events. Examples of such approaches include the Swiss Cheese model (Reason, 2008), Accimap (Svedung & Rasmussen, 2002) and HFACs (Shappell & Wiegmann, 2000). These organisational factor analyses identified improvements in safety management systems that might deliver further reductions in unwanted events. More detailed information on these risk assessment approaches for organisational management systems is provided in Chapters 3 and 5.

The focus on organisational systems and human behaviour led to the insight that organisational culture – particularly it's safety culture – is an important factor that can contribute to, or avert the occurrence and severity of, unwanted events. Safety culture can be defined as the shared safety or risk-related perceptions, beliefs and behaviour shaping norms held by people within the workplace (Casey et al., 2017; Glendon & Stanton, 2000). The importance of culture was highlighted in the Baker Panel review of the BP Texas City Refinery explosion which found *'BP did not instill a common, unifying process safety culture ... The Panel found instances of a lack of operating discipline, toleration of serious deviations from safe operating practices, and apparent complacency toward serious process safety risks'* (Baker et al., 2007, p. 60).

However accidents still continued to occur and analyses has highlighted that these were often due to a failure to implement and/or maintain the known risks controls required to ensure safety (Noetic Solutions, 2014). This lead to approaches being developed to improve the effectiveness of risk controls. These include STAMP-based approaches (Leveson et al., 2003; Leveson, 2011) and Risk Control Management approaches (ENFORM Canada, 2016; Hassall & Harris, 2017; Hassall et al., 2015; ICMM, 2015a, 2015b). More details on Risk Control Management approaches are described in Chapter 5.

All of the above-mentioned approaches have helped to reduce the number of unwanted events. To reduce unexpected, and beyond-design unwanted events, additional approaches

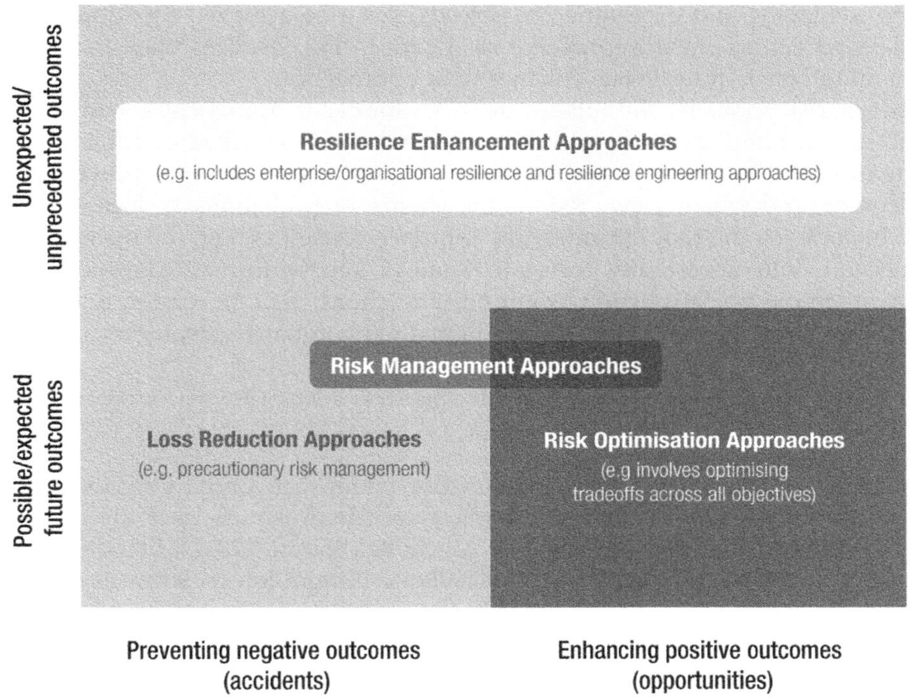

FIGURE 2.9 Scope of risk and uncertainty management.

are being developed to enhance organisational resilience as shown in Fig. 2.9. One technique is the Functional Resonance Analysis Method (FRAM) (Hollnagel, 2012) which seeks to identify how normal variation within a system can lead to unexpected outcomes. Another technique is Blended HAZID that combines function-driven and component-driven approaches to develop detailed structured representations of failure causality in process systems (Seligmann, Németh, Hangos, & Cameron, 2012). A third approach is Strategies Analysis for Enhancing Resilience (SAfER) (Hassall et al., 2014) which seeks to identify ways to improve system designs so humans can better manage industrial operations across both normal and abnormal situations. More details on these different approaches are provided in Chapter 5.

To gain and sustain leading edge operational performance, organisations need to consider both loss prevention and risk optimisation approaches to risk management also shown in Fig. 2.9. These two approaches will be described next.

2.6 Two approaches to modern risk management

Most organisations have two broad objectives: (1) to preserve their existence by preventing disasters and adverse events from happening, and (2) to continue to improve and to capture opportunities in order to realise their potential. The achievement of these objectives can be thwarted by known and unknown threats, incomplete knowledge, variability or change. Achieving and exceeding the objectives can be achieved by adopting risk and resilience-based management approaches as shown in Fig. 2.9. This book focusses on risk management rather than resilience enhancement approaches.

In modern risk management, there are two dominant approaches; namely the 'loss reduction' mindset and the 'risk optimisation' mindset as described in Table 2.4. The **loss reduction** approach is the dominant historical view that focusses on the prevention of negative outcomes, and which views risk as the chance or probability of loss or an adverse outcome. In contrast, the **risk optimisation** approach considers both the upside and downside associated with uncertainty across a range of key performance areas (cost, safety, environment, employee satisfaction, community relations, etc.). In reference to Table 2.4, a leading HSE senior engineer from a global oil and gas company recently commented:

> My company would appoint people who think like the left side of the table, but we would promote the people that think like the right side.

It is recognised that a sole focus on loss prevention can undermine the competitive advantage of an organisation. This point was raised in Australia over the installation of home solar system batteries. According to Macdonald-Smith (2017), Standards Australia were proposing that households should 'incarcerate lithium-ion batteries in concrete bunkers separate from the home, accessible only to authorised persons.' Some in the industry described the proposed standard as 'overkill' that 'would add thousands of dollars on to the cost of installing power storage systems, wiping out the economic gains solar households can make from batteries over their lives as electricity bills soar'. The batteries had a history of occasionally spontaneously catching on fire but manufacturers state they have fixed the problem, and after receiving 3000 public comments, Standards Australia withdrew the draft proposal (Ludlow, 2018).

TABLE 2.4 Two approaches to modern risk management.

Loss reduction mindset (Cameron & Raman, 2005)	Risk optimisation mindset (Hillson, 2010)
What can go wrong? What hazards and threats exist?	What are we trying to achieve? What are our key objectives?
What are the consequences if things go wrong?	What is the 'uncertainty that matters'? Including both downside threats and upside opportunities.
What is the likelihood that things might go wrong?	Acknowledge that risk management is affected by perception and 'zero risk' is unachievable and undesirable, so what is the appropriate level of risk to aim for?
Is the risk low enough to be acceptable, or is action required to lower the risk?	What actions are required to manage risks?
Have enough controls been implemented to prevent the unwanted events from occurring, or to mitigate the consequences if it does occur?	How are we going? What has changed? What have we learned?

Employing a risk optimisation mindset, we are able to state the following six key concepts of risk management (Hillson, 2010, p. 153):

1. **Risk is 'uncertainty that matters'** – but different things matter to different people to a different extent in different circumstances.
2. **Risk includes both downside (threats) and upside (opportunities)** – both types of risk need to be addressed proactively, in order to minimise threats and maximise opportunities.
3. **'Zero risk' is unachievable and undesirable** – all aspects of life (including business and projects) involve risk, so some degree of risk-taking is inevitable, but we should only take appropriate risks in relation to the level of return we expect or require.
4. **Risk has two key dimensions** – uncertainty can be expressed as 'probability' or 'frequency', and how much it matters can be called 'impact' or 'consequence'.
5. **Risk management requires an understanding of both dimensions** – if the uncertain event is very unlikely or it would have negligible effect, it requires less attention.
6. **Risk management is affected by perception** – answers to the questions 'How uncertain is it?' and 'How much does it matter?' are subjective.

A fundamental basis of these six key concepts is that ***risk is intricately linked with human decision-making processes*** which are discussed in detail in Chapter 3.

2.7 Case studies illustrating two approaches to risk management

2.7.1 Case study 1: process plant operation

A processing plant receives raw ore from the mine then crushes it, conveys, washes and sizes it through vibrating screens before transferring it to product stockpiles. From ore receival to product stockpile involves 10 different conveyors transporting the ore over more than a kilometre. Often, due to mine supply issues, there are times when there is no ore supply for the plant. This leaves operators with the question – *Should the plant be kept running or should it shutdown if there is no ore for a specified period of time?*

The answer to such a question could be determined using the loss prevention approach or risk optimisation approach. If the plant is focussed on availability – being able to process ore as soon as it arrives from the mine – then the loss prevention approach would probably lead to the decision to keep the plant running so it is more likely to be available when the ore arrives from the mine. Such an approach would minimise production losses associated with situations where operators are unable to restart the plant in time to process ore when supplies resume.

If a risk optimisation approach was employed, what sort of factors should be considered?

Some factors that might be worth considering include:

- Time: Time before ore supply resumes, time it takes to shutdown and startup plant.
- Cost: Cost of running the plant versus cost of stopping and restarting plant.
- Risk of not being available when ore supply resumes: Risk of breaking down without ore versus risk of not being able restart in time to process ore when supplies resume.

2.7.2 Case study 2: the availability of safety PPE

At present most commercial airlines are equipped with life jackets and require the attendants to run through how to use the life jacket as part of the pre-takeoff safety talk and demonstration. This is an example of loss prevention risk management approach as life jackets on airplanes are intended to save lives if the aircraft ends up in the water. However, when US Airways Flight 1549 landed on the Hudson river, only approximately 10 of the 150 passengers retrieved and evacuated with a life vest (Rosenkrans, 2010). In an article title *'Do planes really need life vests?'* published the Wall Street Journal (McCartney, 2016) included the following quotes:

- In most crash landings, safety experts say, it's more important to get out of the plane as quickly as possible to avoid being trapped by a fire rather than take precious seconds to find a life vest and try to put it on.
- Laboratory tests and actual emergencies have both shown that passengers will give up and flee before actually finding a vest under their seat.
- Life vests are really only useful when there is advance warning of a water landing... But for at least several decades, water landings have universally been the sudden type, not the planned variety.
- There's also a psychological benefit: Passengers would think it ridiculous to travel over an ocean without some type of emergency flotation device beyond life rafts.
- Vests weigh a little more than 1 pound each, so a medium-size jet has about 200 pounds of vests onboard. That weight increases fuel burn and emissions.... Eliminating life vests might save more than 1 million gallons a year at a large airline like American, United or Delta.
- Another cost:... [is that the] vests disappear regularly—passengers steal them as souvenirs, airlines say.

Assuming the Wall Street Journal is correct, what factors would you consider when seeking to determine the optimum trade-offs between the different risks?

2.8 Summary

This chapter introduced the risk management process (ISO31000) and core risk management terms and definitions. It also provided an overview of the history of risk management, illustrating how the development of risk management techniques has evolved from having a solely technical focus, to expanding the scope to consider human and organisational factors. The future of risk management in the process industries should involve two main risk management approaches – the traditional loss prevention approach and the value-adding risk optimisation approach.

Everyone in an organisation is responsible for making risk-based decisions, with upper management having the additional responsibilities of establishing, monitoring and improving an organisation's risk management system. The material covered in this chapter provides the important foundations for the effective management of risk – common definitions, a common and appropriate risk framework, and defined roles and responsibilities.

Review questions

1. What is the ISO Standard for risk management?
2. What is a *hazard*?
3. What is a *threat*?
4. What are the three main steps of Risk Assessment?
5. Sketch the Risk Management Process as presented in ISO 31000, showing the key activities undertaken.
6. What are the stages of Risk Treatment that have been added in the extended risk management process?
7. What is an *unwanted event*?
8. What framework is useful for determining the scope for a risk management activity?
9. What is the meaning of ALARP?
10. What is the dominant factor governing human behaviour (and thus human errors) in industry?
11. Is *zero risk* the ultimate aim for engineers working on projects? Why?
12. What are the differences between the *Risk reduction mindset* and the *Risk optimisation mindset*?
13. Why is it important to adopt human-centred approaches to managing risk?
    Which of the following are either a *hazard, threat, unwanted event* or *consequence*?
    a. A 50 t tank of pressurised LPG
    b. A leak of pressurised propane
    c. An explosion
    d. A worker unfit for work
    e. Incorrect fitting of a flange
    f. An uncontrolled release of steam
    g. A worker injury resulting from an unwanted event

References

AIChE Centre for Chemical Process Safety. (1994). *Guidelines for preventing human error in process safety*. Centre for Chemical Process Safety of the American Institute of Chemical Engineers.

Baker, J. A. I., Leveson, N. G., Bowman, F. L. S., Priest, S., Erwin, G., Rosenthal, I. I., Gorton, S., Tebo, P. V., Hendershot, D., Wiegmann, D. A., & Wilson, L. D. (2007). *The report of the BP U.S. refineries independent safety review panel*.

Bellamy, L. J., Geyer, T. A., & Wilkinson, J. (2008). Development of a functional model which integrates human factors, safety management systems and wider organisational issues. *Safety Science*, *46*, 461–492.

Borys, D., Else, D., & Leggett, S. (2009). The fifth age of safety: The adaptive age? *Journal of Health & Safety Research & Practice*, *1*(1), 19–27.

Cameron, I. T., & Raman, R. (2005). Process systems risk management. *San Diego. CA*. Elsevier.

Casey, T., Griffin, M. A., Flatau Harrison, H., & Neal, A. (2017). Safety climate and culture: Integrating psychological and systems perspectives. *Journal of Occupational Health Psychology*, No Pagination Specified.

Committee on Science and Technology. (1986). *Investigation of the Challenger accident: Report of the Committee on Science and Technology House of Representatives-Ninety-ninth congress-Second session*.

Embrey, D. (1986). *SHERPA: A systematic human error reduction and prediction approach*. International Meeting on Advances in Nuclear Power Systems, Knoxville, TN.

ENFORM Canada. (2016). *A Barrier Focused Approach-How to get started with process safety* (Vol 2). ENFORM Canada.

Glendon, A. I., & Stanton, N. A. (2000). Perspectives on safety culture. *Safety Science, 34*(1–3), 193–214.

Hale, A. R., & Hovden, J. (1998). Management and culture: The third age of safety. A review of approaches to organizational aspects of safety, health and environment. In A. M. Feyer, & A. Williamson (Eds.), *Occupational Injury. Risk Prevention and Intervention*. London: Taylor & Francis.

Hassall, M. E. (2014). MINE4200 humans and risk-Cognitive human factors lecture slides. *March 31, 2014*. Brisbane Australia: The University of Queensland.

Hassall, M.E., & Harris, J. (2017). *Risk controls knowledge: Determining leading practice from case study analysis (ACARP report C25036).*

Hassall, M. E., Joy, J., Doran, C., & Punch, M. (2015). *Selection and optimisation of risk controls (ACARP report C23007).*

Hassall, M. E. (2015). Improving human control of hazards in industry. *19th Triennial Congress of the IEA 9–14 August 2015 Melbourne, Australia 2015.*

Hassall, M. E., Sanderson, P. M., & Cameron, I. T. (2014). The development and testing of SAfER: A resilience-based human factors method. *Journal of Cognitive Engineering and Decision Making, 8*(2), 162–186.

Hillson, D. (2010). *Exploiting future uncertainty creating value from risk*. Farnham, Surrey, England.

Hollnagel, E. (1998). *Cognitive reliability and error analysis method: CREAM*. Elsevier.

Hollnagel, E. (2011). When things go wrong: Failures as the flip side of successes. In D. A. Hofmann, & M. Frese (Eds.), *Errors in organizations*. New York, NY: Routledge.

Hollnagel, E. (2012). *FRAM: the functional resonance analysis method-modelling complex socio-technical systems*. Surrey, England: Ashgate.

ICMM. (2015a). *Health and safety critical control management good practice guide*. Published by the International Council on Mining and Metals (ICMM).

ICMM. (2015b). *Critical control management: Implementation guide*. Published by the International Council on Mining and Metals (ICMM).

ISO 31000 (2018). *ISO 31000:2018 Risk management-Guidelines*. In. Geneva: International Organization for Standardization

Leveson, N. G. (2011). *Engineering a safer world: Systems thinking applied to safety*. The MIT Press.

Leveson, N. G., Daouk, M., Dulac, N., & Marais, K. (2003). *Applying STAMP in accident analysis*. Workshop on the Investigation and Reporting of Accidents, September.

Ludlow M. (2018). *Push for new performance standards to help uptake of battery storage*. Australian Financial Review, July 9, 2018.

Macdonald-Smith, A. (2017). *Battery suppliers battle 'overkill' on installation rules*. Australian Financial Review, April 1, 2017.

McCartney, S. (2016). Do planes really need life vests? The Wall Street Journal, January 20, 2016.

National Transport Safety Board. (1990). *Grounding of the U.S. tankership Exxon Valdez on Bligh Reed, Prince William Sound near Valdez, Alaska, March 24, 1989* (Marine Accident Report).

Pocock, S., Wright, P., & Harrison, M. (1999). *THEA–A technique for human error assessment early in design*. RTO HFM Workshop on the Human Factor in System Reliability–Is Human Performance Predictable?, December 1–2, Siena, Italy.

Noetic Solutions. (2014). *MSAC fatality review 2013–14: Report for NSW safety advisory council*. Noetic Solutions.

Reason, J. T. (2008). *The human contribution: Unsafe acts, accidents and heroic recoveries*. Farnham, England: Ashgate.

Renn, O. (1992). Concepts of risk: A classification. In S Krimsky, & D Golding (Eds.), *Social theories of risk*. (pp. 53–79). London: Praeger, In press.

Renn, O. (2008). *Risk Governance: Coping with uncertainty in a complex World (1st ed.)*. Earthscan, In press.

Rosenkrans, W., Survival on the Hudson. version. August 5. 2010. https://flightsafety.org/asw-article/survival-on-the-hudson/ [Accessed 06 Feb 2023].

Seligmann, B. J., Németh, E., Hangos, K. M., & Cameron, I. T. (2012). A blended hazard identification methodology to support process diagnosis. *Journal of Loss Prevention in the Process Industries, 25*(4), 746–759.

Shappell, S. A., & Wiegmann, D. A. (2000). *The human factors analysis and classification system-HFACS*. Report Number DOT/FAA/AM-00/7. Office of Aviation Medicine, Washington DC.

Shorrock, S. T., & Kirwan, B. (2002). Development and application of a human error identification tool for air traffic control. *Applied Ergonomics, 33*(4), 319–336.

Stanton, N. A., Salmon, P. M., Walker, G. H., Baber, C., & Jenkins, D. P. (2005). *Human factors methods: A practical guide for engineering and design*. Aldershot, UK: Ashgate.

Svedung, I., & Rasmussen, J. (2002). Graphic representation of accident scenarios: Mapping system structure and the causation of accidents. *Safety Science*, *40*, 397–417.

Tuncel, S., Lotlikar, H., Salem, S., & Daraiseh, N. (2006). Effectiveness of behaviour based safety interventions to reduce accidents and injuries in workplaces: Critical appraisal and meta-analysis. *Theoretical Issues in Ergonomics Science*, *7*(3), 191–209.

U.S. Chemical Safety and Hazard Investigation Board. (2007). Investigation report-Refinery explosion and fire - BP, Texas City, March 23, 2005. U.S. Chemical Safety and Hazard Investigation Board.

Whalley-Lloyd, S. (1998). Reducing the impact of human error. *The Safety & Health Practitioner*, *16*(5), 20.

Williams, J. (1986). *HEART-A proposed method for assessing and reducing human error*. Paper presented at the 9th Advances in Reliability Technology Symposium, University of Bradford.

CHAPTER

3

Humans and risk

OUTLINE

3.1 Introduction 42

3.2 The changing role of the human operator in industry 43

3.2.1 Era 1 – localised, direct and physical control 43

3.2.2 Era 2: regional, centralised and more cognitive control 45

3.2.3 Era 3: global, collaborative, cognitive 47

3.2.4 What have we learnt from the changing nature of industrial work? 50

3.3 The role of humans – risk perceivers, analysers and controllers 50

3.3.1 Risk perceivers 51

3.3.2 Risk analysers 52

3.3.3 Risk controllers 52

3.4 Risk communication 53

3.5 The human decision-making process 56

3.5.1 Situation assessment 56

3.5.2 Response strategies 58

3.6 Improving human performance 61

3.6.1 An example of designing for improved human performance – overfilling vessels 64

3.7 Summary 65

Review questions 65

References 67

Fundamentals of Risk Management for Process Industry Engineers
DOI: https://doi.org/10.1016/B978-0-12-820320-0.00001-5

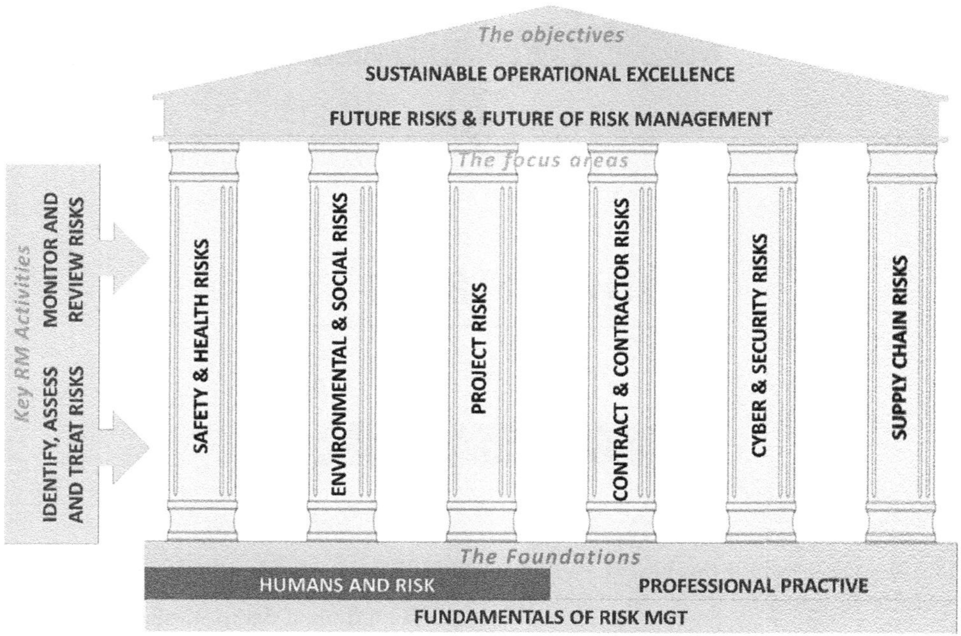

There are risks and costs to action. But they are far less than the long range risks of comfortable inaction
John F. Kennedy.

3.1 Introduction

In Chapter 2, we saw that human behaviour is a critical factor that can contribute to, or avert the occurrence and severity of, unwanted events. Human and organisational factors are commonly attributed as key causal factors in process incidents, but it is also human behaviour also averts incidents, and moreover, ensures operational targets are achieved or exceeded. The humans in our process systems are not only causes of risk, they are also perceivers and managers of the risk.

In this chapter, we explore in more detail how human decisions and actions at various organisational levels directly impact, and are directly impacted by, risk as shown in Fig. 3.1. It is these decisions and actions that dictate the performance of an organisation and how it interacts with the external world. We will see that human performance dominates risk in high hazard industries, but that humans are also the adaptable resource that control risk in real time. This paradox means that organisations operating in high hazard industries are often challenged with how best to approach the role of the humans in the risk zone.

At the end of this chapter, we hope that you will have developed a clearer appreciation of the critical role that humans play in managing risk, and also a commitment to adopting human-centred approaches to risk management. It is thus a central tenet of this book that *adopting a human-centred approach to risk management is needed to deliver real improvements in risk identification, understanding, analysis, control, communication and governance.*

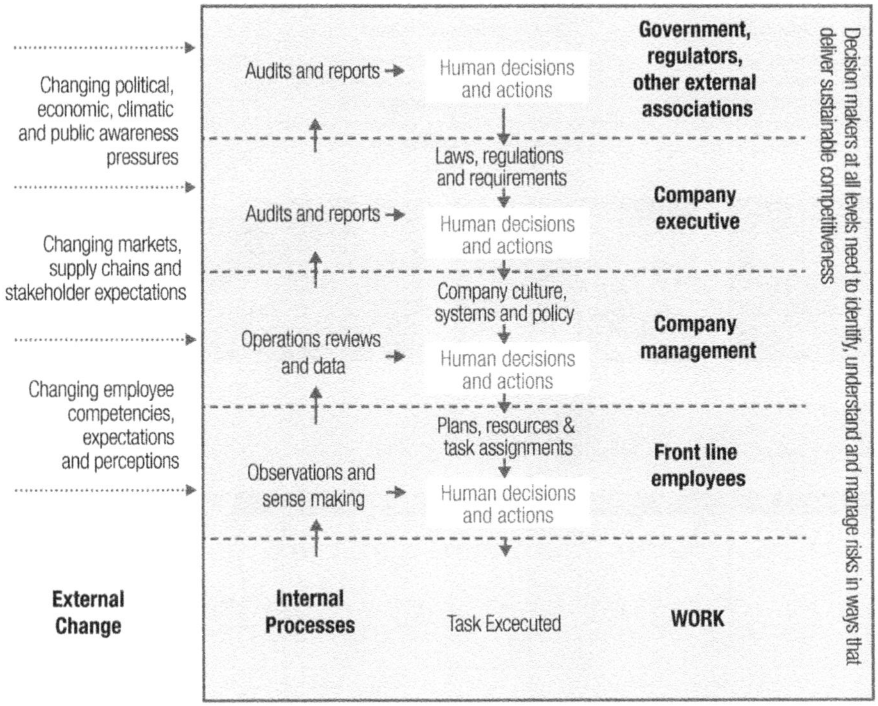

FIGURE 3.1 Organisational decision-making (Rasmussen, 1997). *Source: Adapted from Rasmussen, J. (1997). Risk management in a dynamic society: a modelling problem. Safety Science, 27(2–3), 183.*

3.2 The changing role of the human operator in industry

In the process industries, humans have always played an important role in controlling the operations. Humans are often required to oversee the process and to make the adjustments necessary to keep the process under control. However, the nature of human control and the nature of industry have changed over time as shown in Fig. 3.2. This section provides a brief history of industrial work, and examines the evolving role of the human operator. This is critical for understanding the contemporary context within which operational staff work, and for appreciating the key factors that will inform human-centred approaches to managing risk.

3.2.1 Era 1 – localised, direct and physical control

Era 1 began with the introduction of electricity into industry. It ran from the 19th century to after the Second World War. Many industrial companies that operated in this era tended to be locally owned and operated companies that used local suppliers and distributed to local markets. The control of technology with these companies tended to

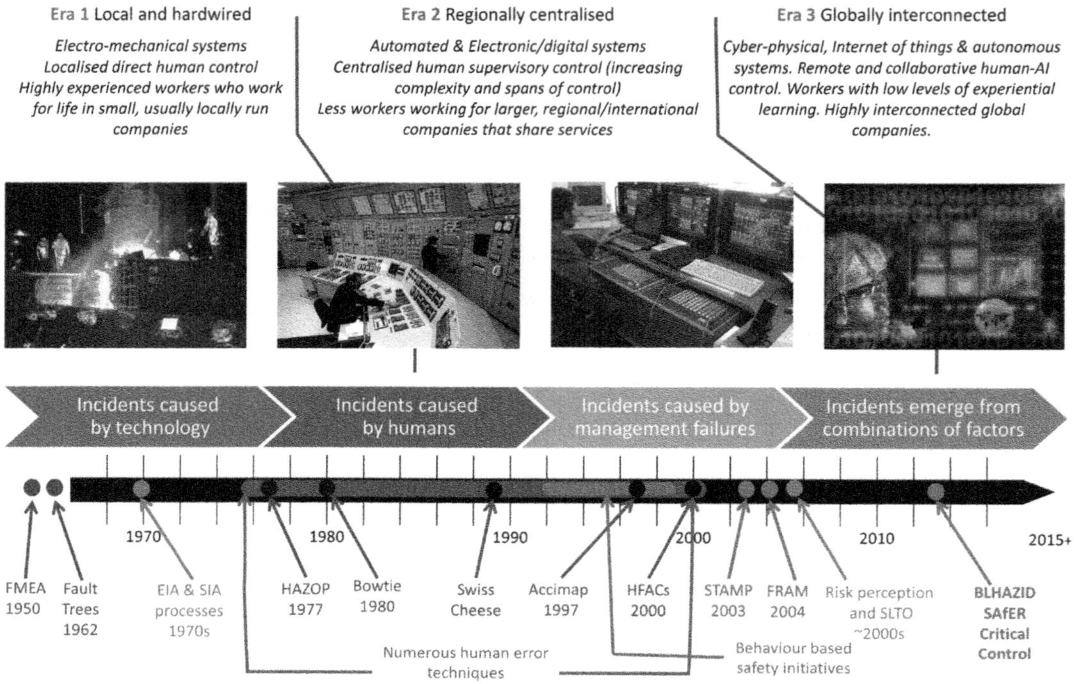

FIGURE 3.2 History of industrial work (Hassall, 2015). *Source: Adapted from Hassall, M. E. (2015). Improving human control of hazards in industry. 19th Triennial Congress of the IEA 9-14 August 2015, Melbourne, Australia.*

be performed by human operators who were physically located with the equipment they managed. Their span of control was often constrained to the equipment, levers and dials that were within their sight and reach. Often, the technology was unreliable and operators had to regularly manually intervene to keep production going. This regular interaction, and the fact that at this time workers often stayed with the same company or type of work for their whole career, meant that human controllers often developed good mental models of the fundamental processes and good tacit knowledge on how to keep these processes operating. However, regular manual interventions combined with pressure to keep processes operating led to high fatigue levels and unacceptable injury rates. Specialist boards and committees were developed to improve human productivity and wellbeing (Zionchenko & Munipov, 2005).

During this era, scientific studies into how to improve safety and reduce accidents were introduced. These studies tended to focus on improving machine guarding, stopping explosions and preventing structural failures (Hale & Hovden, 1998). Tools and techniques were developed to help identify, assess and address risks associated with technology failures. Such techniques included Failure Mode Effects Analysis (FMEA) and Hazard and Operability Studies (HAZOPs). Scientific studies were also developed to study workers time and motion when conducting tasks. The aim of these time and motion studies was to identify the most efficient and least stressful way that workers should be performing tasks (Hassall et al., 2015).

Toward the end of this era, environmental and social impacts started to be considered. The initial focus was on identifying and reporting the potential economic, environmental and social impacts of a proposed project.

3.2.2 Era 2: regional, centralised and more cognitive control

The second era of industrial control began after the Second World War. Military technologies developed during the war included radar and sonar as well as nuclear weapons. After the war, these technologies were converted into civilian applications along with their screen-based monitoring jobs (Chapanis, 1996). This resulted in the emergence in industry of centralised control rooms from which operators controlled large areas of more complex industrial systems such as nuclear power plants. Thus these human controllers were removed from being colocated with the equipment in the plant where they were able to see, hear, smell, touch and directly interact with the process into rooms often located away from most equipment and often insulated from the plant sounds, vibrations, temperatures and visual appearance. This led to the development of large-scale, complex work systems with centralised control rooms. Such industries often were managed by large corporations and they employed experienced people to run the operations. Examples include nuclear power plants, oil refining, petrochcemicals and civil aviation systems. This transformation meant that human controllers were more isolated from directly interacting with, or being able to use their senses to assess and control, the systems. They were physically isolated from directly observing the system and had to rely more on instrumentation readings and alarm systems to diagnose and control system states.

During this period, operators in the control room were required to be supervisory controllers rather than direct controllers. Supervisory controllers monitor the process and provide control instructions and receive feedback from the process through intermediary interfaces instead of directly interacting with the system (Sheridan, 2011). These human controllers were reliant on instrumentation, alarm systems and screen-based technologies to detect, diagnose and control processes. Initially, the control room instrumentation was electrical and mechanical in nature so operators could see and hear if these were malfunctioning. However, the transformation to digital technologies (commencing in the 1970s) meant that the instrumentation and alarms were removed from operators with only the outputs being displayed on computer screens. This reduced the physical layouts of information sources but increased the difficulty in detecting incorrect control information or malfunctioning controllers or instrumentation.

As technology became more reliable, and as the human controls became more removed from the process, the accidents that continued to occur were often attributed to ***human error***. For example, the U.S. DOE found that human error caused 80% of incidents (U.S. Department of Energy, 2009) and Reason (2008) found that human error caused 70% of U.S. nuclear power plant incidents and 80%–90% of chemical industry incidents. It is interesting to note that in a somewhat pioneering work of its time, Heinrich (1941) found that 88% of industrial accidents resulted from workers' unsafe actions.

In addition, the humans involved in many incidents were experienced operators. For example, the Three Mile Island nuclear power plant incident occurred when experienced operators did not correctly interpret the state of the reactor from the instruments and as a result they made control actions that worsened the situation (Booth, 1987). This led to the realisation that systems should be designed to fit people rather than trying to change people to fit the system, which in turn led to the establishment of the field of human factors and ergonomics (Shaver, 2009).

This second era of industrial control occurred in conjunction with what Hale and Hovden (1998) referred to as the 'Age of ***Human Factors***'. The focus of this era was to resolve safety problems by redesigning work to better match humans (Hale & Hovden, 1998). To help analysts identify ways to improve design, many human factors techniques were developed during this period. Examples include Cognitive Reliability Analysis Method (CREAM) (Hollnagel, 1998), Human Error Assessment and Reduction Technique (HEART) (Williams, 1986), Systematic Human Error Reduction and Prediction Approach (SHERPA) (Embrey, 1986), The Technique for Human Error Assessment (THEA) (Pocock et al., 1999), and Technique for the Retrospective and Predictive Analysis of Cognitive Errors (TRACEr) (Shorrock & Kirwan, 2002). See Stanton et al. (2013) for more details.

As this era progressed, there was growing realisation that errors in human control actions could be attributed back to issues with ***organisational factors***. An analysis of human error incidents revealed that 30% were caused by individual mistakes and the remaining 70% result from latent weaknesses in the organisation as shown in Fig. 3.3.

Examples of these organisational weaknesses include issues with 'management decisions, core organisational processes - designing, building, maintaining, scheduling, budgeting, and the like - along with the corporate safety culture' (Reason, 2008, p. 98). Disasters including Space Shuttle Columbia and Fukushima Nuclear Power Plant accident were

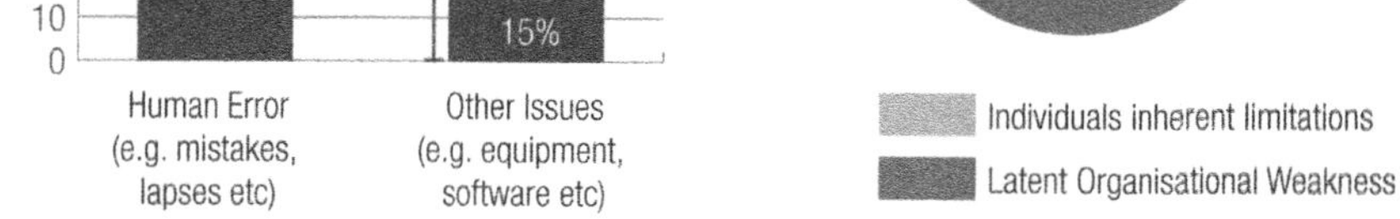

FIGURE 3.3 Types of error (Kletz, 2009; Reason, 2008; U.S. Department of Energy, 2009). *Source: Kletz, T.A. (2001). An engineer's view of human error (3rd ed.). Institution of Chemical Engineers; Reason, J.T. (2008). The human contribution: Unsafe acts, accidents and heroic recoveries. Ashgate. http://library.uq.edu.au/record = b2443566; U.S. Department of Energy (2009). DOE Standard: Human performance improvement handbook - Volume 1: Concepts and principles (DOE-HDBK-1028-2009). Washington D.C: U.S. Department of Energy, Article DOI: https://www.standards.doe.gov/standards-documents/1000/1028-BHdbk-2009-v1/@@images/file.*

attributed to organisational factors. The Columbia Accident Investigation Board (2003) found organisational causes of the accident which included:

> Cultural traits and organizational practices detrimental to safety were allowed to develop, including: reliance on past success as a substitute for sound engineering practices (such as testing to understand why systems were not performing in accordance with requirements); organizational barriers that prevented effective communication of critical safety information and stifled professional differences of opinion; lack of integrated management across program elements; and the evolution of an informal chain of command and decision-making processes that operated outside the organization's rules (p. 6).

The investigation into the Fukushima Nuclear Power Plant disaster found that it

> Was the result of collusion between the government, the regulators and TEPCO, and the lack of governance by said parties ... we therefore conclude that the accident was clearly "manmade." We believe that the root causes were the organizational and regulatory systems that supported faulty rationales for decisions and actions (The National Diet of Japan, 2012, p. 16).

To identify and address the organisational contributions to accidents, numerous approaches were developed. These include the Swiss Cheese Model (Reason, 2008), human factors analysis and classification system (HFACS) (Shappell & Wiegmann, 2000), and Accimap (Rasmussen & Svedung, 2000) and Systems-Theoretic Accident Modelling and Processes (STAMP) (Leveson, 2011).

This era also highlighted the importance of considering public perceptions of risk. Significant nuclear power plant accidents including Three Mile Island, Chernobyl and Fukushima raised public awareness of the risks of nuclear technologies, and this has subsequently had significant impacts on energy policy in many countries. Research into public perceptions revealed that the public often use distinctly different bases than technical experts to assess risk and impacts (e.g. OECD, 2010). These insights highlight the importance of employing sociotechnical, and not just technical, analysis of risk to understand all of the uncertainties that could impact the achievement of objectives.

3.2.3 Era 3: global, collaborative, cognitive

In modern industries, companies are adopting automation, autonomous devices, and internet-connected technologies. Some of these companies are exploring nanotechnologies, the use of cyber-physical systems (e.g. 3-D printing), and wearable technologies to improving the operational performance and safety of a facility. Some of these technological advances have further removed humans from the work face. Modern control rooms now can be thousands of miles from the plants they are controlling. For example, the oil and gas industry is developing increasing numbers of normally unmanned or unattended installations which are operated remotely (Maslin, 2016). Similarly, Rio Tinto runs its operations in the Pilbara, Western Australia, from a control room hundreds of kilometres away in Perth. The use of autonomous devices such as drones and robots are controlled by coders and maintainers who are separated from real-time operations not only by distance but also by time. This evolution to internet-connected technology means that organisational resource allocation and locations are changing. In addition, there is a growing dependency on internet and telecommunication companies to ensure safe and secure operations. These dependencies introduce new risks both in terms of threats and opportunities.

In this era, industrial control transitions from supervisory control type towards more interactive, collaborative or adaptive control where humans and digital technology share and exchange control duties (Parasuraman & Wickens, 2008; Sheridan, 2002). With this form of control humans and/or preprogrammed computers monitor, diagnose and adjust the control system parameters (Sheridan, 2011). The allocation of control functions between humans and computers can be static or dynamically changed to match the operator's capability and the current plant state. The ability to change function allocation to incorporate human capabilities will be important because, as stated by Hassall (2015), this generation of workers may not have the experiential learning background or stay with the same job for the same length of time that was experienced by workers in previous eras. For example, previous generations of workers often grew up in environments where there was a risk of getting physically hurt (while playing on see-saws or monkey bars in playgrounds or playing in wilderness areas, etc.). They were also often exposed to, and got experience with, trying to repair mechanical and electrical devices (radios, motor vehicles, TVs, etc.). In previous generations there was also more exposure to serious workplace accidents and occupational injuries. Thus previous generations had more extensive, direct, personal and physical experiences that help them understand the fundamental operation of equipment and cause-effect-consequence relationships. In contrast, current and future generations of workers are growing up in a digital world where cause-effect-consequence is learned from electronic gaming that rewards finding the 'cheat' or shortcuts and where being reborn, often more than once, is possible. They are also learning more by watching videos and viewing social media posts than by reading detailed text. Moreover, in this era it is commonplace to dispose and replace, rather than to repair and reuse. This in turn reduces the learning opportunities that come from pulling apart an item, trying to diagnose and fix the fault, then rebuilding in a manner to restore full functionality. These step changes in learning experiences may affect workers' ability to understand fundamental process operations and their perception, assessment and management of hazards and risks (Hassall, 2015). In addition, current and future generations of workers are less likely to stay in one job or company or even industry for life. These generational differences can in turn affect workers' learning and compliance preferences as well as their operational decision-making, and control execution strategies.

Thus this era is characterised by significant changes in technology, organisations and human factors. The interaction effects caused by these dynamic changes can produce novel, emergent nonlinear systems and external stakeholder responses for which there is no precedent (Hollnagel, 2012; Leveson, 2011). These responses can positively and/or negatively impact on human wellbeing as well as overall system performance and sustainability (Hassall, 2015). For example, the development of interconnected industrial networks (aka the Industrial Internet of Things) is affecting not only the way industrial and commercial enterprises run, but also having disruptive impacts on personal lives, community knowledge and risk perceptions and societal behaviour (Lee et al., 2019). Therefore the linear risk assessment techniques that rely on past experience and learning from incidents to prevent reoccurring events (Woods & Hollnagel, 2006) will not necessarily deliver the risk identification and management requirements in the third era of industrial control.

New techniques are needed to help industry practitioners identify and address risks in Era 3. Approaches that help to prospectively identify risks that matter and that design interventions to help workers successfully detect, diagnose and manage deviations are needed, especially in complex systems where unexpected or even unimagined situations might occur (Hassall, 2015). These approaches also need to enhance learning from both successful and unsuccessful performances so that beneficial practices can be promoted and potentially harmful practices can be prevented (Borys et al., 2009). To address this need, there are a couple of types of approaches being developed that are applicable to the process industries.

The first category of approaches is based on Resilience Engineering. Resilience engineering, according to Hollnagel et al. (2006), focusses on developing tools and techniques that help people cope with complexity under pressure to diagnose, assess, understand the situation, and anticipate consequences in a manner that delivers success. Examples of engineering-based techniques that aim to enhance organisational resilience include Functional Resonance Analysis Method (FRAM), Systems Theoretical Accident Model and Processes (STAMP) and Strategies Analysis for Enhancing Resilience (SAfER).

FRAM, developed by Hollnagel (2012), helps analysts describe work systems in terms of functions and the interconnections between the functions. Each function is in terms of inputs, outputs, preconditions, resources or execution conditions, control and time. This information is visually represented in a manner that analysts can then use to assess how normal variation might propagate through the system to produce unexpected outcomes.

STAMP and its accompanying Systems-Theoretic Process Analysis (STPA) and Causal Analysis using System Theory (CAST) were developed by Leveson (2011). These approaches are based on the premise that systems involve '*interactions among people, societal and organizational structures, engineering activities, and physical system components*' (Leveson, 2011, p. 90) that are dynamically adapting and continually reacting to internal and external change. These approaches aim to identify the appropriate design and operational constraints on system and component behaviour that are needed to sustain safe operations (Leveson, 2011).

SAfER was developed by Hassall and colleagues (Hassall et al., 2014, 2016). It firstly seeks to identify the range of ways that system controllers might respond in different operating conditions. It then identifies the potential impacts that the range of controllers' responses might have on the system when it is operating in different states (e.g. normal, abnormal, startup and shutdown states). These impacts are then used to determine whether a response should be prevented because it will lead to adverse outcomes, or promoted because it will lead to beneficial outcomes. Responses that could lead to adverse outcomes but cannot be prevented need to be tolerated – that is the system needs to be designed so that if the response is performed then the design will tolerate it in a manner that does not lead to an unwanted event. More information with example applications of the SAfER approach are provided later in this chapter.

The second category of approaches for prospectively identifying risks that emerge from complex nonlinear systems is the use of dynamic models and digital twin technology. Dynamic models and digital twins allow decision makers to simulate and assess different operating modes, scenarios, faults and failures (Lee et al., 2019). These models can be used to simulate how disturbances from different process deviations propagate through a system both in space and time. They can also be used to forecast the impact that potential changes to process inputs can have on processing parameters and outputs by processing scenarios at speeds many times faster than real time.

3.2.4 What have we learnt from the changing nature of industrial work?

Why does this history matter? In this book we argue that it is critical to adopt human-centred approaches to managing risk. It is thus important to understand the factors and conditions that will affect how humans perceive and manage risks. In this section, we have seen how the role of operations staff in industry has evolved, and that the next generation of operators will have little, if any, tactile experiences with mechanical or electrical equipment, that their predominant experiences will come from gaming and media, and they will likely be responsible for operating remote autonomous equipment. It is within this context that we will have to identify, assess, evaluate and treat risks in the future. Our approaches will clearly have to evolve to be appropriate for this new world.

3.3 The role of humans – risk perceivers, analysers and controllers

When it comes to risk in industry, humans adopt a number of roles:

- ***Risk perceiver***: Stakeholders who consider and hold a view about a risk.
- ***Risk analyser***: People who identify and assess risks and determine the controls needed to manage risk.
- ***Risk controller***: People who are exposed to the risk and have to manage it.
- ***Risk communicator***: People who disseminate information about risk and risk management processes.

How people perform each of these roles is shaped by the attributes of the person, the attributes of the risk and the attributes of the situation. The attributes of the person include their experience, motivation, preferences, abilities and cognitive biases. The attributes of a person shape how they perceive, assess, control and communicate risk. Such attributes can also lead a person to be sensitised or desensitised to the risk. The attributes of the risk can also affect how people perceive, assess, respond to and communicate about the risk. Such attributes include the observability of risk, controllability of risk, immediacy, likelihood, severity and irreversibility of outcomes, as well as whether the persons affected will be voluntarily and involuntarily impacted (Slovic, 1987; Slovic et al., 1982). These factors can be summarised into the degree that the risk is known and the degree to which the risk is dreaded (Slovic et al., 1982).

The attributes of the situation that can affect risk perception, analysis, control and communication include organisational factors and contextual factors. From an organisational perspective, factors such as the culture and risk appetite of the leadership can shape risk perceptions, analysis, controls and communications. For example, organisations can be risk averse or risk takers. They can have centralised top-down decision-making or decentralised bottom-up decision-making. The contextual factors that can shape risk perceptions, analysis, controls and communications are quite extensive. They include attributes such as the social complexity, the technical complexity, the normality/novelty of context, the time pressure of the situation, plus the perceived severity of the consequences if things do go wrong. Other attributes might include the number and seniority of people involved in the situation, as well as whether the risk is observable or not. In creating, using and refining risk management systems, it is necessary to understand all of these factors as well as the range of roles that humans will perform to deliver acceptable and possible optimal levels of organisational risk which is discussed next.

3.3.1 Risk perceivers

Risk perceivers are stakeholders who consider and hold a view about a risk or a number of risks. Their point of view is often referred to as their 'risk perception'. A person's perception of risk can be influenced by a number of factors such as those shown in Fig. 3.4.

There are conscious, subconscious and affective factors that can impact risk perceptions. The conscious, subconscious and affective factors may include anticipated consequences, costs, rewards, past experiences, skills and training, individual propensity and cultural norms. In turn, these combine with company and societal factors to shape a persons' risk perceptions, risk assessment and motivation to respond. Such individual influencing factors can be moderated by the situation, organisational objectives, and the exposure of the person. These combinations of factors can influence risk perceptions, the assessment of risk and the motivation to act as shown in Fig. 3.4.

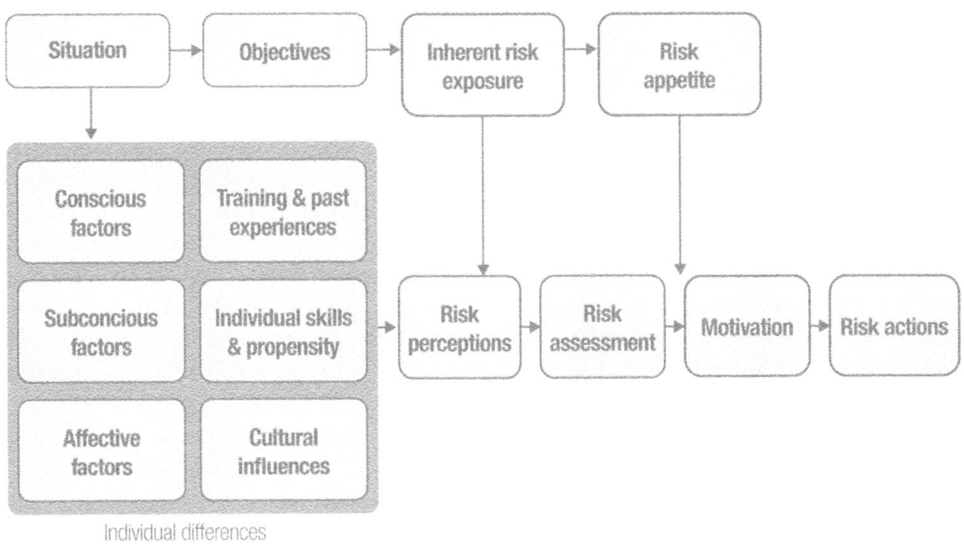

FIGURE 3.4 Risk perception inputs and outputs (Hillson & Murray-Webster, 2012). *Source: Adapted from Hillson, D., & Murray-Webster, R. (2012). A short guide to risk appetite. Burlington, VT: Gower Pub.*

3.3.2 Risk analysers

Risk analysers are the people who apply risk management principles and processes to identify, assess and determine ways of addressing the uncertainties that matter to a business. Some organisations have dedicated risk professionals, some organisations leave risk analysing to subject-matter specialists (e.g. environmental engineers, safety specialists), some organisations make risk analysis a core competency for all key management and technical roles, some organisations outsource the risk analysis and some organisations use a combination of approaches. Other stakeholders can also be important risk analysers for a company. Examples include financial analysts, shareholders, insurance agents, suppliers and customers. Some of the approaches and techniques used for risk analysis in process industry contexts are described in detail in Chapters 5 and 6.

The analysis of risk is influenced by an entities risk appetite. Risk appetite is a measure of how much risk an individual or entity is willing to accept. Some of the differences between high-risk appetite entities and low-risk appetite entities are illustrated in Fig. 3.5.

Risk appetite is the foundation on which risk analysers determine whether a risk is tolerable, too low, or too high. If the risk is too low then an entity might seek more risk to ensure they are not missing out on the opportunities that come from taking risks. If the risk is tolerable then the entity should monitor it to see if it remains tolerable or increases or decreases. If the risk is too high then an entity may seek to transfer it to another entity, manage or control it, or if they can't control it then insure against it. In process industries most risks are inherent to the process operations and therefore need to be managed or controlled by humans. Humans as risk controllers are discussed next.

3.3.3 Risk controllers

Risk controller refers to the role that humans can take in controlling risks in real time. There are two perspectives of humans and risk. There is the perspective that humans are the source of risk within industry. This is evident in findings that the majority of incidents are caused by human error as highlighted in (Pasquale et al., 2015) who stated that the percent failures attributed to human error was approximately 70% for nuclear power plants, 85% for chemical industry and road transport and 90% for air traffic control. These findings have been extended beyond individual humans to organisational error as shown in Fig. 3.3. The key point from this information is that ***human performance dominates risk in high hazard industries***.

The other perspective on humans and risk is that humans are the adaptable resource that controls risk in real time (Hollnagel, 2014). Reason (2008) provides some great studies that illustrate the role humans can play in preventing disasters.

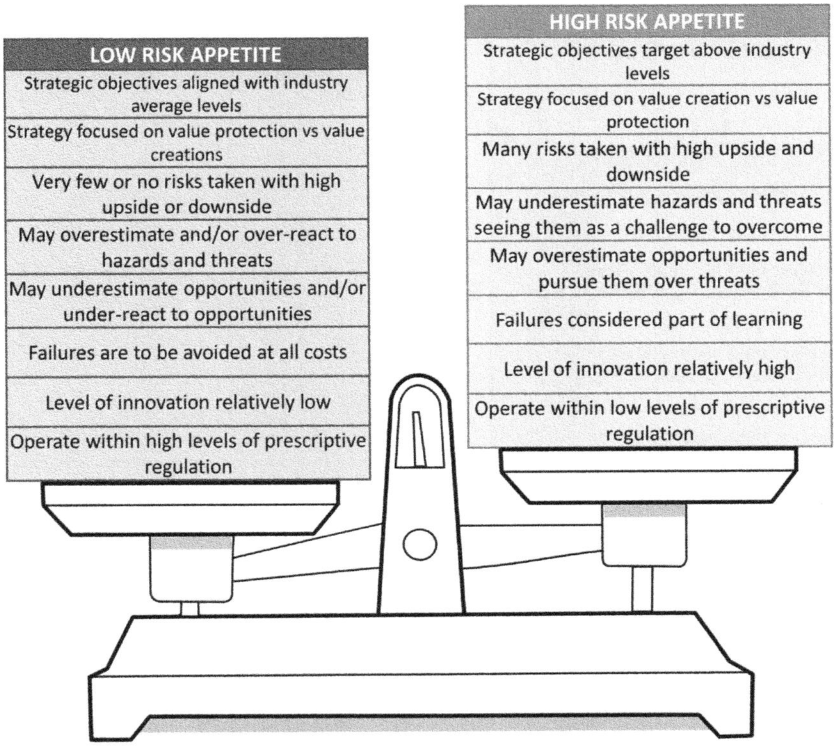

FIGURE 3.5 Factors that characterise high and low organisation risk appetites (Hillson, 2010; Indicators of High/Low level of Risk Appetite for an organisation). *Source: Adapted from Riskspotlight (n.d.). Indicators of High/Low level of risk appetite for an organisation. Retrieved from https://www.riskspotlight.com/risk-infographics/ and Hillson (2010). Exploiting future uncertainty creating value from risk. Farnham, Surrey, England, Burlington, VT.*

3.4 Risk communication

Risk communication involves the exchange of information about what risks exist, their prevalence, causes, consequences and the assessment and treatment of them. It is a prescribed step in the ISO31000 standard for risk management as shown in the left box of Fig. 3.6. According to the ISO31000 standard *'Communication seeks to promote awareness and understanding of risk, whereas consultation involves obtaining feedback and information to support decision-making'* (ISO 31000, 2018).

Risk communication can be formal or informal. It is a human-centred process that involves senders, messages and receivers as shown in Fig. 3.7.

There is also growing recognition that risk communication needs to extend beyond solely technical information such as probability, consequence and exposure type, to incorporate the risk perception and risk appetite aspects of risk. It also needs to extend beyond internal company documents to reach external stakeholders. Risk communication has

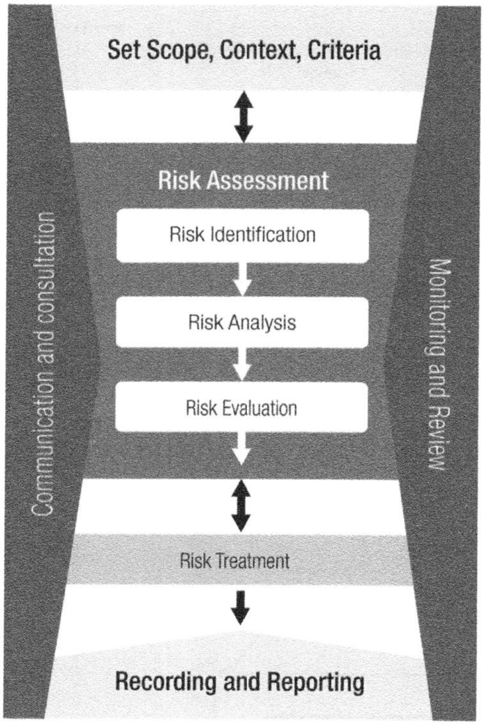

FIGURE 3.6 The risk management process (ISO 31000, 2018). *Source: From ISO 31000:2018 Risk management-Guidelines. Geneva: International Organization for Standardization.*

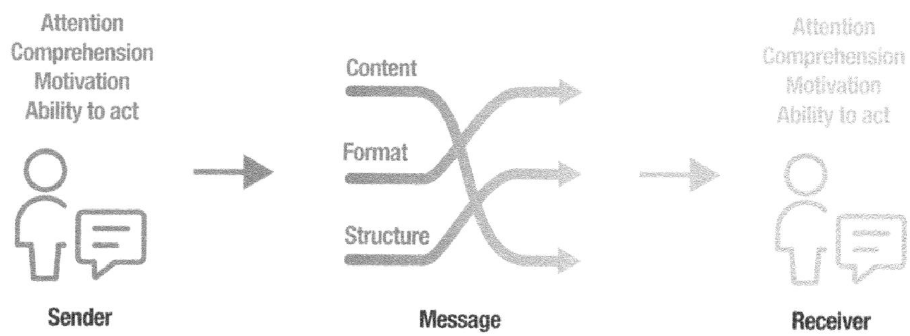

FIGURE 3.7 Risk communication model.

evolved over time. The evolution phases of risk communication as identified by Fischhoff (1995) are shown in Fig. 3.8.

Risk communication can occur across the many modes of communication including face-to-face discussions, printed material, online materials and local/social media as shown in Fig. 3.9. In communicating risk, consideration needs to be given to how risk information is presented to the intended audience so that they can comprehend the significance of the risk and its associated uncertainties in a manner that will help them make timely and

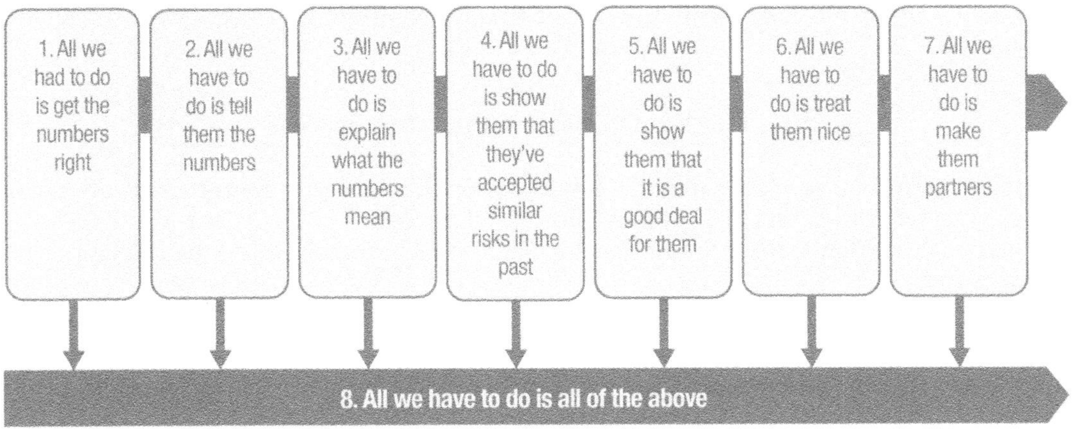

FIGURE 3.8 The evolution of risk communication (Fischhoff, 1995). *Source: Adapted from Fischhoff, B. (1995). Risk Perception and Communication Unplugged: Twenty Years of Process. Risk Analysis 15(2): 137-145.*

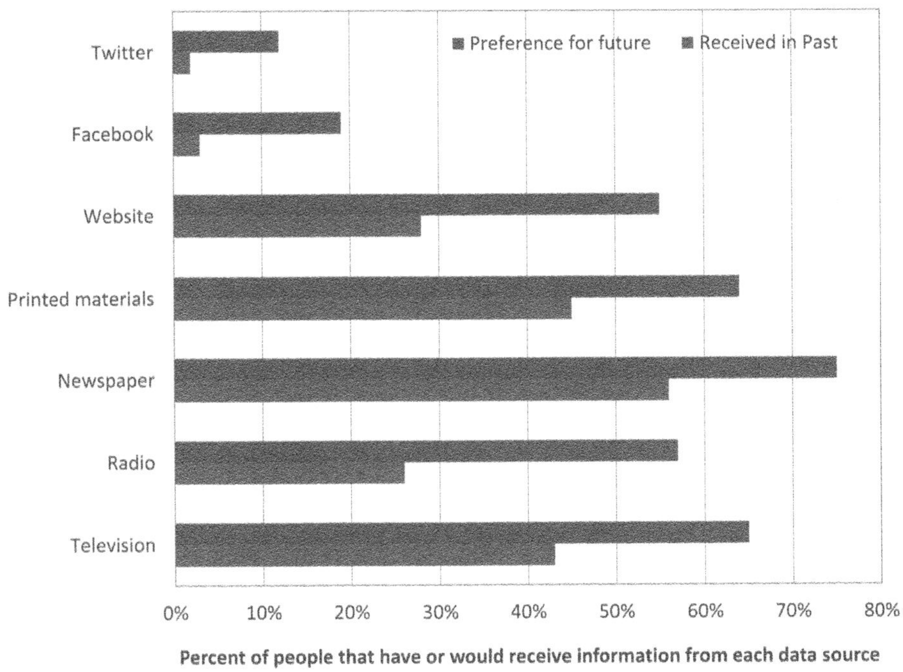

FIGURE 3.9 Methods of risk communication – results from flood research (Feldman, 2016). *Source: From Feldman, D., et al. (2016). Communicating flood risk: Looking back and forward at traditional and social media outlets. International Journal of Disaster Risk Reduction 15: 43–51.*

good quality decisions. Visualisation of risk is becoming a more prevalent way of communicating risk.

3.5 The human decision-making process

Human performance dominates the successful or unsuccessful management of risk in high hazard industries. But, as stated by Kletz (2001, p. 2):

To say that accidents are due to human failing is not so much untrue as unhelpful, for three reasons:

1. Every accident is due to human error
2. Saying an accident is due to human failing is about as helpful as saying that a fall is due to gravity
3. The phrase "human error" lumps together different sorts of failure that require different actions to prevent them happening again.

More recent research (Woods et al., 2010) recognises that the human error approach is limited for the following reasons:

- It is only with hindsight that actions are labelled errors. The same or similar actions might be considered either effective or even heroic if they lead to success, but erroneous if they lead to failure.
- Humans rarely aim to fail. If an accident occurs it is often because issues with the human-system or work design have induced the failure.

Humans can effectively identify and manage risks leading to what some refer to as Highly Reliable Organisations (HROs) (Weick & Sutcliffe, 2015). Therefore to effectively identify and manage risk, we need to design systems that allow humans to accurately perceive, assess, control and communicate risks. This involves understanding the two sides to human performance as shown in Fig. 3.10. In order to create designs that optimise both human wellbeing and overall system performance, designers need to balance the defensive and offensive approaches and design:

- Tolerant systems that prevent and protect against adverse outcomes should deviations in desired performance occur.
- Adaptive systems that support and promote human performance that leads to successful detection and management of unexpected situations.

3.5.1 Situation assessment

At a fundamental level, human decision-making comprises the following steps as shown in Fig. 3.11:

- The ongoing assessment of the situation
- The selection and execution of a response strategy

Correctly understanding the situation that a person is confronted with is key to making correct decisions about how to respond. Situational Assessment entails *'being aware of what*

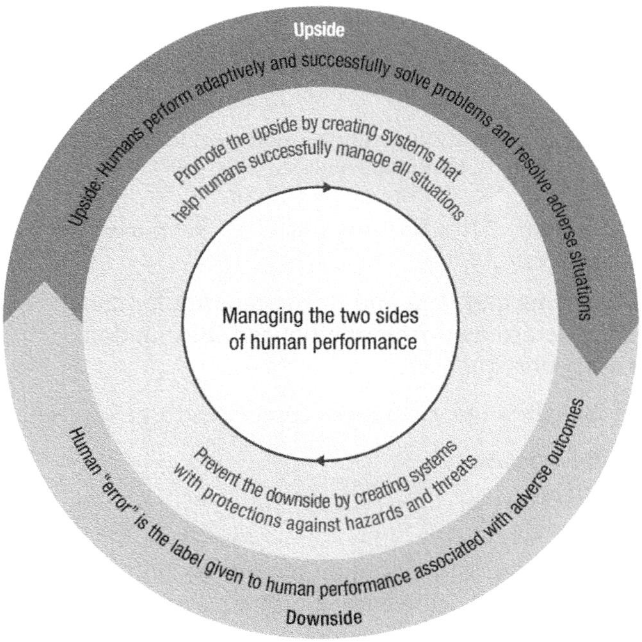

FIGURE 3.10 Designing for humans.

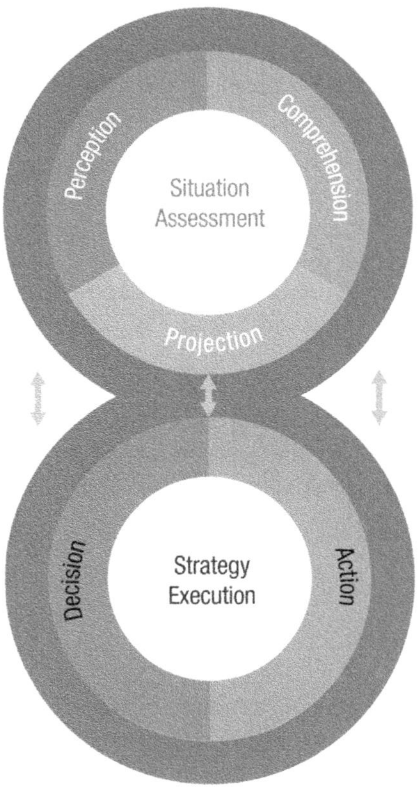

FIGURE 3.11 Components of human performance.

is happening around you and understanding what that information means to you now and in the future' (Endsley & Jones, 2012, p. 12). *Situation assessment can also be described as situational awareness which has been defined as:*

> the perception of the elements in the environment within a volume of time and space, the comprehension of their meaning, and the projection of their status in the near future (Endsley, 1988, p. 97).

Rasmussen and others (Lintern, 2010; Naikar, 2010; Rasmussen, 1976; Rasmussen et al., 1994) specify that situation assessment involves:

- information that is noticed from observations and data and from scanning for cues.
- assessment of current situation which is determined from the noticed information.
- prediction of potential system states and consequences.

Similarly, Endsley (2012, p. 13) states that situation awareness comprises three levels:

- *Level 1: Perception of the elements in the environment*
- *Level 2: Comprehension of the current situation*
- *Level 3: Projection of future status*

In risk management, accuracy in the higher levels of situation awareness will help produce better identification and assessments of risk. However, building and maintaining accuracy in situation awareness can be challenging. Some factors or mechanisms that can confound accurate situation assessment include attentional tunnelling, memory traps, stress (e.g. due to workload, anxiety, fatigue and other stressors), data/information overload, misplaced salience, complexity, erroneous mental models and being out-of-the-loop (Endsley & Jones, 2012). Examples of incidents that were caused, at least in part, by situational assessment issues include:

- Texas City Refinery explosion: In the 2005 Texas City refinery operators were unaware that splitter tower was being overfilled. Through a sequence of events the operators made decisions based on misleading indicators. As a result, they failed to realise that they actually overfilled the column. This resulted in a geyser-type release of hydrocarbon vapours and liquids from a vent stack which ignited killing 15 people and injuring 180 others (U.S. Chemical Safety and Hazard Investigation Board, 2007). It also caused major damage to the refinery.
- The UK Buncefield oil storage depot fire was caused when a fuel tank overflowed and the escaping petrol was ignited and exploded. Employees monitoring the filling operation were not aware of the hazardous situation as the level gauge on a tank was stuck and did not show the high level or activate the high level alarm (COMAH, 2011).

Thus improving designs to enhance situational assessment is crucial for identifying and managing risks within the process industries.

3.5.2 Response strategies

When people make decisions and take actions they can use a number of strategies for doing so. The most commonly known strategies are the fast, intuitive strategies and the slower, deliberative strategies. However, in the high hazard industries it can be useful to think in more detail about the categories of strategies used in order to create designs that promote the response strategies that will lead to success, prevent the response strategies that will lead to failure and tolerate the use of other strategies so that they will not lead to adverse outcomes (Hassall, 2013).

Different people (e.g. experts vs novices, thinkers versus doers, conformists versus mavericks) can have different preferences for decision-making strategies. Different tasks (e.g. complex vs simple, new/novel vs well-practiced and routine) can elicit different decision-making strategy preferences. Different contexts (e.g. high risk vs low risk, easy to predict vs unpredictability, high stress vs low stress, significant time pressure vs little to no time pressure) can also lead people to use different decision-making strategies. The range of strategies that might be employed in risk management activities are described in Table 3.1.

TABLE 3.1 Range of possible decision-making strategies.

Types of strategies	Description
Avoidance strategy	Strategies associated with not doing a task that includes delaying or deferring the task or forgetting to perform the task. Avoidance strategies may increase risk if a situation requires action and no action is given (e.g., response is required to address an emergency but not action taken). They may also decrease risk if the safest option is not to respond (e.g., situation unsafe to start up a plant because control system is down)
Intuitive strategy	Strategies associated with performing a task with little to no explicit or deliberate thought. This includes automatic and habitual responses typically used on familiar tasks such as opening a door at home, opening and shutting commonly used valves in a processing plant, and experienced drivers changing gears when driving their own manual vehicle. Intuitive strategies can be a mentally efficient and effective way for experienced people to deal with familiar situations. However, if the person is not experienced and/or the situation is not exactly the same as the one they are familiar with, then the intuitive strategy deployed might be wrong (e.g., openly a valve an incorrect amount because the valve had been replaced with a different one)
Arbitrary-choice strategy	Strategies associated with a guessed, panicked, random, ad hoc, scrambled, or haphazard response where there is no consideration of the available response options or cues. Arbitrary-choice strategies can be deployed when people are distracted, under stress, fatigued, inexperienced, rushed, or they assess that there is no differences in outcomes so choice of which way to do task does not matter. These strategies can get used to deal with situations like alarm floods, unprecented, beyond-design, situations, and in these circumstances this can increase risks if the choice selected makes the situation worst. Arbitrary-choice strategies can also get used in situations where the response does not matter, because there is negligible risk involved. In these cases, arbitrary choice strategies can be an efficient but possibly not effective way of executing a task
Imitation strategy	Strategies that involve coping another person or the way a task was previously and successfully executed. Training workers in process industries is often done by leveraging the imitation strategies. Operators are taught with training simulators how to successfully response to abnormal situations so they can then copy this response when they are face with similar situations while overseeing the real operations. Tradespersons are often shown how to tackle maintenance issues by first watching an experienced person execute a task then being asked to copy the actions. Imitation strategies can be successful, if the activity steps are easy to remember, the actions required easy to copy and the situation faced is the similar enough to previously

(Continued)

TABLE 3.1 (Continued)

Types of strategies	Description
	observed situation that copying previous actions is appropriate. If these three factors are not presented, then using the imitation strategy can lead to an undesirable outcome
Option-based strategy	Strategies to a task where the worker selects a possible action that meets a minimum requirement, based on a heuristic, rule-of-thumb, or another rule set that does not involved considering cues from system or environment are referred to as option-based strategies. An example of such a strategy is a person choosing to read the first-level measurement found when tasked to read a tank level without checking whether it was right gauge on right tank. It can be difficult to discern whether a given strategy is option-based, intuitive or an imitation strategy. In these circumstances, it matters more that the strategy has been captured in a manner that allows analysis rather than which category of generic strategy it has been recorded against
Cue-based strategy	Approaches to tasks that involve people actively seeking and considering information from the system and environment to determine how to response are cue-based strategies. These strategies work well when the cues are easy to perceive, comprehend, and project, and they are valid. If cues are not readily available, or are misleading or incorrect, then this could result in wrong response selections
Compliance strategy	Strategies that involve following authorised procedures and rules. *'Compliance strategies require workers to make the effort to find, read, understand and execute procedures as they are written'* (Hassall and Sanderson, 2012). This can be beneficial for complex, unfamiliar activities like plant startups if the procedures are correct. However, if the procedures are not correct or are misinterpreted then this can lead to undesirable outcomes. Also the strict compliance to procedures prohibits the adoption of effective "work arounds" or improvements that often are identified by workers
Analytical reasoning strategy	Analytical reasoning strategies involve thinking through and reasoning about the best approach to a given situation or task. These types of strategies may involve using the fundamental principles of the work system, undertaking mental simulations or trial and error thinking to identify and compare options, and/or inductive, deductive or another type of reasoning process to select the best option. As such analytical reasons can be time consuming and mentally demanding and it requires an in-depth and actuate knowledge of system and options to identify an appropriate response. However, it can be the best option available for addressing novel and beyond design type situations

Source: Adapted from (Hassall et al., 2014; Hassall & Sanderson, 2012).

The choice of which decision strategy to use will depend on the human making the choice, the circumstances within which the choice is made and the perceived consequences associated with the choice. Variability in people, circumstances and perceived consequences will exist so we should expect there will be a range of decision-making strategies that will be used and we need to understand and support this range of strategies so that good decisions will be made. Examples of response strategies that impact the risk of a system include:

- Captain Sullenberger seemed to deploy a range of strategies including compliance and intuition and by swapping strategies to respond to the situation was able to land U.S. Airways Flight 1549 on the Hudson in 2009 saving the passengers, crew and airplane.
- The decisions made by NASA to launch Space Shuttle Challenger seem to include a combination of avoidance and imitation which lead to the loss of the spacecraft and crew.

3.6 Improving human performance

Sometimes decisions are made and actions are taken that turn out to be the incorrect ones. This is can happen when it comes to the identification, assessment, treatment and ongoing management of risks. Decision-making bias is one reason for incorrect decisions. As humans, the decisions we make are all subject to bias. There are many types of bias that influence our ability to perceive, comprehend and forecast a situation and there are many types of bias that influence our decision-making processes. Decision-making bias was found to be a contributing factor in the Deepwater Horizon drilling rig disaster in 2010 (Hopkins, 2012).

Improving human performance requires a human-centred approach to design whereby industrial systems are designed to support and enhance human performance, rather than designing the system with the expectation that humans can be trained to deal with any operational complexity thrown at them. Conversely, poor designs do not consider human–system interactions. They do not provide the right people with the correct and timely information needed to diagnose the systems operating state. They do not equip the human and technology system with the capability to successfully control the full range of operating states.

Good designs use human factors approaches to optimise human–system interactions. Human factors (also called ergonomics) is 'the scientific discipline concerned with the understanding of interactions among humans and other elements of a system, and the profession that applies theory, principles, data and methods to design in order to optimise human well-being and overall system performance' (http://www.HFES.org). The term 'human factors' is commonly misused and often misunderstood. The following facts and fictions as quoted from (Russ et al., 2013) can help to better define it:

Fact #1: Human factors is about designing systems that are resilient to unanticipated events.
Fiction: Human factors is about eliminating human error (p.803).
Fact #2: Human factors addresses problems by modifying the design of the system to better aid people.
Fiction: Human factors addresses problems by teaching people to modify their behaviour (p.803).
Fact #3: Human factors work ranges from the individual to the organisational level.
Fiction: Human factors is focussed only on individuals (p.804).
Fact #4: Human factors is a scientific discipline that requires years of training; most human factors professionals hold relevant graduate degrees.
Fiction: Human factors consists of a limited set of principles that can be learnt during brief training (p.804).
Fact #5: Human factors professionals are bound together by the common goal of improving design for human use, but represent different specialty areas and methodological skills sets.
Fiction: Human factors scientists and engineers all have the same expertise (p.805).

Human factors work can include:

- **Physical ergonomics** which focusses on designing jobs, tools, equipment, and workspaces to fit the physical attributes of workers. Physical ergonomics professionals

draw on human anthropometrical, biomechanical and physiological information to design physical environments that accommodate and enhance human sensation and perception, human physical postures, and human musculoskeletal performance (Kroemer & Grandjean, 2009; Marras & Karwowski, 2006).
- **Cognitive ergonomics** which focusses on human mental processing abilities, and limitations. Following the introduction of computerised technology, work has become less physically demanding, and more cognitively demanding. Cognitive ergonomists seek to design tools and systems based on a scientific understanding of attention, perception, memory, mental models, expertise and mental workload in order to enhance human situation awareness, problem solving, and decision-making capabilities (Harris, 2013).
- **Macroergonomics** (also known as organisation ergonomics) which focusses on the optimisation of organisational and work systems design to ensure the organisation as a whole is designed to facilitate safe and effective interactions between technological subsystems, personnel subsystems and the external environment (Hendrick, 2005).

Therefore when we are considering the design of inherently risky systems, it is worth engaging with human factors practitioners in order to develop a human-centred design. Such a design will help people to more successfully perceive, assess, control, manage and communicate about risk. Some common considerations for design interventions include:

- Forcing functions: Forcing functions are aspects of a design that prevents a user from executing an action without consciously completing a previous action or considering important information. An example forcing function is not being able to remove vehicle keys from the ignition until the vehicle is in park and handbrake applied. Forcing functions work well when standardised. However, humans can become reliant on them, which can induce them to make errors in scenarios where the forcing function doesn't exist or has failed.
- Automation and computerisation: Automation and computerisation is when human actions have been allocated to computer-based hardware and software systems. This can help standardise performance but can create other risks associated with human–system interaction or when system drifts beyond design.
- Simplification and standardisation: Simplification and standardisation can help humans monitor and assess situations and choose successful strategies. For example, safety signage and hazardous chemical labelling is standardised and simplified to improve identification and management of risks.
- Reminders, checklists, double checks: The provision of reminders, checklists and double checks is often done to prompt humans to the safety critical elements that need to be checked before, during and after activities. These can help ensure comprehensive monitoring but if not treated seriously they can result in 'tick and flick' activity where the sheets are ticked without the items being properly checked.
- Rules and policies: Example of rules and policies often seen in the process industry are 'Golden Safety Rules' or 'Life Saving Rules' as shown in Fig. 3.12. Other examples exist in written procedures and instructions.
- Training and education: Training and education involves activities were people are given the skills and knowledge associated with their tasks. Training and education can

FIGURE 3.12 Example of life saving rules. (IOGP, 2022). *Source: IOGP, 2022. Life-saving rules. Available from http://www.lifesavingrules.org.*

be effective if the skills learned are practiced and the knowledge taught is retained and correctly recalled in practice.

This list above is referred to as the hierarchy of intervention effectiveness, with the intervention effectiveness reducing from the top (forcing functions) to the bottom (rules, policies and training). A major reason for the reducing effectiveness is that the interventions at the top are system focussed, whereas the interventions at the bottom are people focussed. Interventions that are system focussed are more effective.

3.6.1 An example of designing for improved human performance – overfilling vessels

Overfilling vessels is a recurring cause of accidents in industry (Waite, 2013). In this section, we will us1e the preceding information about *situational awareness* and *decision-making strategies* to inform a human-centred design to assist operational staff when filling vessels such as storage tanks, operational columns, drums, reactors, road tankers, railroad tankers and other transportation containers. You will be required to 'walk in the shoes of the operators', and consider the human–technology interactions.

Consider the case of an operator who is responsible for filling a 40,000 L petrol road tanker at a depot. The tanker is parked at the filling bay, and the operator is responsible for filling the tank with fuel (petrol/gasoline) prior to distributing it to fuel stations. What does a human-centred approach to designing this filling operation look like?

We have just seen that human decision-making comprises *awareness and assessment of the situation*, selecting *a response strategy* to decide how to respond, and then *execution* of that strategy (Fig. 3.11). Given that situational assessment comprises perception, comprehension and projection, in the case of an operator supervising the filling of a tanker, consider the following questions:

1. What elements need to be perceived?
2. How do they need to be comprehended?
3. What types of projection would be useful for ensuring safe operation?
4. What design features could help to enhance the situation awareness for the operator?

What elements need to be perceived? We would want the operator to perceive when fuel is flowing or stopped, the liquid level in the tank, when the tank is full, when there is a loss of containment of fuel and when there is a loss of control of flow (e.g. it goes too high). Various ways of enhancing level 1 perception could be to include flashing lights to indicate fuel is flowing, level indicators on the tanker, vapour monitoring and alarms around the work area.

How do the elements need to be comprehended? We want the operator to determine whether the system is within specified operating parameters, whether it is outside of them but safe, or whether it is outside of them and needs actioning. Flow, pressure, temperature alarms and shut off systems could be used to help the operator comprehend safe versus unsafe situations.

What types of projection would be useful for ensuring safe operation? When the system is safe, we want the operator to anticipate and prepare for shutdown in a timely manner. A display of the planned 'time-to-fill' versus the actual could enhance level 3 awareness.

We now need to consider the decision strategies that the operator may make in response to perceiving, comprehending and projecting a situation that needs an action. The range of decision strategies available to operators are presented in Fig. 3.12 and Table 3.1. Your task *is now to*:

1. Select a strategy that the operator may choose
2. Identify actions that might be associated with this generic strategy
3. Specify if the design should promote, prevent or tolerate this strategy
4. Recommend design improvements

Table 3.2 shows some possible answers to the above questions. This small case study hopefully illustrates that human-centred design can contribute to practical outcomes that can assist to achieve effective and safer task performance.

3.7 Summary

The aim of this chapter was to highlight the importance of humans in risk perception, risk assessment, risk appetite, risk control, risk communication and risk management activities. Human performance dominates the successful or unsuccessful management of risks in high hazard industries. While humans at an individual or organisational level were seen to be a major cause of accidents, we have also seen that human capacity to adapt and solve problems can also be critical for resolving adverse situations. It is thus important to adopt human-centred approaches to risk management, thus designing procedures, systems and plants that enable the human operators to optimise performance.

Identifying and managing risk is about being aware of the situation and the risks associated with it, and then making decisions about how to manage those risks. Understanding decision-making and its components, as well as factors that influence decision makers like the choice of decision-making strategies, bias and safety culture, are crucial for ensuring that good risk-based decisions are made.

Much of the material in this chapter has been quite theoretical and conceptual in nature. This is because our understanding of these areas of risk is still emerging. A lot of contemporary research aims to understand these concepts, and to develop useful techniques and tools to help people identify and manage these aspects of risk in the process industries. The tanker filling exercise, however, has hopefully demonstrated that human-centred design tools can have a genuinely practical application, and that by adopting a different perspective, we may identify beneficial design improvements.

A simple take-home message from this chapter is that *engineers should stop thinking that they can get humans to do things 'the right way', and start designing systems where 'normal/typical' human performance will lead to successful outcomes.* We are not suggesting that this is a trivial task, but it is a good mindset to adopt when designing plant and systems.

Review questions

1. Humans perform a number of roles in the management of risk. What are they?
2. What factors affect how people perceive risks?

TABLE 3.2 Designing improvements in response to possible operator strategies for road tanker filling.

Generic strategy	What decision/actions might be associated with this generic strategy	Should design promote, prevent or tolerate strategy?	Design improvements for both safe and unsafe operations
Avoidance – Not done, defer, or forget to do	Do not fill the tanker	Tolerate	None
	Person does not action or delays actioning of spill	Tolerate	Install automated response, e.g. design pad to collect spills, install deluge system to treat spills
Intuitive – Automatic response, done without explicitly or deliberately using thought processes	Fill truck without checking on the status	Tolerate	Install alarms and automated shutdown for loss of control/containment situations
	Assume others know status of operation and will respond correctly	Tolerate	Install exclusion zones to warn others of hazardous nature of tank filling operations
Arbitrary-choice – Guessed, scrambled haphazard or panicked response	Random choice of the fuel to fill the truck	Prevent	Use unique nozzles to ensure only correct fuel goes into tanker
	Guess how to treat a spill	Prevent	Provide regular spill response training/drills
Imitation strategies – Copy how others do it or copy what has worked in the past	Use same filling process without considering changes (in tankers, in equipment and/or with fuels)	Tolerate	Use tankers equipped with RFID or barcode scanning that is used to automatically set up fill parameters within the control system
Cue-based strategies – Select chosen option using the observed info/cues and predict consequences	Operator tracks fill rate and fill level throughout the entire filling process	Promote	Provide operator console with plan versus actual fill display with forecast completion time
	Operator monitors truck and pad for signs of leaks	Promote	Provide cameras and fume monitors with feedback showing on operator console
Compliance-based strategies – Following procedures as they are written/practiced	During normal and expected conditions, operator follows planned fill procedures and planned emergency response if required	Promote	Make process simple and standard and provide easy to follow checklists/procedures
	During unexpected conditions, where operations are beyond scope of design and procedures	Tolerate	Provide alarms and automatic shutdown to indicated beyond design conditions and operator should call for backup
Analytical Reasoning strategies – Using analytical thinking to reason out the best way to perform task	For use during fault finding operations	Tolerate	As above

3. How has the nature of the work of human operators in industry changed over time?
4. What does good situation awareness require?
5. We can use a variety of 'interventions' to prevent users from making mistakes. For example, education and training is a common method employed to help people to employ the right behaviours. List the following interventions in order of effectiveness (LEAST effective first and finishing with the MOST effective).
   - Forcing functions
   - Education and training
   - Checklists
   - Automation
   - Rules and policies
   - Standardisation
6. State three approaches that address *organisational factors* for accidents.
7. Approximately, what % of chemical industry failures are attributed to human error?
8. State four characteristics for organisations with a high-risk appetite.
9. What roles do humans adopt in regard to risk in industry?
10. What are the three stages of Situation Assessment?

References

Booth, W. (1987). Postmortem on Three Mile Island. *Science*, *238*(4832), 1342–1345.

Borys, D., Else, D., & Leggett, S. (2009). The fifth age of safety: The adaptive age? *Journal of Health & Safety Research & Practice*, *1*(1), 19–27.

Chapanis, A. (1996). *Human factors in systems engineering*. Wiley. Available from http://uq.summon.serialssolutions.com/2.0.0/link/0/eLvHCXMwY2BQAGb_FHODZKPUVEvwzsdEY4MU00TTRNCpmOYWSQYoYx1IpbmbEANTap4og7Sba4izh25pYTx0CCM-CVgLA5uywNpPjIEF2C1OFWdgTQNGD5AGFpniQO3iDBwRluGhFpHefhCuEIyrVwzevqRXWCIOLKHBsatrrGcAAGFXJwA.

Columbia Accident Investigation Board. (2003). *Columbia accident investigation board report volume 1*. https://history.nasa.gov/columbia/Troxell/Columbia%20Web%20Site/CAIB/CAIB%20Website/CAIB%20Report/Volume%201/introduction.pdf.

COMAH. (2011). *Buncefield: Why did it happen?* UK: UK Health and Safety Executive on behalf of the Competent Authority for the Control of Major Accident Hazards.

Embrey, D. (1986). *SHERPA: A systematic human error reduction and prediction approach* International Meeting on Advances in Nuclear Power Systems, Knoxville, TN.

Endsley, M. R. (1988). Design and evaluation for situation awareness enhancement. In *Proceedings of the 23rd Annual Meeting of the Human Factors and Ergonomics Society* (pp. 97–101). Human Factors and Ergonomics Society.

Endsley, M. R. (2012). *Situation awareness. Handbook of human factors and ergonomics* (pp. 553–568). John Wiley & Sons, Inc. Available from https://doi.org/10.1002/9781118131350.ch19.

Endsley, M. R., & Jones, D. G. (2012). *Designing for situation awareness: An approach to user-centered design*. CRC Press.

Feldman, D., et al. (2016). Communicating flood risk: Looking back and forward at traditional and social media outlets. *International Journal of Disaster Risk Reduction*, *15*, 43–51.

Fischhoff, B. (1995). Risk perception and communication unplugged: Twenty years of process. *Risk Analysis*, *15*(2), 137–145. Available from https://doi.org/10.1111/j.1539-6924.1995.tb00308.x.

Hale, A. R., & Hovden, J. (1998). Management and culture: The third age of safety. A review of approaches to organizational aspects of safety, health and environment. In A. M. Feyer, & A. Williamson (Eds.), *Occupational Injury. Risk Prevention and Intervention*. Taylor & Francis.

Harris, D. (2013). *Engineering psychology and cognitive ergonomics: understanding human cognition: 10th International Conference, EPCE 2013, held as part of HCI International 2013, Las Vegas, NV, USA, July 21-26, 2013, Proceedings*. (8019-8020). Springer.

Hassall, M. E. (2013). *Methods and tools to help industry personnel identify and manage hazardous situations*. Queensland, Australia: The University of Queensland.

Hassall, M. E. (2015). *Improving human control of hazards in industry*. 19th Triennial Congress of the IEA 9-14 August 2015, Melbourne, Australia.

Hassall, M. E., & Sanderson, P. M. (2012). A formative approach to the strategies analysis phase of cognitive work analysis. *Theoretical Issues in Ergonomics Science*, 1–47. Available from https://doi.org/10.1080/1463922X.2012.725781.

Hassall, M. E., Sanderson, P. M., & Cameron, I. T. (2014). The development and testing of SAfER: A resilience-based human factors method. *Journal of Cognitive Engineering and Decision Making*, *8*(2), 162–186. Available from http://edm.sagepub.com/content/8/2/162.abstract.

Hassall, M. E., Sanderson, P. M., & Cameron, I. T. (2016). Incident analysis: A case study comparison of traditional and SAfER methods. *Journal of Cognitive Engineering and Decision Making*, *10*(2), 197–221. Available from https://doi.org/10.1177/1555343416652749.

Hassall, M. E., Xiao, T., Sanderson, P. M., & Neal, A. (2015). Human factors and ergonomics. In (2nd ed. J. D. Wright (Ed.), *International Encyclopedia of the Social & Behavioral Sciences* (Vol. 11 Elsevier. Available from http://doi.org/10.1016/B978-0-08-097086-8.22025-4.

Heinrich, H. W. (1941). *Industrial accident prevention: A scientific approach*. McGraw-Hill book company, inc. Available from http://uq.summon.serialssolutions.com/link/eJxjYLALDvX19ffTDXYNDvYE0rnlqUmGegVF-Sl6xaW5ufl5esWpRZmJOcXF-TmlJZn5ecV6yfm5uubGBha6RmamjAxczm6mAb4WwVFGPkwMGhklJQVW-vo5mUlFiUWVeqWFeqkppXqJpfpFqcn5RSm2SYYmZhamxsYsDMxGhgHcDEylhTwMjIa8QMzXsSzkZ7izd3ZXqPWhzyee1gMA7g8w7A.

Hendrick, H. W. (2005). An overview of macroergonomics. In N Moray (Ed.), *Ergonomics: Major writings*. In: (Vol. 1 - The history and scope of human factors). Taylor and Francis.

Hillson, D. (2010). *Exploiting future uncertainty creating value from risk*. Burlington, VT.

Hillson, D., & Murray-Webster, R. (2012). *A short guide to risk appetite*. Burlington, VT: Gower Pub.

Hollnagel, E. (1998). *Cognitive reliability and error analysis method: CREAM*. Elsevier. Available from http://uq.summon.serialssolutions.com/link/0/eLvHCXMwY2BQME5JSzNPMzZOSjEA5h5gAWhhbpSYbJAMdHSyWWpqI-spYB1Jp7ibKIO3mGuLsoVtaGA8dvYhPAu0ENQO2YwzFGHgTQau-80rAu8NSAILRG8M.

Hollnagel, E. (2012). *FRAM: the functional resonance analysis method - modelling complex socio-technical systems*. Ashgate.

Hollnagel, E. (2014). *Safety-I and safety-II: The past and future of safety management*. Ashgate Publishing Limited. Available from http://uq.summon.serialssolutions.com/2.0.0/link/0/eLvHCXMwdV1LT8JAEJ4IJEZOIFIRTPoH2uyr2-XgCSQQuRg18UZKO3s0QcrBf-_stk0qwdPm28tuNjvPzHwDIEXMojOdIBIrMiXyTCEaxXNbpCRqOZPIhJ37kUvt9EczZdEVWX68bv8WTP9HXvvEydaQxelAx3gZbUieq3SLU8Tct3alzodgWjSMTzXmrU5Ib11WA-ihazkYwhV-3UK_RRE4gvFbZrH8iTYhhfzhsQabO5iunt8X6-h02NUZmN3exYFMJlqOoUtRPQbQs_S7aCWNH9BpAVx_zrdLs35ZVHDYwPjou6_iQxmQgfGfM9Jxeg8hl0rnezSolVGJYxnMVFEYx7CmXJQzgdGlOzxc3p7CDfkCqsouzKBbfp_w0T3IL34wc_A.

Hollnagel, E., Woods, D., & Leveson, N. G. (Eds.), (2006). *Resilience engineering concepts and precepts* (Vol. 1). Ashgate.

Hopkins, A. (2012). *Disastrous decisions: The human and organisational causes of the Gulf of Mexico blowout*. CCH Australia Limited.

IOGP (2022). Life-saving rules. https://www.lifesavingrules.org/ [Accessed 22 Sep 2022].

ISO 31000 (2018). *ISO 31000:2018 Risk management-Guidelines*. Geneva: International Organization for Standardization.

Kletz, T. A. (2001). *An engineer's view of human error* (3rd ed.). Institution of Chemical Engineers.

Kletz, T. A. (2009). *What went wrong? Case histories of process plant disasters and how they could have been avoided*. Houston. Texas, USA.: Gulf Publishing Company.

Kroemer, K. H. E., & Grandjean, E. (2009). *Fitting the task to the human: a textbook of occupational ergonomics* (5th). Taylor & Francis.

Lee, J., Cameron, I., & Hassall, M. (2019). Improving process safety: What roles for digitalization and industry 4.0? *Process Safety and Environmental Protection*, *132*, 325–339. Available from https://doi.org/10.1016/j.psep.2019.10.021.

Leveson, N. G. (2011). *Engineering a safer world: Systems thinking applied to safety*. The MIT Press. Available from http://uq.summon.serialssolutions.com/link/0/eLvHCXMwY2BQAFaaqRaWhpZpSSYWyYkmicAqM8XYyDg1LSkxMTUlyQJlrAOpNHcTZZB2cw1x9tAtLYyHjl7EJxkDazQjU2MjQzEG3kTQqu-8EvDusBRxBtY0YBSlioMmLcWBRogzcERYeriYOPr4QrhcMK5eojgDs7GOCQCJ7yQK.

Lintern, G. (2010). A comparison of the decision ladder and the recognition-primed decision model. *Journal of Cognitive Engineering and Decision Making*, 4(4), 304–327. Available from http://edm.sagepub.com/content/4/4/304.abstract.

Marras, W. S., & Karwowski, W. (2006). *The occupational ergonomics handbook*. CRC/Taylor & Francis.

Maslin, E. (2016). PDA on the drillfloor. *Offshore Engineer, April 1, 2016*. Available from https://www.oedigital.com/news/449820-pda-on-the-drillfloor.

Naikar, N. (2010). *A Comparison of the Decision Ladder Template and the Recognition-Primed Decision Model*. Air Operations Division Defence Science and Technology Organisation DSTO-TR-2397.

OECD. (2010). *Radioactive waste in perspective*. OECD Publishing. Available from https://www.oecd-nea.org/ndd/pubs/2010/6350-waste-perspective.pdf.

Parasuraman, R., & Wickens, C. D. (2008). Humans: Still vital after all these years of automation. *Human Factors: The Journal of the Human Factors and Ergonomics Society*, *50*(3), 511–520. Available from https://doi.org/10.1518/001872008x312198.

Pasquale, V. D., Miranda, S., Iannone, R., & Riemma, S. (2015). A simulator for human error probability analysis (SHERPA). *Reliability Engineering & System Safety*, *139*, 17–32. Available from https://doi.org/10.1016/j.ress.2015.02.003.

Pocock, S., Wright, P., & Harrison, M. (1999, 1–2 December). *THEA – A technique for human error assessment early in design*. RTO HFM Workshop on the Human Factor in System Reliability – Is Human Performance Predictable? Siena, Italy.

Rasmussen, J. (1976). *Data, models, and strategies in human information processing (Report No: N-8-76)*.

Rasmussen, J. (1997). Risk management in a dynamic society: A modelling problem. *Safety Science*, *27*(2–3), 183.

Rasmussen, J., Pejtersen, A. M., & Goodstein, L. P. (1994). *Cognitive systems engineering*. Wiley. Available from http://library.uq.edu.au/record = b2441055.

Rasmussen, J., & Svedung, I. (2000). *Proactive risk management in a dynamic society*. Risk & Environmental Department, Swedish Rescue Services Agency.

Reason, J. T. (2008). *The human contribution: Unsafe acts, accidents and heroic recoveries*. Ashgate. Available from http://library.uq.edu.au/record = b2443566.

Riskspotlight.com. Available from https://www.riskspotlight.com/risk-infographics/ [Accessed 7 Feb 2023].

Russ, A. L., Fairbanks, R. J., Karsh, B.-T., et al. (2013). The science of human factors: separating fact from fiction. *BMJ Quality & Safety*, *22*(10), 802–808. Available from https://doi.org/10.1136/bmjqs-2012-001450.

Shappell, S. A., & Wiegmann, D. A. (2000). *The human factors analysis and classification system - HFACS* (DOT/FAA/AM-00/7).

Shaver, E. (2009). A short history of human factors and ergonomics. *The Human Factor Advocate* (January).

Sheridan, T. B. (2002). Some musings on four ways humans couple: Implications for systems design. *Systems, Man and Cybernetics, Part A: Systems and Humans, IEEE Transactions on*, *32*(1), 5–10. Available from https://doi.org/10.1109/3468.995525.

Sheridan, T. B. (2011). Adaptive automation, level of automation, allocation authority, supervisory control, and adaptive control: Distinctions and modes of adaptation. *IEEE Transactions on Systems, Man, and Cybernetics - Part A: Systems and Humans*, *41*(4), 662–667. Available from https://doi.org/10.1109/TSMCA.2010.2093888.

Shorrock, S. T., & Kirwan, B. (2002). Development and application of a human error identification tool for air traffic control. *Applied Ergonomics*, *33*(4), 319–336.

Slovic, P. (1987). Perception of risk. *Science*, *236*(4799), 280–285.

Slovic, P., Fischhoff, B., & Lichtenstein, S. (1982). Why study risk perception? *Risk Analysis*, *2*(2), 83–93. Available from https://doi.org/10.1111/j.1539-6924.1982.tb01369.x.

Stanton, N. A., Salmon, P. M., Rafferty, L. A., Walker, G. H., Baber, C., & Jenkins, D. P. (2013). Human factors methods (2nd ed.). Ashgate Publishing Ltd. Available from https://doi.org/10.1201/9781315587394.

The National Diet of Japan. (2012). *The official report of the Fukushima nuclear accident independent investigation commission: Executive summary*. The National Diet of Japan. http://warp.da.ndl.go.jp/info:ndljp/pid/3856371/naiic.go.jp/en/report/.

U.S. Chemical Safety and Hazard Investigation Board. (2007). *Investigation report - Refinery explosion and fire - BP, Texas City, March 23, 2005*. U.S. Chemical Safety and Hazard Investigation Board. Available from http://www.csb.gov/investigations/completed-investigations/.

U.S. Department of Energy. (2009). *DOE Standard: Human performance improvement handbook - Volume 1: Concepts and principles (DOE-HDBK-1028-2009)*. Washington D.C: U.S. Department of Energy. Available from https://www.standards.doe.gov/standards-documents/1000/1028-BHdbk-2009-v1/@@images/file.

Waite, P. (2013). Recurring accidents: Overfilling vessels. *The Chemical Engineer* (861), 40–44.

Weick, K. E., & Sutcliffe, K. M. (2015). *Managing the unexpected: Sustained performance in a complex world* (3rd ed.). Hoboken, New Jersey: Wiley.

Williams, J. (1986). HEART - A proposed method for assessing and reducing human error. Ninth Advances in Reliability Technology Symposium, University of Bradford.

Woods, D. D., Dekker, S., Cook, R., Johannesen, L., & Sarter, N. (2010). *Behind human error*. Surrey, England: Ashgate.

Woods, D. D., & Hollnagel, E. (2006). Prologue: Resilience engineering concepts. In E. Hollnagel, D. Woods, & N. Leveson (Eds.), *Resilience engineering concepts and precepts* (pp. 1–6). Ashgate.

Zionchenko, V., & Munipov, V. (2005). Fundamentals of ergonomics. In N. Moray (Ed.), Ergonomics: Major writings *(Vol. 1 - The history and scope of human factors* (pp. 17–37). Taylor and Francis.

CHAPTER

4

Professional practice

OUTLINE

4.1 Introduction 72

4.2 What is professional practice? 72

4.3 What is a professional engineer? 73

4.3.1 Ethics 74

4.3.2 Competence 75

4.3.3 Performance 78

4.4 Obligations, accountabilities and responsibilities 78

4.4.1 Legal and regulatory obligations 78

4.4.2 Accountabilities 80

4.4.3 Responsibilities 81

4.5 Case study: dieselgate – VW emissions scandal 82

4.6 Summary 82

Review questions 83

References 83

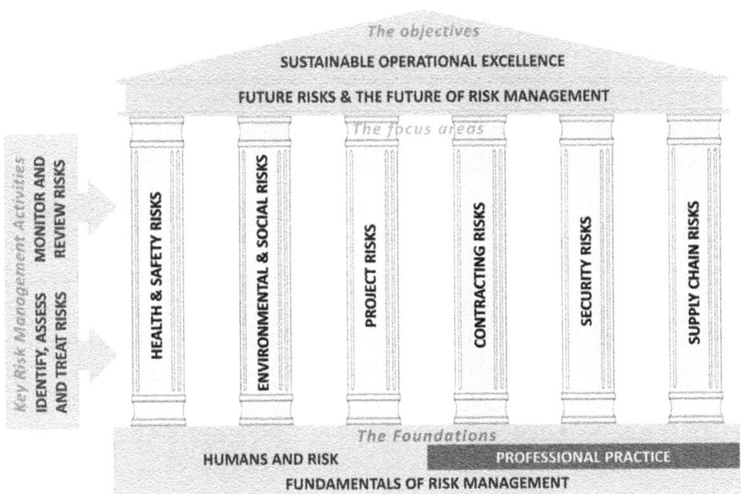

The aim of a chemical engineer is to be of service to the community and society expects the highest professional standards. Ethics lies at the heart of our discipline (Institution of Chemical Engineers, 2023)

Fundamentals of Risk Management for Process Industry Engineers
DOI: https://doi.org/10.1016/B978-0-12-820320-0.00007-6

4.1 Introduction

In the previous chapter we discussed the importance of the role of humans in managing process industry risk. Engineers, in particular, are crucial for facilitating the risk identification, analysis and control of process industry risks. As a result, and as mentioned in Chapter 1, risk management is a core competency of all engineers. Engineers are employed to *facilitate continual improvement of the organisation.* To do so, professional engineers are responsible for the identification, assessment and management of risks associated with all aspects of a business. This will include the planning and execution of projects, the operation of processing plants and management of other aspects of the business including its environment and community interactions required to obtain and sustain operational excellence.

This chapter examines the concept of process engineering professional practice. Specifically, this chapter should help you answer the following questions:

- What does it mean to be a professional engineer?
- Why is it important in managing process industry risk?

At the end of this chapter, you should be able to describe what it means to be a professional engineer, and why your professional practice is important for how you will manage risk. You should also be able to discuss your own professional risk, and how you plan to manage it. We also hope to encourage you to adopt a professional attitude of self-development.

4.2 What is professional practice?

We can use Aristotle's terms and framework, discussed in Chapter 1 to describe engineering practice. As shown in Fig. 4.1, the three tiers of knowledge comprise:

1. Underpinning knowledge and understanding (i.e. *episteme* and *sophia*)
2. Engineering tools (i.e. *techne*)
3. Professional practice (i.e. *phronesis and praxis*)

Undergraduate engineering education predominantly focuses on *episteme, sophia* and *techne*. In many ways, this makes sense, as it is critical that engineers do have the fundamental knowledge and tools to enable them to execute their tasks. However, Fig. 4.1 helps to illustrate that being a professional engineer requires much more as it also relies on *praxis* and *phronesis*. That is, professional engineers should be able to demonstrate *prudent understanding of what should be done in a practical situation*, and should also demonstrate *practical, thoughtful doing*. Of course, this is knowledge and behaviour that is not acquired in a classroom, but necessitates many years of development and mentoring in the workplace.

In presenting this knowledge framework, we aim to impart a broader understanding of what it is to be professional, and acknowledge that it will take many years of post-university training and development to get there.

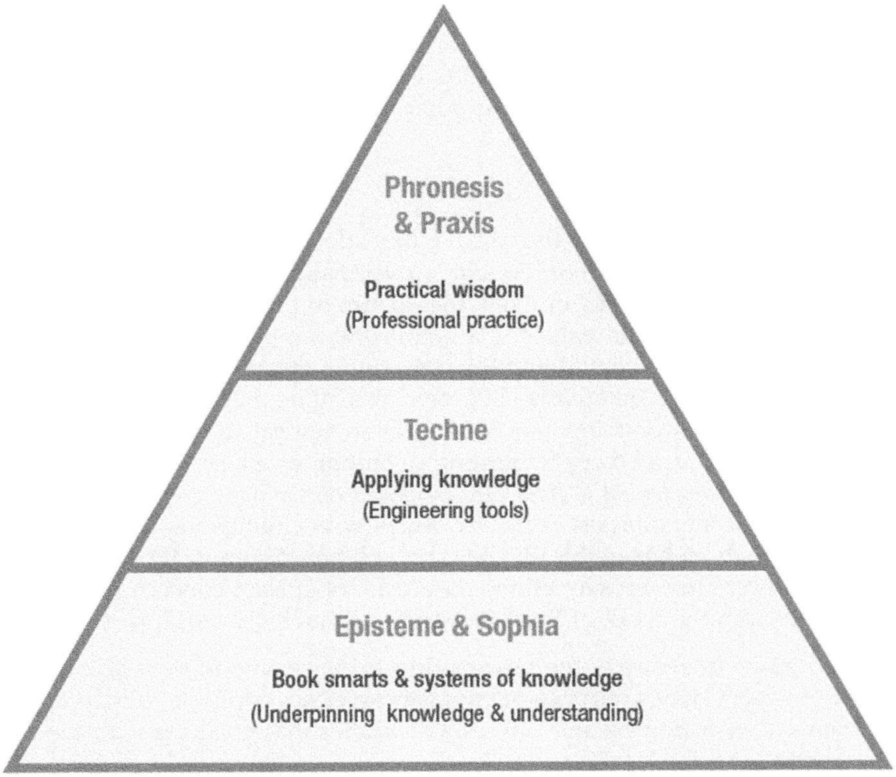

FIGURE 4.1 Three-tiered system of knowledge.

4.3 What is a professional engineer?

Typically, professional engineers are those that:

- have passed a 4-year accredited engineering degree.
- completed 4–5 years of engineering practice under the supervision or mentorship of a professional engineer.
- work within the bounds of their competency to high ethical, quality and transparency standards.
- are able to demonstrate that they are continuously maintaining and improving their professional competencies and skills.

The Warren Centre for Advanced Engineering (2009) proposed a concise framework of professional engineering that comprised three dimensions: *ethics*, *competency* and *performance* (Fig. 4.2).

Increasingly around the world, laws are being enacted that require engineers to become licensed or registered to practice as a professional engineer (e.g. Professional Engineers

Act 1990, Ontario Canada (Government of Ontario., 2017); Professional Engineers Act 2002, Qld Australia (Queensland Government., 2013); Professional Engineers Act 1991, Singapore (Government of Singapore., 2020). The laws are often enforced by regulatory bodies to protect public welfare, safety and other public interests.

4.3.1 Ethics

A standard dictionary will define the word 'ethics' along the lines of being 'the moral system of a particular writer or school of thought', or 'the rules of conduct recognised in certain limited departments of human life' or even 'the science of human duty in its widest extent'. In essence, ethics is about moral values – a moral philosophy or set of moral principles – that express in a formal way what 'doing the right thing' means. However, because ethics is about moral values, it is also inherently subjective. It may be that the appropriate prevailing moral values in one social or religious context are somewhat different in another. Clancy and Luegenbieghl (2017) use a very interesting definition of ethics, which is that *ethics concerns actions that have the potential to have serious impact on the lives of others*. This broad view of ethics makes it clear that ethics is implicitly linked with engineering decision-making.

Shouldn't everybody behave ethically, so why is it any different for engineers? The following compelling reasons for why ethics should be of special concern to engineers have been summarised from the work of Rockwell and Luegenbieghl (2017, p. 8):

- Engineers are often unaware of the powers they exercise, and of their abilities to affect the lives of others. A major purpose of engineering ethics could be to help engineers to become aware of their powers and the responsibilities that go along with the exercise of those powers.
- Engineers have specialised knowledge and skills that are acquired through long periods of intensive study and training. The public does not understand what engineers do, and

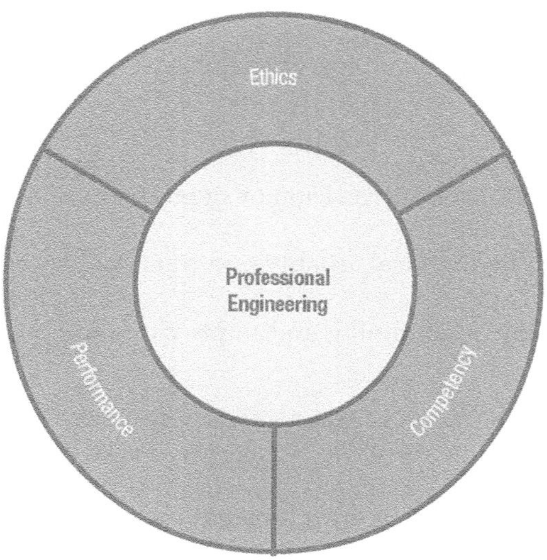

FIGURE 4.2 The three dimensions of professional engineering (The Warren Centre for Advanced Engineering). *Source: Adapted from The Warren Centre for Advanced Engineering (2016).* PPIR: a program for professional performance. *The Warren Centre for Advanced Engineering. https://www.engineersaustralia.org.au/sites/default/files/2022-06/ppir-program-for-professional-performance.pdf.*

thus exists in a relationship of dependence on engineers. It must trust the work engineers do on its behalf.
- The specific nature of engineering means that engineers will face ethical challenges that are not applicable to other people. For example, engineering activities may directly impact on the well-being of society and the environment. Society expects engineers to act in their best interests while performing their duties.
- Engineers often work in business environments within which they can be exposed to undue pressures to which they may respond inappropriately.

It is thus critical that ethics in engineering is given special consideration. When we describe ethics in a professional context, we usually express these professional values as a 'higher duty' that transcends differences in social or religious values, yet responds to the generic interests of the community and pays proper respect to the inherent dignity of the individual. It is usual for a profession to set out its version of these higher duty professional values in a document such as the American Institute of Chemical Engineers Code of Ethics (American Institute of Chemical Engineers, 2015) or the National Society for Professional Engineers Code of Ethics (National Society of Professional Engineers, 2019).

The preamble in the USA National Society of Professional Engineers (2019) Code of Ethics states:

> Engineering is an important and learned profession. As members of this profession, engineers are expected to exhibit the highest standards of honesty and integrity. Engineering has a direct and vital impact on the quality of life for all people. Accordingly, the services provided by engineers require honesty, impartiality, fairness, and equity, and must be dedicated to the protection of the public health, safety, and welfare. Engineers must perform under a standard of professional behaviour that requires adherence to the highest principles of ethical conduct.

Often risks emerge where there are ethical dilemmas that need resolving. Figs. 4.3 and 4.4 illustrate some of the ethical decision considerations. It is worth using these figures to consider the Volkswagen 'dieselgate' scandal (see Section 4.5 later in this chapter). In that case, the behaviour of some people was both unethical and illegal, and we can surmise that the external pressures and personal benefits created a situation where there was a very high risk of this behaviour occurring.

As an engineer working in industry, you have the power to seriously impact on society, and you have the skills to act in societies best interests. Ethics is clearly implicit in almost everything that you will do as an engineer, and risk management provides a framework within which you can professionally (ethically) perform your engineering tasks.

4.3.2 Competence

> Competence is 'a professionals ability to carry out engineering tasks successfully and safely within their field of practice. This includes having the individual skills, knowledge and understanding, personal behaviour and approach, to be able to work collaboratively with others to achieve the intended outcomes. Competence includes the ability to make professional judgments and an awareness of the limits of one's own ability and knowledge in order to seek assistance when required (Engineering Council., 2020, p. 7). Competence is developed by a combination of formal and informal learning, and training and experience, generally known as initial professional development. (Engineering Council, 2014, p. 6).

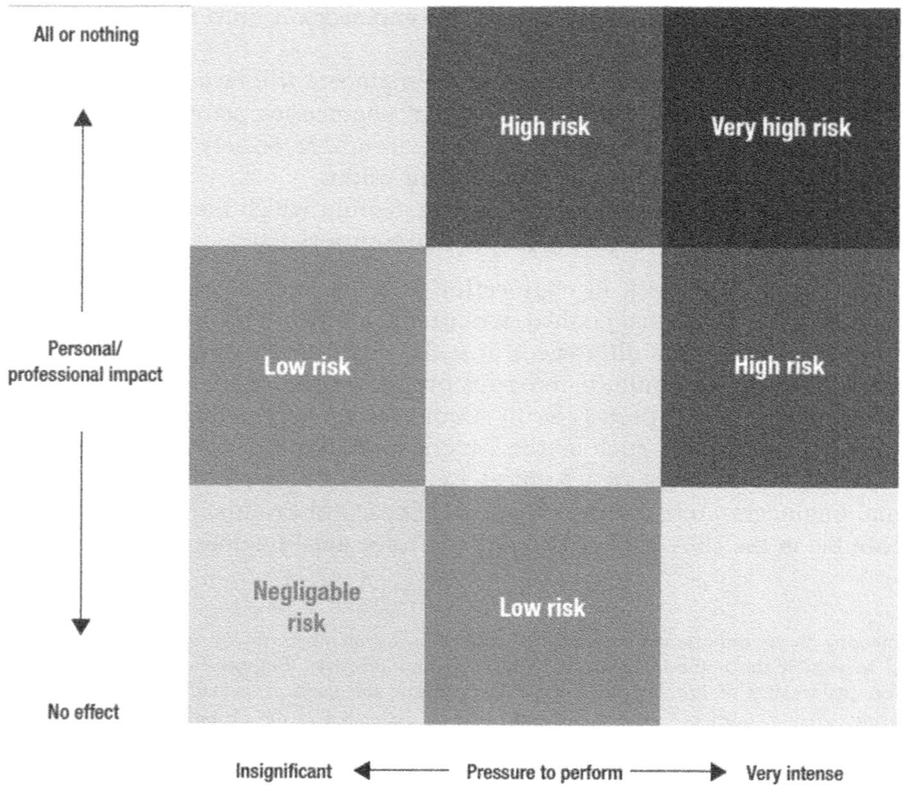

FIGURE 4.3 Ethical dilemma grid. (nparkhaev, 2013). *Source: Adapted from nparkhaev. (2013). Product Marketing In Hospital. Retrieved from http://blogs.ubc.ca/nparkhaev/2013/03/24/product-marketing-in-hospital/.*

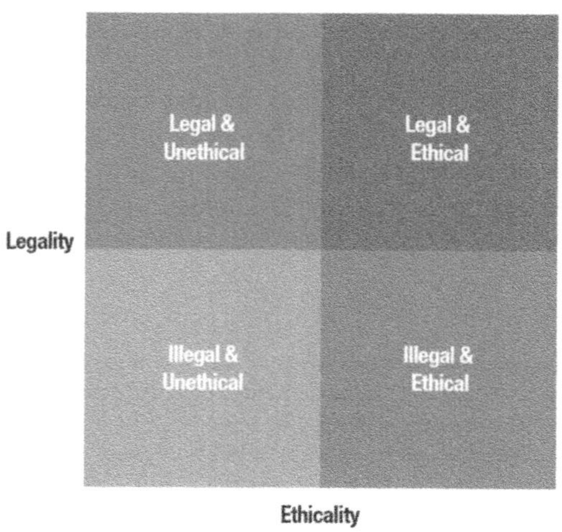

FIGURE 4.4 Ethical decision grid (Dowling, 2020). *Source: Adapted from Dowling, D. (2020).* Engineering your future an Australasian guide *(Fourth edition.). John Wiley & Son Australia, Ltd.*

The Warren Report (The Warren Centre for Advanced Engineering, 2009) distinguishes between two levels of competence:

1. Gateway competence
2. Task-specific competence

Gateway competence refers to acquisition of the specified (accredited) university degree, with associated workplace practice, which is required in order to become a *qualified* engineer. There are also higher levels of gateway competence such as Chartered Professional Engineer status.

Task-specific competence refers to competence required to perform a particular task. Even if you are a qualified engineer, you need to continuously assess your level of competence for particular tasks, and continue to develop your skills and knowledge. The management of risk is recognised as a core competence for engineers. For example, the UK Engineering Council lists the following as examples of risk-related competencies that could be employed by engineers:

- Contribute to the design and development of engineering solutions which includes Identifying operational risks and evaluating possible engineering solutions, taking account of cost, quality, safety, reliability, accessibility, appearance, fitness for purpose, security (including cyber security), intellectual property constraints and opportunities, and environmental impact (Engineering Council., 2020, p. 26)
- Plan the work and resources needed to enable effective implementation of engineering tasks and projects which includes carrying out holistic and systematic risk identification, assessment and management (Engineering Council., 2020, p. 27)
- Understand the safety implications of their role and manage, apply and improve safe systems of work which includes Identifying and taking responsibility for your own obligations for health, safety and welfare issues; Managing systems that satisfy health, safety and welfare requirements; Developing and implementing appropriate hazard identification and risk management systems and culture (Engineering Council., 2020, p.29)
- Take an active role in the identification and definition of project requirements, problems and opportunities which includes carrying out technical risk analysis and identifying mitigation measures (Engineering Council, 2020, p.33)
- Can implement engineering tasks and evaluate the effectiveness of engineering solutions which includes "Implementing design solutions, taking account of critical constraints, including due concern for safety, sustainability and disposal or decommissioning; Identifying and implementing lessons learned; Evaluating existing designs or processes and identifying faults or potential improvements including risk, safety and life cycle considerations; Actively learning from feedback on results to improve future design solutions and build best practice (Engineering Council., 2020, p. 34)

Similarly, Engineers Australia recognises risk management as a core competency for experienced professional engineers in the following ways (quoted from Engineers Australia, 2012):

- *ensuring that costs, risks and limitations are properly understood in the context of the desirable outcomes*
- *managing risk as well as sustainability issues*

- *minimise engineering outputs only on the basis of an informed understanding of the costs, risks, consequences and limitations*
- *identify, assess and manage risks* [which] *means that you develop and operate within a hazard and risk framework appropriate to engineering activities [which includes]:*
- *identify, assess and manage product, project, process, environmental or system risks that could be caused by material, economic, social or environmental factors;*
- *establish and maintain a documented audit trail of technical and operational changes during system or product development, project implementation or process operations;*
- *follow a systematic documented method and work in consultation with stakeholders and other informed people to identify unpredictable events (threats, opportunities and other sources of uncertainty or missing information) that could influence outcomes;*
- *assess the likelihood of each event, and the consequences, including commercial, reputation, safety, health, environment, regulatory, legal, governance and social consequences;*
- *devise ways to influence the likelihood and consequences to minimise costs and undesirable consequences, and maximise benefits;*
- *help in negotiating equitable ways to share any costs and benefits between stakeholders and the community, manage projects effectively, including scoping, procurement and integration of physical resources and people; control of cost, quality, safety, environment and risk; and*
- *monitoring of progress and finalisation of projects.*

4.3.3 Performance

It is interesting to consider what it means to perform as an engineer. The Warren Centre for Advanced Engineering (2009) attempts to define performance and actually proposes a protocol for engineering performance. The protocol which describes 'the essentials of performance for Professional Engineers acting in a professional capacity' is summarised in Table 4.1.

4.4 Obligations, accountabilities and responsibilities

As a professional engineer, you will have obligations, accountabilities and responsibilities. It is important to understand the difference, as your required level of conformance is very different for each. To assist you, the definitions for obligation, accountability and responsible have been provided in Table 4.2.

4.4.1 Legal and regulatory obligations

Professional engineers are obligated to comply with the law at all times. This raises two key questions:

1. What laws are out there that you have to comply with?
2. What are the specific requirements within each law that you have to comply with?

TABLE 4.1 Summary of the Warren Centre PPIR Protocol for Professional Performance (The Warren Centre for Advanced Engineering, 2009).

Entity	Description of professional performance
1. Relevant Parties and Other Stakeholders	The Professional Engineer should develop a clear understanding of the Relevant Parties to and Other Stakeholders in the Engineering Task and the relationships between them.
2. The Engineering Task	The Professional Engineer should consult and agree with the Responsible Person for the objectives and extent of the Engineering Task.
3. Competence to Act	The Professional Engineer should assess and apply the competencies and resources appropriate to the Engineering Task.
4. Statutory Requirements and Public Interest	The professional Engineer should identify and respond to relevant statutory requirements and public interest issues.
5. Risk Assessment and Management	The Professional Engineer should develop and operate within a Hazard and Risk Framework appropriate to the Engineering Task.
6. Engineering Innovation	The Professional Engineer should seek to use engineering innovation to enhance the outcomes of the Engineering Task.
7. Engineering Task Management	The Professional Engineer should apply appropriate engineering task management protocols and related standards in carrying out and accomplishing the Engineering Task.
8. Contractual Framework	The Professional Engineer should ensure that any contract or other such evidence of agreement governing or relevant to the Engineering Task is consistent with the provisions of this Protocol.

Have you ever thought about these legal obligations? It can be quite confronting to consider this for the first time. This subsection introduces some concepts to help you to start navigating through this complex topic.

Legal and regulatory regimes can be broadly categorised as compliance or risk-based systems. In compliance systems, the governing body (Government and/or Regulator) specify mandatory guidelines and standards that a company must abide by in order to operate legally and to avoid civil and criminal penalties. In risk-based systems the emphasis is on the operator to demonstrate *duty of care* to ensure health and safety of workers and to demonstrate systematic processes are used to identify and manage potential risks to an acceptable/tolerable level. In the mining industry, different countries have adopted different approaches. For example, the United States uses a compliance-based regulatory system. Australia, on the other hand, officially adopted risk-based regulations in 1999 for Queensland and in 2002 for NSW (Poplin et al., 2008).

Most jurisdictions have laws relating to the management of risks associated with health and safety, environment and social impact and executing business activities. In addition, there can be legislation aimed at a specific industry. Further, there are other legal requirements that a company and/or individual should abide by. These include an environment authority/license/permit, commitment agreements with stakeholders, contractual agreements with employees, vendors and suppliers.

Therefore there are a myriad of laws that an industry professional needs to be aware of especially when you consider the OHS, environmental, financial, contractual, human resources laws that typically apply to the activities of companies operating high hazard facilities. Thus you are encouraged to examine the legislation and regulations that govern your activities. Finally, as mentioned earlier, competent professional practitioners know the scope of their competencies, and work within that scope. Engineers are not lawyers. Do not assume otherwise. As a junior engineer, always seek advice from suitably competent colleagues and/or advisors. And remember that wherever you are working, your work and that of your company, will be subject to a broad range of laws and regulations. Not knowing them is not an acceptable defence.

As mentioned above, **in risk-based legislation and regulation the emphasis is on a person to demonstrate duty of care**. According to the Warren Centre (The Warren Centre for Advanced Engineering, 2009, p. 5):

> 'Duty of care' is designed to ascertain whether there is a duty to apply a standard of reasonable care in the particular circumstances; and given that duty, the 'standard of care' is about the appropriate standard in those circumstances. The standard of care owed by professionals is determined by what can reasonably be expected of a person professing the professional skill, taking into account all the relevant circumstances at the time – that is, the appropriate professional performance in that particular situation.
>
> Notwithstanding these important changes, if a particular profession does not have a generally applicable and widely shared view of standard professional practice, the professional's duty and standard of care is defined by default by the view of performance formed by the court in retrospect, in the course of each particular litigation proceeding.

The difference between professional practice and legal requirements is shown in Fig. 4.5.

4.4.2 Accountabilities

Accountability involves being answerable for given activities, actions, performance and/or behaviour (derived from Merriam-Webster.com dictionary). Engineers are often accountable for their own work performance, professional advice and output produced. If they are leaders, project managers or leaders, they may also be accountable for the work performance of people reporting to them or working on their projects. As mentioned in Table 4.2, accountability is established within an authority system. An authority system is the formal organisational rules that state the power or rights and limitations of individuals within specific roles to issue orders, to make decisions and direct or control other people or things (derived from Merriam-Webster.com dictionary). Authority systems for engineers can be created by their organisations, by their professional bodies or by regulators. Examples of engineering accountabilities set by organisations might include performing statutory inspections, managing a project team, redesigning a part of the plant, collecting, analysing and reporting production data. Examples of engineering accountabilities set by external organisations might include:

- Professional associations hold engineers accountable for adhering to their code of ethics
- Companies and Regulators in most jurisdictions hold engineers accountable for the health and safety of themselves, their subordinates and others in their workplace.

FIGURE 4.5 A comparison of professional standards and legal requirements. (Hannah, 2015). *Source: Adapted from Hannah, R. (2015).* Course notes for CHEE4002: Impact and risk in the process industries. *The University of Queensland.*

TABLE 4.2 Meaning of obligation, accountability and responsibility (Hannah, 2015).

Element	Meaning	Level of conformance
Obligation	Established within a system of 'law' by organisations that have been legally constituted. Requirements should be specific (spelt out).	**MUST** conform
Accountability	Established within an authority system. Requirements set by your team leader and external bodies. Requirements open to interpretation, room for personal judgement.	You need to be able to account for your decisions, judgements and actions. Behaviours driven by professional and social ethics.
Responsibility	Standards and actions that you impose upon yourself.	Responsible to your own personal ethics.

4.4.3 Responsibilities

Responsibility has been variously defined in dictionaries, literature and guidelines. For our purposes, we refer to responsibilities as the requirements individuals place on themselves to be an upstanding person, member of the community and professional engineer. Responsibilities may overlap with obligations and accountability but they also can include additional requirements you choose to adopt that are not specified by the law, companies or professional bodies. Examples of such responsibilities might include donating certain amount of time and/or money to volunteer and charitable activities, ensuring you are aware of and understand causes behind contemporary catastrophic accidents, staying abreast of the latest global trends that could impact your and your organisation's objectives, and to publicly advocate for specified causes.

4.5 Case study: dieselgate – VW emissions scandal

In 2014 it was revealed that VW, the world's number one manufacturer of cars at that time, had used a 'software fix' to enable diesel cars to pass emissions tests when under test conditions. Under normal driving conditions, the cars emissions exceeded regulated levels in the United States by up to 40 times. This 'cheat device' affected 11 million vehicles worldwide. After initially denying that this was happening, VW's senior executive in the United States testified in a public enquiry that this was not due to senior management, but that it was the fault of two software engineers. Criminal investigations and legal proceedings have since commenced in jurisdictions around the world. By the end of 2018, VW had paid more than US$35bn in costs and several senior executives faced criminal charges (Stumf, 2019). To date, two VW executives have been imprisoned in the United States.

Many articles have been written about the 'dieselgate' saga, many of which rightly focus on the role of the senior executives, and the corporate culture that allowed this illegal behaviour to occur at such a massive scale. However, some articles did also interrogate the role of the engineers working on this project. The following quote from Merkel (2015) is particularly pertinent in the context of this chapter.

> ... as well as the senior decision-makers, there is very likely to have been a much larger group of engineers who knew of the illegal deception, understood the consequences and chose not to reveal it to authorities or the media. The ethical duties of software engineers in these circumstances are, theoretically, quite clear ......The Software Engineering Code of Ethics, agreed jointly by the Association for Computing Machinery (ACM) Institute of Electrical and Electronics Engineers (IEEE), states that a software engineer should:
>
> Disclose to appropriate persons or authorities any actual or potential danger to the user, the public, or the environment, that they reasonably believe to be associated with software or related documents.

How could this happen at a company like VW? This is a company with a proud engineering pedigree. How do you think you would behave if you worked in such a project in a multinational company? Why is it so difficult to speak out?

4.6 Summary

The main aim of this chapter was to highlight how critical your own professional practice is in guiding how you will, and should, make decisions. This is, of course, pivotal in how you will manage risk.

You must make a lifelong commitment to your own development as a professional, and you must take responsibility for this. The reality of modern professional life is that you, and only you, must take control and manage your own professional career.

We have covered a broad range of issues, and merely scratched the surface. You should now understand what it means to be a professional engineer. Professional practice relies on *praxis* and *phronesis*. That is, professional engineers should be able to demonstrate *prudent understanding of what should be done in a practical situation*, and should also demonstrate *practical, thoughtful doing*. We presented a simple framework for professional engineering, based on ethics, competence and performance.

Being a professional engineer means that you have an obligation to comply with the law at all times. The law can be extremely complex, and there is a lot of legislation that is relevant to practicing professional engineers. Whilst you do not need to be a lawyer, you should understand how the law works, how you will interact with the law in your professional life, know which laws are important in your specialist area, and have the professional judgement to seek advice from an expert (work within your competence).

Review questions

1. What are the three dimensions of engineering professionalism?
2. What are obligations, accountabilities and responsibilities, and how are they different?
3. What is the difference between something illegal and something unethical? Can you describe an activity that is unethical but legal, and another that is illegal but not unethical?
4. Identify an ethical challenge that you have encountered in an engineering or work environment. What should the person/people involved have done? Remember, ethics is about actions that have the potential to seriously impact the lives of others.
5. Consider an industry of specific interest to you (e.g. energy, food, pharmaceutical, chemical, mineral processing). Can you identify some key ethical challenges that the industry is confronting? For example, should large food process companies be producing food with high-fat content and/or high sugar in a world facing obesity and diabetes crises?
6. You are a junior engineer working at a large factory. Your factory has a wastewater stream which is discharged to the sewer (this is called trade waste). Your company pays a fee to discharge the trade waste. They paid more than $1 million last year in trade waste charges. The trade waste specifications are regulated by the trade waste agreement that your factory has with the local council, and the trade waste is monitored by trade waste officers. It is your job to manage the operations of the wastewater treatment plant. You have noticed that the trade waste officers collect their samples on regular days each month, and on those days the effluent is within the specifications, largely because the operators are modifying the operations on those particular days to improve the water quality (e.g. by blending with higher-quality effluent that has been stored in a spare storage tank). You are also aware that the factory tends to run batches that result in lower-quality effluent on the days when the trade waste officers are not in attendance. The factory has very few licence breaches. What would you do?

References

American Institute of Chemical Engineers. (2015). AIChE Code of Ethics. Available from https://www.aiche.org/about/governance/policies/code-ethics. (2015) Accessed 12.02.23.

Clancy, R., & Luegenbieghl, H. (2017). *Global engineering ethics*. Elsevier Science & Technology.

Dowling, David. (2020). *Engineeering your future an Australasian guide* (Fourth). John Wiley & Son Australia, Ltd.

Engineering Council. (2014). *UK-SPEC: UK Standard for professional engineering competence* (3rd ed.). Engineering Council. Available from https://www.engc.org.uk/engcdocuments/internet/Website/UK-SPEC%20third%20edition%20(1).pdf.

Engineers Australia. (2012). *Australian engineering competency standards stage 2 - Experienced professional engineer*. Engineers Australia. Available from https://www.engineersaustralia.org.au/sites/default/files/content-files/2016-12/competency_standards_june.pdf.

Engineering Council. (2020). *The UK standard for professional engineering competence and commitment (UK-SPEC)* (Fourth). Engineering Council. Available from https://www.engc.org.uk/media/3877/uk-spec-v12-web.pdf.

Government of Ontario. (2017). *Professional Engineers Act, R.S.O. 1990, c. P.28*. Ontario, Canada: Government of Ontario. Available from https://canlii.ca/t/5568z. Accessed 12.02.23.

Government of Singapore. (2020). *Professional Engineers Act 1991* (2020 revised). Government of Singapore. Available from https://sso.agc.gov.sg/Act/PEA1991. Accessed 12.02.23.

Hannah, R. (2015). *Course notes for CHEE4002: Impact and risk in the process industries*. The University of Queensland.

Institution of Chemical Engineers. Professional ethics. 2023. Available from https://www.icheme.org/about-us/governance/code-of-professional-conduct/professional-ethics/. Accessed 09.02.23.

Merkel. (2015). Where were the whistleblowers in the Volkswagen emissions scandal? *The Conversation*, September 2015.

National Society of Professional Engineers. (2019). *Code of ethics. National Society of Professional Engineers*. Available from https://www.nspe.org/sites/default/files/resources/pdfs/Ethics/CodeofEthics/NSPECodeofEthicsforEngineers.pdf.

nparkhaev. (2013) Product Marketing In Hospital. Available from https://blogs.ubc.ca/nparkhaev/2013/03/24/product-marketing-in-hospital/.

Poplin, G. S., Miller, H. B., Ranger-Moore, J., Bofinger, C. M., Kurzius-Spencer, M., Harris, R. B., & Burgess, J. L. (2008). International evaluation of injury rates in coal mining: A comparison of risk and compliance- based regulatory approaches. *Safety Science*, *46*(8), 1196–1204.

Queensland Government. (2013). *Professional Engineers Act 2002*. Queensland Government. Available from https://www.legislation.qld.gov.au/view/pdf/inforce/2013-09-23/act-2002-054. Accessed 12.02.23.

Rockwell, C., & Luegenbieghl, H. (2017). *Global engineering ethics*. Elsevier.

Stumf, R. VW Executives From Dieselgate Era Formally Charged With Stock Market Manipulation. 2019. Available from https://www.thedrive.com/news/29996/vw-executives-from-dieselgate-era-formally-charged-with-stock-market-manipulation. Accessed 12.02.23.

The Warren Centre for Advanced Engineering. (2009). *Professional performance, innovation and risk in Australian engineering practice*. The Warren Centre for Advanced Engineering Limited. http://thewarrencentre.org.au/wp-content/uploads/2012/02/PPIR_full_report.pdf.

The Warren Centre for Advanced Engineering. (2009). *Professional performance, innovation and risk in Australian engineering practice*. Sydney, Australia: Sydney University.

The Warren Centre for Advanced Engineering (2016). PPIR: a program for professional performance. The Warren Centre for Advanced Engineering. Available from https://www.engineersaustralia.org.au/sites/default/files/2022-06/ppir-program-for-professional-performance.pdf.

CHAPTER

5

Identify, assess and treat risks

OUTLINE

5.1 Introduction 86

5.2 Establishing the context 88

5.3 Risk assessment 90

5.3.1 Risk identification 90

5.3.2 Risk analysis 95

5.3.3 Risk evaluation 99

5.4 Risk treatment and management 103

5.4.1 Overview of risk treatment 103

5.4.2 Bowtie analysis 105

5.4.3 Unwanted event characterisation 108

5.4.4 Unwanted event identification 108

5.4.5 Determine the scope of the analysis 109

5.4.6 Identify the threats that can cause the unwanted event 111

5.4.7 Identify the possible consequences that could result from the unwanted event 112

5.4.8 Control analysis 114

5.4.9 Management of controls 118

5.5 Summary 122

Review questions 122

References 123

DOI: https://doi.org/10.1016/B978-0-12-820320-0.00008-8

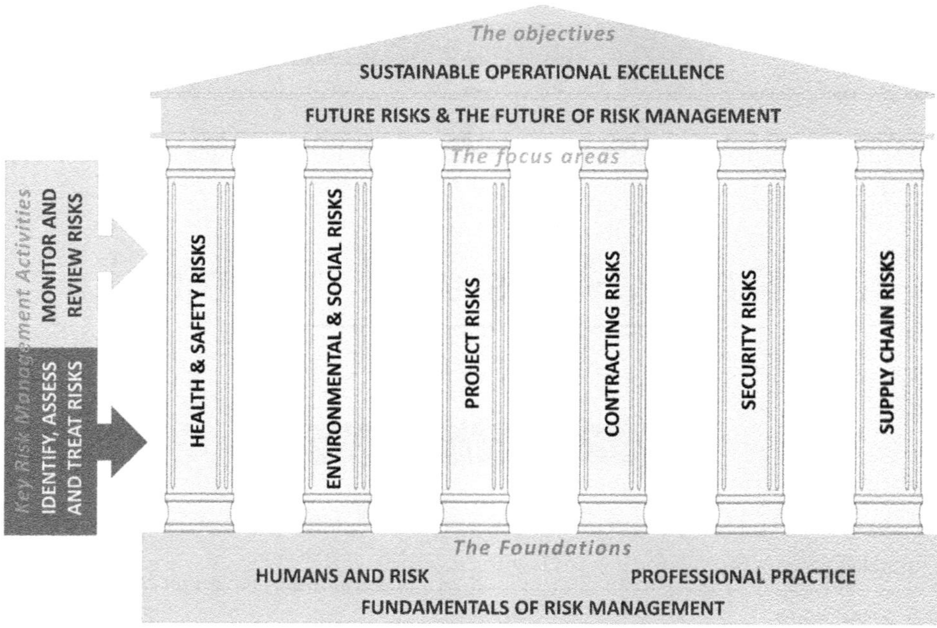

Risk comes from not knowing what you're doing *Warren Buffett*. (Synder, 2017)

5.1 Introduction

This is the first of two chapters which focus on the key risk management activities. In this chapter we will introduce knowledge and tools to assist with performing risk identification, assessment and treatment as shown in Fig. 5.1. Managing risk first requires setting scope, context and criteria, identifying and assessing the risks, then determining how to treat the unacceptable risks. This embellished version of the ISO31000 framework (as introduced in section 2.3) puts equal emphasis on risk assessment and risk treatment, which is crucial for assuring effective management of process industry risks.

Note that risk management not only covers the management of controls, but also the monitoring and review, and communication and consultation processes required to effectively manage the risk as shown in Fig. 5.1.

We may also consider the ***risk management process as a continuous and infinite process involving risk identification, assessment, monitoring, control, review and evaluation, followed by learning and adaptation*** as shown in Fig. 5.2. This representation is very powerful because it emphasises that risk management is an ongoing process, and that it is never completed. ***This is a key message to remember.***

The first part of the chapter focuses on establishing the context. The second part of the chapter focuses on risk assessment. This encompasses the key activities of risk identification, risk analysis and risk evaluation. In the third part of the chapter, we will look at how

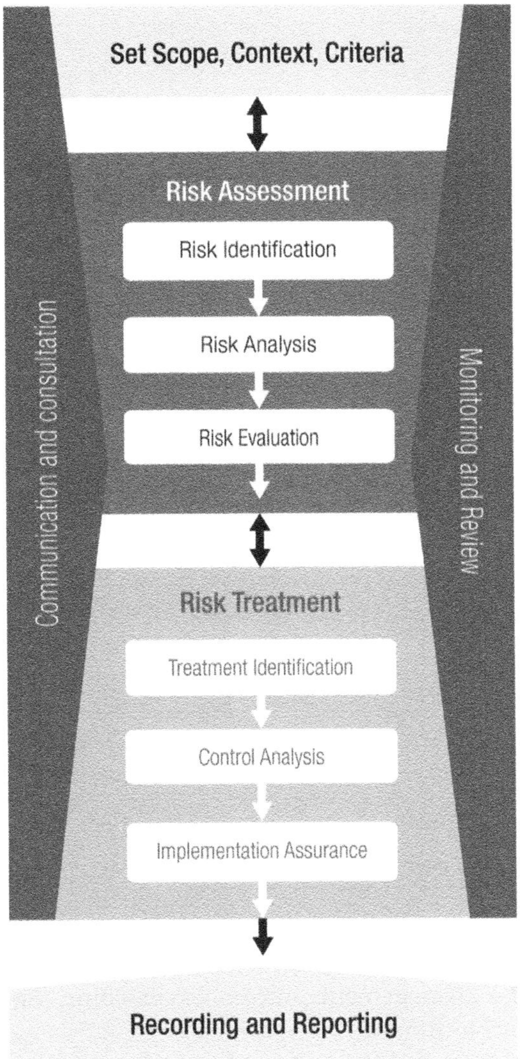

FIGURE 5.1 Extended version of ISO31000 risk management process.

we can then treat the risks that we assess as requiring further action. This encompasses the key activities of scenario identification, control selection and analysis and control management and evaluation.

This chapter aims to equip you with appropriate knowledge and tools to enable you to identify, assess and treat risks in industry. There are a number of tools that can be used in each step of the risk management process. These will become more evident as we go through this chapter. On completion of this chapter you should be able to describe and understand how to apply some contemporary tools used to identify, assess and treat risks. This chapter will also further reiterate the importance of good professional practice for performing rigorous, robust, practical and useful risk assessment and treatment work.

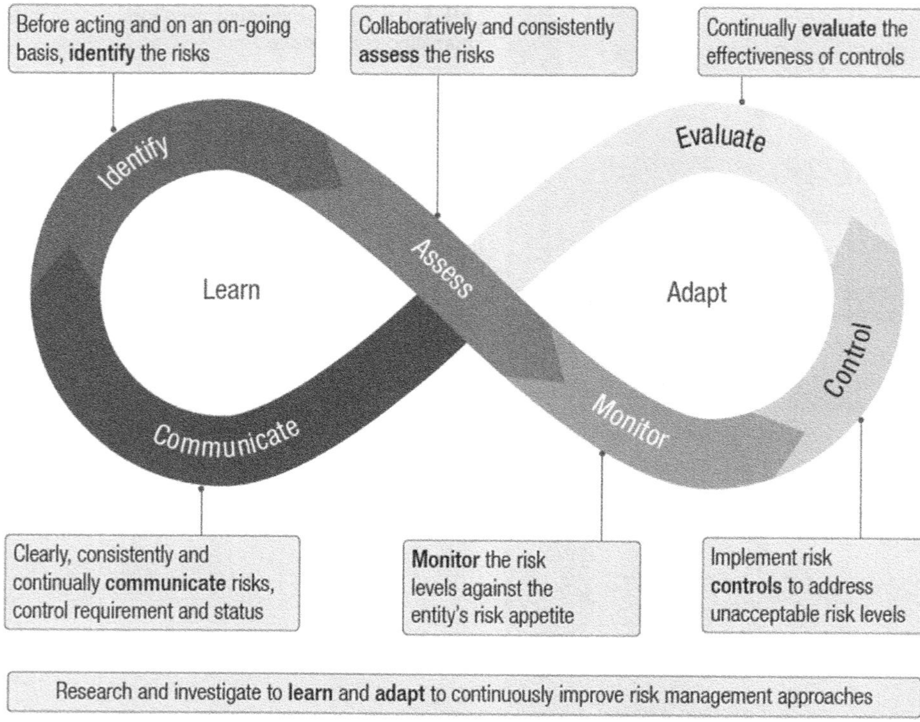

FIGURE 5.2 Risk management process (Solicitors Regulation Authority, 2014). *Source: Adapted from Solicitors Regulation Authority (Ed.). (2014). SRA regulatory risk framework. http://www.sra.org.uk/risk/risk-framework.page.*

5.2 Establishing the context

There are a number of important steps when establishing the context for a risk management activity. These include, determining the objective, identifying the stakeholders who will be involved and consulted as part of the risk management process, establishing the scope and selecting which risk assessment approaches to use.

Documenting the context helps ensure that those involved in performing, using or reviewing the risk work will understand the objective, focus areas, assumptions and limitations or boundaries of the risk management work conducted.

To establish the context for a risk management activity it is important to first determine the objective of the risk assessment. Will you adopt a 'loss reduction' mindset or a 'risk optimisation' mindset? Is the objective focused on identifying and addressing hazards that could cause adverse incidents? Is it to determine the deviations and uncertainties that need to be proactively managed in order to minimise adverse outcomes and maximise beneficial outcomes? Or is the objective just to identify and address safety and health risks? It is clearly important to define the objective.

The second step in setting the context is identifying the stakeholders who will be involved and consulted as part of the risk management process. In some regions, the need to consult relevant stakeholders is a legislative requirement for certain risk management activities. Typically, in the process industries, the stakeholders will include the frontline people exposed to the risk (including employees and contractors), subject matter experts who have a deep understanding of the risk, and managers responsible for ensuring the risk is managed. Sometimes the process can involve external stakeholders who might be members of the community, government representatives, external experts and others. Stakeholder management is discussed in detail in Chapter 8.

Establishing the scope of a risk management activity involves clarifying what is included and excluded from consideration in the risk assessment and risk treatment processes. Using a table of categories can help ensure completeness of the scope. Table 5.1 uses the categories of People, Locations, Equipment, Activities, Timeframes, Environmental conditions, Scenarios

TABLE 5.1 Scope table populated with examples from driving in a processing plant site.

Categories	In scope	Out of scope
People	Employees, contractors, visitors.	People off-site.
Locations	Processing plant site.	Off-site locations.
Equipment	Vehicle and its components including passenger vehicles, light vehicles, freight vehicles and specialist vehicles such as cranes, forklifts, vacuum trucks etc. Road, road equipment (signs, barriers, traffic lights etc.).	Vehicle maintenance equipment. Equipment carried as loads on vehicles. Aerial vehicles (e.g. drones).
Activities	Vehicles while in transit, e.g. – Transporting people around site. – Transporting goods around site. – Conducting inspections.	Maintenance activities. Activities done while vehicle is stationary e.g. unloading, lifting, loading etc.
Timeframes	Driving around site can take place any hour of the day or night and on any day of the year with activities occurring across shift changes and when shutdowns are occurring.	
Environmental conditions	Adverse weather conditions including lightning, cyclones/hurricanes and flood events should be consider.	Seismic events. Landslide events.
Scenarios	Loss of control of vehicle leading to crashes that could result fatalities of vehicle occupants and/or pedestrians. Loss of awareness of pedestrians resulting in them being struck and severely injured by vehicle. Loss of control of vehicle or situation awareness resulting in vehicle crashing into plant causing loss of containment of flammable/toxic substances which could result in fire, explosion and/or poisoning that results in fatality or sever injuries	Vehicle breakdowns with no safety impacts Loss of containment of load resulting in environmental impacts only
Other assumptions	Vehicles are driven by a person. Vehicles are powered by traditional petrol, diesel and battery sources. Vehicles comply with road registration safety standards. Plant contains hydrocarbon that is volatile and toxic to humans.	Autonomous and remote controlled vehicles Lithium ion battery and hydrogen powered vehicles.

and Other assumptions to list what is included and excluded from the scope. The table is populated with examples of what might be included and excluded from a risk assessment for driving a vehicle on around a processing plant. It shows that in-scope considerations include employees, contractors and visitors, as well and the plant and equipment within the processing plant site. Other offsite people, plant and locations are out-of-scope for this analysis. The activities being considered are the driving of vheicles to transit people and goods around the processing plant. The second last row of the table shows that for this example, the risk management exercise was focused on identifying and addressing loss of containment and loss of control of vehicle scenarios and loss of driver situation awareness scenarios that would adversely impact safety. The last row states the assumptions.

5.3 Risk assessment

Risk assessments entail risk identification, risk analysis and risk evaluation. How each of these can be performed in process industry contexts are explained in more detail in the following subsections.

5.3.1 Risk identification

ISO31000 defines risk identification as the process of finding, recognising and describing risks. Risk is uncertainty that matters because it can have an impact on objectives. Risk identification is thus about trying to identify the uncertainties that matter. Uncertainty can be caused by variability, known and unknown threats and opportunities and incomplete knowledge. Uncertainty can be associated with the present and the future. Risk identification should involve looking for uncertainty that derives from current and future variability, threats and potential threats, opportunities and potential opportunities and incomplete knowledge. Risk identification activities should also refer to the past. We can learn from history to identify the variability, threats, opportunities and knowledge issues that have impacted the objectives, so these risks can be managed into the future in ways that prevent reoccurrences of unwanted events.

Risk identification activities can be formal, informal or a combination of both as shown in Fig. 5.3. The activity of risk identification should be regular and ongoing and it should be linked to learnings from past events as well as current and anticipated hazards, opportunities and uncertainties that span the range of risks (e.g. health, safety, projects, contractors, supply chain, environmental, social, cybersecurity, financial areas as shown in the pillar diagram).

There are numerous risk identification techniques. Informal or unstructured techniques include brainstorming and blank sheet approaches. Formal techniques provide guidance to help people identify the risks, as well as the uncertainties that can cause deviations in expected outcomes.

In the process industries, commonly used formal risk identification techniques include:

- **Hazard identification and analysis techniques** such as Hazard Identification (HAZID), Process or Preliminary Hazard Analysis (PHA), and Job Hazard Analysis (JHA). HAZID

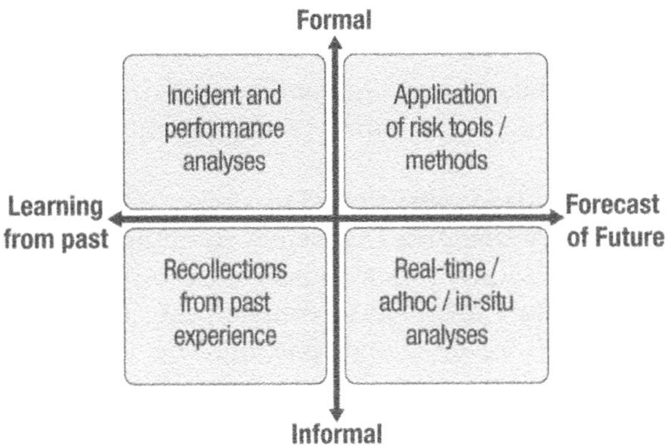

FIGURE 5.3 Approaches to risk identification.

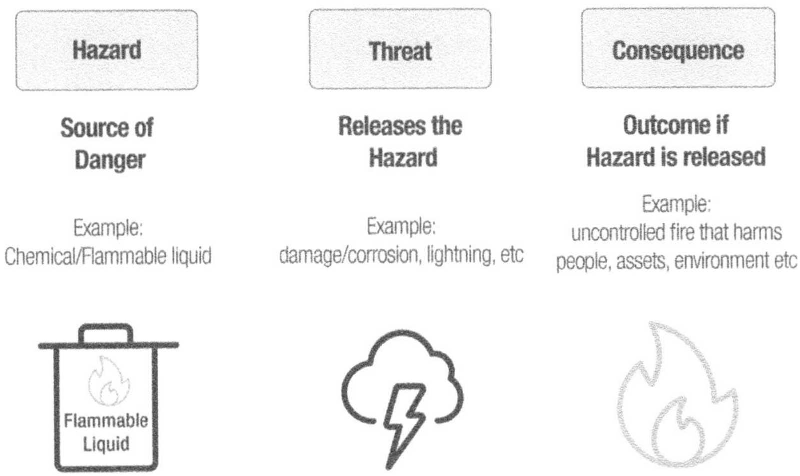

FIGURE 5.4 Example of a hazard, threat and consequence.

techniques involve [semi-]structured brainstorming type processes. The aim of HAZID is to determine what could go wrong – identify the things that could harm people, the environment, assets and the economic situation of an organisation in ways that would prevent them achieving their objectives – and to identify and assess the severity of the consequences that might result if something does go wrong as shown in Fig. 5.4. Thus HAZID and analysis techniques include:

1. Identify the hazards (e.g. chemical and/or flammable liquid) which could potentially cause harm to the assets that matter (e.g. people, environment, plant, and company financial and reputation status).
2. Identify the threats which are scenarios that can release a hazard (e.g. damage, corrosion, lightning).
3. Describe the consequences which are the possible outcomes that might result from an unwanted scenario where a threat has released a hazard (e.g. uncontrolled fire and loss of assets, personal injury and air and water pollution).

4. Assess the level of harm (impact) that might result if these hazards are not eliminated, substituted or controlled.

– **Process deviation analysis techniques** such as Hazard and Operability study (HAZOP). HAZOPs are a systematic and structured examination of an operational process or procedure. They comprise:
  1. Decomposing the process or procedure into nodes (subsystems) or steps.
  2. Identifying the safety critical ***parameters*** and ***guidewords*** that when combined describe deviations in the operations. For operational process analyses, parameters typically include 'temperature', 'flow', 'pressure', 'level' and 'composition'. The guidewords typically include 'no', 'low', 'high', 'reverse' and 'part of'. The guidewords can also include decision and action terms such as 'none', 'wrong', 'more', 'less', 'too fast', 'too slow' and 'incorrect order'.
  3. Using the various combinations of parameter-guideword to determine potential deviations in the operations that could lead to unwanted events/scenarios.
  4. Determining the causes and consequences of each unwanted event scenario.
  5. Identifying current safeguards (risk controls) and any additional actions/recommendations needed to manage the risk.

– **Plant failure analysis techniques** including Failure Modes Effects and Criticality Analysis (FMECA). Failure analysis involves:
  1. Separating the system into subsystems, equipment and/or components depending on the level of detail required from the analysis.
  2. Brainstorming is used to identify the different ways these entities might fail and the potential effects (consequences) of each failure.
  3. Assessing the criticality of the effects in terms of severity of the consequences.
  4. Identifying the potential causes of failures that have critical effects/consequences and estimating the likelihood of these causes.
  5. Describing any current controls and other additional actions/recommendations that need to the deployed to manage the risk.

An overview of the steps involved in the three types of risk identification approaches described above is shown in Fig. 5.5. These techniques can differ in how thoroughly and systematically they are applied. It is preferable that HAZID processes are performed by a team and appropriately reference historical incident data. The team should comprise both people familiar with the system being analysed and external people who can bring a different perspective to the process. Guidelines for performing these techniques are widely available in literature and on the internet from government and industry sites.

Hazard-focused risk identification tools focus on identifying negative risks – that is risks that can cause adverse deviations from the objectives. They typically also focus on deviations in operational performance. However, as discussed in Chapter 1, process industries are subject to both upside and downside risks from a number of internal and external sources. Identification of these types of risks can be done using techniques such as:

- Strengths, Weaknesses, Opportunities and Threats analysis (SWOT) that covers Political, Economic, Social, Technology, Legal and Environmental impacts (PESTLE analysis)
- Human factors' analyses
- Social and environmental impact assessments

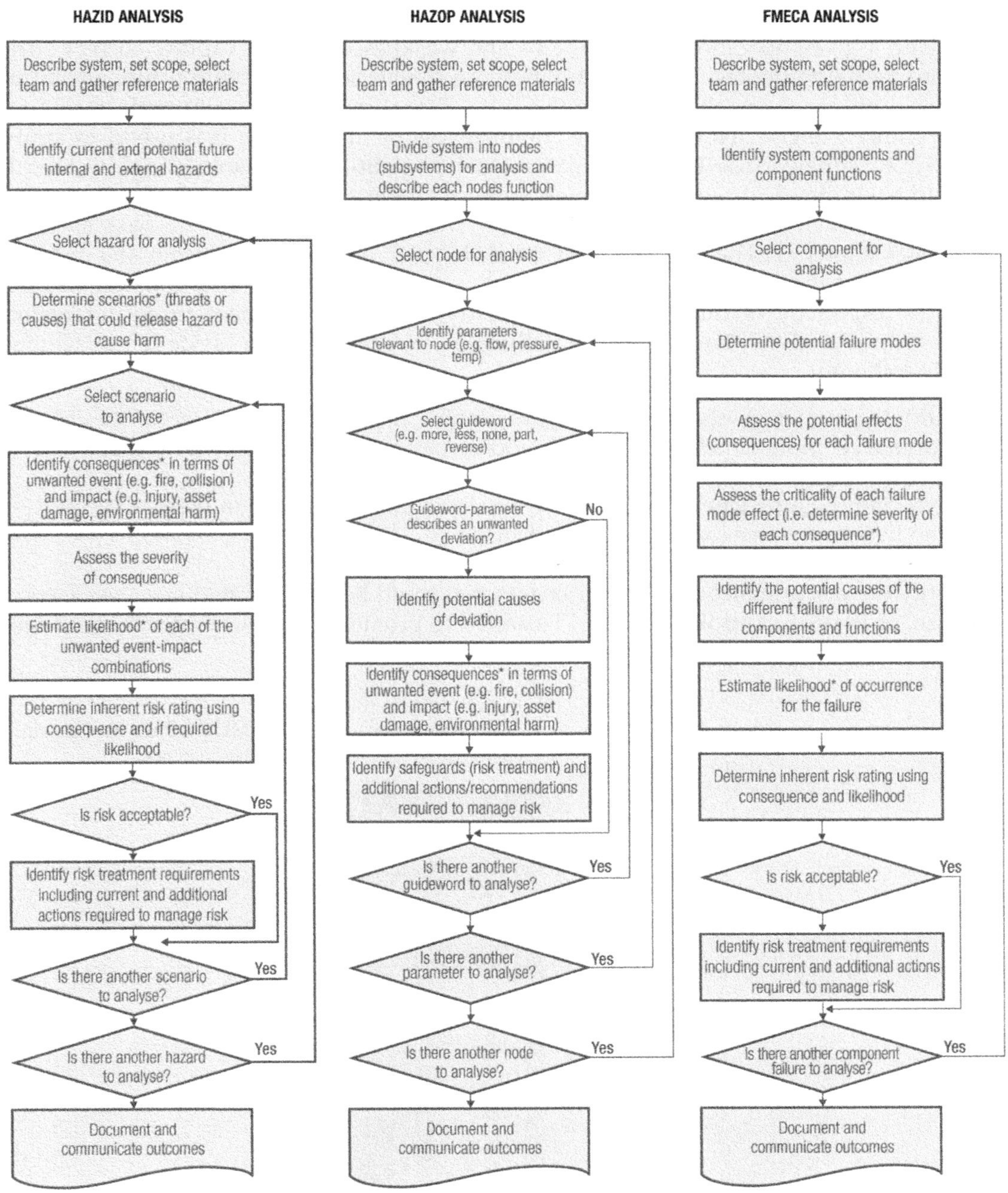

FIGURE 5.5 Generic method of different risk identification techniques in the process industries.

SWOT analysis involves listing the strengths, weaknesses, opportunity and threats associated with a given operation. The strengths and weaknesses seek to capture current attributes or internal attributes. The opportunities and threats seek to capture future attributes or external influences as shown in Fig. 5.6. SWOT analyses are typically performed as a brainstorming exercise. To help ensure completeness of the analysis it is useful to apply the PESTLE framework so that participants consider political, economic, social, technical, legal and environmental aspects that might help or hinder the achievement of objectives.

Human interaction with systems can create both upside and downside risks as mentioned in Chapter 3. The identification of such risks might emerge from the HAZID approaches and the SWOT analysis mentioned above. However, often in the high hazard industries, especially in the oil and gas industry, it is a requirement to do a targeted human factors' analysis. To identify the risks associated with human–system interactions, an analyst should:

- Identify the human–system interactions to identify the full range of interactions. At the frontline level more detailed analysis should be performed to identify critical risk-related tasks.
- Understand the interactions and identify the deviations or failures in them that could lead to better or worse than expected outcomes.
- Assess the risks, consequences and likelihood, of the deviations.
- Identify the performance influencing factors that could be incorporated into the system design, or the risk controls that could be added to promote positive or prevent negative high-risk interactions.

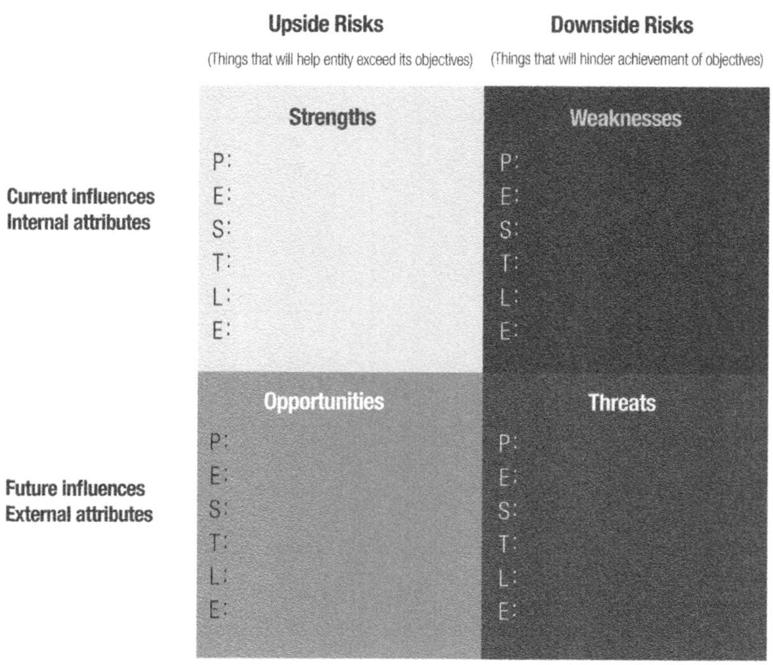

FIGURE 5.6 SWOT analysis framework.

Specific techniques for understanding human factor risks associated with those that control process industry operations include SAfER and STPA (introduced in Chapter 3 and demonstrated in Chapter 7). Further guidance on human factors in risk assessments can be found on the UK HSE website (http://www.hse.gov.uk) and on the International Association of Oil & Gas Producers (IOGP) website (http://www.iogp.org) which highlights how to *'optimise the human contribution to production and minimises potential for design-induced risks to health, personal or process safety or environmental performance'* (IOGP, 2011).

The other type of risk identification techniques often used in the process industries are social and environmental impact assessments. Impact assessments are a specific type of risk assessment that seek to identify the impact (both positive and negative) that an entity (i.e. company or project) will have on society and the environment, and that society and the environment will have on the entity both during construction, operation, closure and post closure. Impact assessments help identify risks by:

- Identifying the stakeholders and other risk receptors and their objectives (i.e. the 'things that matter to them').
- Baselining current performance and determining potential changes from the project that might impact on stakeholders' ability to achieve their objectives.
- Assessing the significance of changes or impacts and whether stakeholders may perceive them as positive or negative.
- Determining options to mitigate negative and enhance positive impacts.

Impact assessments typically identify risks that are not identified in operational hazard analysis studies. Impact assessments are discussed further in Chapter 8.

5.3.2 Risk analysis

Once risks are identified, they need to be analysed to determine the likelihood and the potential seriousness of the resultant consequence. This analysis is used to rank the risks in terms of significance so they can be evaluated. The two most common risk analysis techniques used in the process industries are the risk matrix and quantitative risk assessment (QRA). In this chapter we will focus on the use of risk matrices. QRA is discussed further in Chapter 7 and Appendix B.

Risk matrices are conceptually very simple. But care is required in firstly designing an appropriate risk matrix for a particular scenario, and secondly in applying it appropriately. Risk matrices are designed to provide a consistent framework to help people to determine and articulate what is tolerable/acceptable or intolerable/unacceptable. The determination can (and is often) used to direct people to the approaches that they need to determine the risk treatments required. So the absolute number isn't the goal. But making the right decision on what sort (if any) of risk treatment is to be applied is crucial.

Very simply, a risk matrix maps the potential impact of an unwanted event against the likelihood of the event occurring. Fig. 5.7 provides an example of a risk matrix. The impacts represent the objectives that matter. In this case, they include occupational health and safety (OH&S), asset damage, environmental impacts and reputational damage. The impact that a risk can have ranges from insignificant to catastrophic. The matrix also

RISK RANKING MATRIX									
IMPACT					LIKELIHOOD				
					Rare	Unlikely	Possible	Likely	Almost Certain
Impact	Injury and Disease (includes workers and community)	Business Cost (assets losses, fines, etc)	Environment	Social/Reputation	Event may occur under exceptional circumstances. (Estimated frequency between 1 every 10,000 years and 1 every 1000 years)	Event could occur at some time. (Estimated frequency between 1 every 1000 years and 1 every 100 years)	Event should occur at some time (May happen in plant life based on industry events. Est frequency between 1 every 100 years and 1 every 10 years)	Event will probably occur in most circumstrances (Probably occur once based on history of near misses. Est freq = 1 every 10 years or 1 per year)	Event expected to occur in most circumstances (e.g. does occur, definitely history of occurrence or estimated frequency = 1-10 times per year)
Catastrophic	Multiple fatality and/or significant irreversible exposure to a health risk that affects greater than 10 people	> $25 m	Permanent effects on the environment (e.g. air, water, soil, plants, animals). Potential loss of licence to operate. Prosecution of company and directors possible.	Very serious widespread/ irrepairable damage to significant relationships. Extremely adverse media internationally. Reputation impacted with majority of key stakeholders. Will not gain approval to operate.	Very High	Very High	Very High	Very High	Very High
Major	Single fatality, permanent disabling injury or exposure to a health risk that is permanently disabling and/or reduces life expectancy	$5m - $25 m	Major effects on environment (e.g. air, water, soil, plants, animals). Substantial cleanup costs. Personal & business prosecution possible	On-going serious social issues with significant relationships. Reputation impacted by adverse media and with significant number of key stakeholders. May lose licence to operate or not gain approval.	Medium	Significant	Significant	Very High	Very High
Moderate	Serious injury/exposure that require hospitalisation with moderate permanent effects (e.g. internal/head injuries, toxic inhalations).	$0.5m - $5m	Pollution with moderate effects on environment (e.g. air, water, soil, plants, animals). Required to report to authorities. Extensive cleanup may be required	Ongoing social issues. Attention from media. Heightened concern by local community. Criticism by local NGOs. Significant difficulties in gaining approvals. Reputation impacted with some key stakeholders.	Low	Medium	Significant	Significant	Very High
Minor	Significant injury. Medical treatment required from which recovery is likely (e.g. burns, broken bones, severe rashes/reactions, cuts).	$50k - $500k	Pollution with minor effects on enivronment. Required to inform authorities. May involve a cleanup. Visible discharge observed offsite.	Minor medium term impacts on local population / items of significance. Mostly repairable. Minor, adverse local public or media attention. Reputation is impacted with a small number of people.	Low	Medium	Medium	Medium	Significant
Insignificant	Minor injury not requiring medical treatment (e.g. cuts, mild skin irritations, no measurable physical effects).	< $50 k	Low Pollution = No observable effect to environment. No requirement to inform authorities. No observable discharges offsite.	Low-level social or cultural impacts. Public concern restricted to local complaints and individual concerns. No discernable impact on reputation.	Low	Low	Low	Medium	Medium

RISK RATING			
	High	Unacceptable Risk	Unacceptable – Operations do not to continue until risk is reduced. Risk treatment considerations should include inherently safe design and defense in depth analyses. Assign to Executive Team Member
	Significant	Tolerable Risk - Reduce risk to ALARP levels only stopping whenf risk reduction is impracticable or it's cost is grossly disproportionate to improvement gained)	Tolerable if actioned as a high priority and industry recognised risk controls measures are identified, implemented, monitored and reported on with respect to their ability to reduce risk. Assign senior manager responsible to action, monitor and review.
	Medium		Tolerable if actioned with industry recognised risk controls measures are identified, implemented, monitored and reported on with respect to their ability to reduce risk. Assign manager responsible to action, monitor and review.
	Low	Broadly Acceptable	Generally Acceptable - Manage with regular monitoring and review.

FIGURE 5.7 Example risk matrix.

considers the likelihoods that the consequence might occur. These range from rare, which equates to not expected to occur but has happened in industry, to almost certain.

As you can see, examples of insignificant risks are shown as first aid/minor injuries, less than 100k in asset damage, low level lasting environment effects, and no legal or reputational issues. At the other end of the scale, examples of catastrophic impacts include fatality events, events where greater than $50 million of asset damage is incurred or there is serious long-term environmental damage and/or widespread irreversible and adverse reputational and legal damage. Assessing the impact and likelihood gives a risk rating which in this matrix ranges from Very High, as shown in the top right, to Low as shown in the bottom left. Once a risk rating is agreed, the matrix also provides guidance on what risk evaluation to make as shown at the bottom. For example, if the risk is rated low (i.e. it is in blue zone) then the risk evaluation is that it is broadly acceptable and the risk treatment is 'continue to monitor'. Conversely, if the risk rating is rated very high (i.e. it is in the red zone) then the risk evaluation is that the risk is unacceptable and opperations must stop until actions taken to reduce it to an acceptable level have been identified and implemented. Identifying what the actions are that must be taken is a risk treatment analysis.

The risk matrix also provides guidance on what is tolerable and intolerable risk. As such, risk matrices are a means by which an organisation can express its risk appetite. The matrices also provide guidance on actions to take when the risk is assessed as having a certain consequence and likelihood rating.

You may have noted that the preceding discussion about risk analysis is predominantly targeted at 'loss reduction'. That is how most risk analyses are typically performed in industrial settings. However, it is worth recalling that risk is not always negative, and that we sometimes not only focus on loss reduction, but also looking to optimise the risk benefits. In such cases, the consequences can be positive. As such, a risk matrix may also be constructed to consider the risk benefits. Fig. 5.8 provides an example of a risk matrix that enables consideration of both positive and negative consequences. In this case, it is being used to analyse the social impacts or opportunities for a project. We will discuss this in more detail in chapter 8.

Risk matrices vary between scenarios and organisations, and that is appropriate. However, whatever the scenario, they need knowledgeable and critical interpretation. Consider the risk matrix in Fig. 5.9. This is an example of a 'standard' risk matrix that has a range of consequences, and provides for assessing the severity of the different consequences as well as specific guidance on assessing likelihood and required focus of risk treatments. Can you see a major problem with using this risk matrix in the process industries?

A key lesson learned from using risk matrices is that high consequence, low likelihood events can often get overlooked. Events such as dust explosions, well blowouts, tailings dam failures, and project closure due to community action, might not get the priority or attention they deserve because some matrices rank them as low[er] risk, because the likelihood is very low. Major disasters have taught us not to look at likelihood for high consequence outcomes, as that often results in them being incorrectly categorised as requiring lower levels of risk treatment. For high hazard industry contexts, we should look at consequence first. If the consequence is intolerable (i.e. a fatality or multiple fatalities) then that should require people to use appropriate and rigorous risk treatment approaches to determine how best to manage the risk. That is, put simply, any risk deemed catastrophic should be 'in the red' (i.e. unacceptable). If the consequence is below the intolerable level

Likelihood that threat/issue or opportunity/benefit will occur	Threats (how would stakeholder rate issue/threat?)					Opportunities (how would stakeholder rate opportunity?)				
	Insignificant	**Minor**	**Moderate**	**Major**	**Severe**	**Huge**	**Material**	**Moderate**	**Minor**	**Small**
	Localised short term negative reversible changes that all risk receptors can easily adapt to	Short term negative reversible changes that most risk receptors can easily adapt to	Medium term negative changes that are recoverable and risk receptor can cope with or adapt to	Long term negative changes that are recoverable and that risk receptors can cope with	Irreversible negative change. Risk receptor has no capacity to cope.	Permanent and significant positive change that addresses high priority wants/needs of risk receptors	Long term positive change that addresses priority wants/needs of risk receptors	Medium term change that is received positively by all of risk receptors	Short term change that is received positively by majority of risk receptors	Localised short term change that is received positively by some risk receptors
Almost certain: Probably will occur during project (>75% chance) at either a specific stage of project or more broadly	No action just monitor	Take 'reasonably practicable' action	Must action	Must action	Must action	Should action	Should action	Should action	Take 'reasonably practicable' action	No action just monitor
Likely: Could occur during project (>50% chance) at either a specific stage of project or more broadly	No action just monitor	Take 'reasonably practicable' action	Take 'reasonably practicable' action	Must action	Must action	Should action	Should action	Take 'reasonably practicable' action	Take 'reasonably practicable' action	No action just monitor
Possible: Might occur during project (>25% chance) at either a specific stage of project or more broadly	No action just monitor	Take 'reasonably practicable' action	Take 'reasonably practicable' action	Take 'reasonably practicable' action	Must action	Should action	Take 'reasonably practicable' action	Take 'reasonably practicable' action	Take 'reasonably practicable' action	No action just monitor
Unlikely: Might occur during project but unlikely (<25% chance) at either a specific stage of project or more broadly	No action just monitor	No action just monitor	Take 'reasonably practicable' action	Take 'reasonably practicable' action	Must action	Should action	Take 'reasonably practicable' action	Take 'reasonably practicable' action	No action just monitor	No action just monitor
Very Unlikely: Not expected to occur (<1% chance) at either a specific stage of project or more broadly	No action just monitor	No action just monitor	No action just monitor	No action just monitor	Take 'reasonably practicable' action	Take 'reasonably practicable' action	No action just monitor	No action just monitor	No action just monitor	No action just monitor

FIGURE 5.8 Threats and opportunities matrix.

		Likelihood				
		Rare	Unlikely	Possible	Likely	Almost certain
Consequence	Catastrophic	MEDIUM	HIGH	CRITICAL	CRITICAL	CRITICAL
	Major	MEDIUM	MEDIUM	HIGH	CRITICAL	CRITICAL
	Moderate	LOW	MEDIUM	HIGH	HIGH	CRITICAL
	Low	LOW	LOW	MEDIUM	HIGH	HIGH
	Negligible	LOW	LOW	LOW	MEDIUM	HIGH

FIGURE 5.9 Common risk matrix.

then the risk treatment approach might be less rigorous. The risk matrix in Fig. 5.7 better reflects this approach.

5.3.3 Risk evaluation

Once a risk has been analysed, an entity (person or organisation) needs to evaluate it. Risk evaluation is conducted to determine whether the risk will be accepted as it is, or whether it needs to be proactively managed so that it becomes acceptable. Risk acceptability is often judged using terms such as tolerability and ALARP (As Low As Reasonably Practicable). Understanding the concepts of ALARP and risk tolerability can be challenging. A good explanation of the concepts is as follows and as shown in Fig. 5.10.

ALARP *'refers to reducing risk to a level that is As Low As Reasonably Practicable. In practice, this means that the operator has to show through reasoned and supported arguments that there are no other practicable options that could reasonably be adopted to reduce risks further'* (NOPSEMA, 2015).

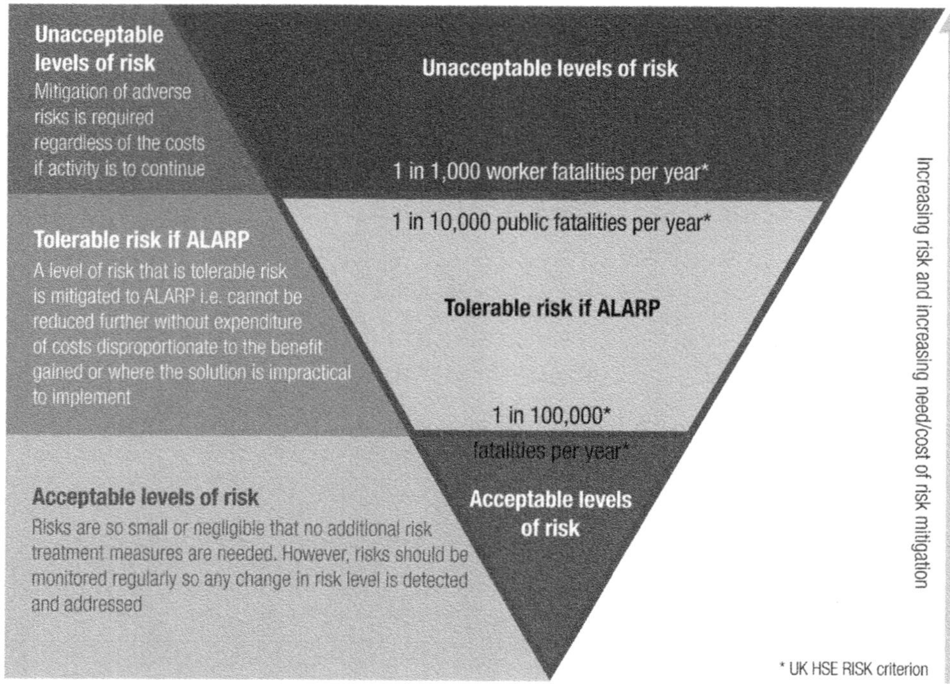

FIGURE 5.10 Risk tolerability and ALARP.

Reasonably Practicable *'is a narrower term than 'physically possible' and seems to me to imply that a computation must be made by the owner, in which the quantum of risk is placed on one scale and the sacrifice involved in the measures necessary for averting the risk (whether in money, time or trouble) is placed in the other; and that if it be shown that there is a gross disproportion between them – the risk being insignificant in relation to the sacrifice – the defendants discharge the onus on them. Moreover, this computation falls to be made by the owner at a point of time anterior to the accident. This English decision has since been confirmed by the Australian High Court'* (NOPSEMA, 2015).

Risks can be **intolerable** when ALARP has not been demonstrated and 'action must be taken to reduce the risk almost irrespective of cost' (UK Health and Safety Executive, n. d.). Risks can also be considered **tolerable** if ALARP has been demonstrated and that the extent of the demonstration is proportionate to the level of risk (UK Health and Safety Executive, n.d.). Lastly, risks can be **broadly acceptable** 'if the ALARP demonstration shows that the risks can be managed by adhering to required codes, standards and established good practice and that demonstration and codes are up-to-date and relevant to the operations' (UK Health and Safety Executive, n.d.).

To demonstrate ALARP, it has to be shown, through reasoned and supported arguments, that there are no other practicable options that could reasonably be adopted to reduce risks further. Alternately, it must be shown that the cost of further risk reduction measures becomes unreasonably disproportionate to the additional reduction in risk that would be achieved. Note that the ALARP point can be below the risk tolerability criteria

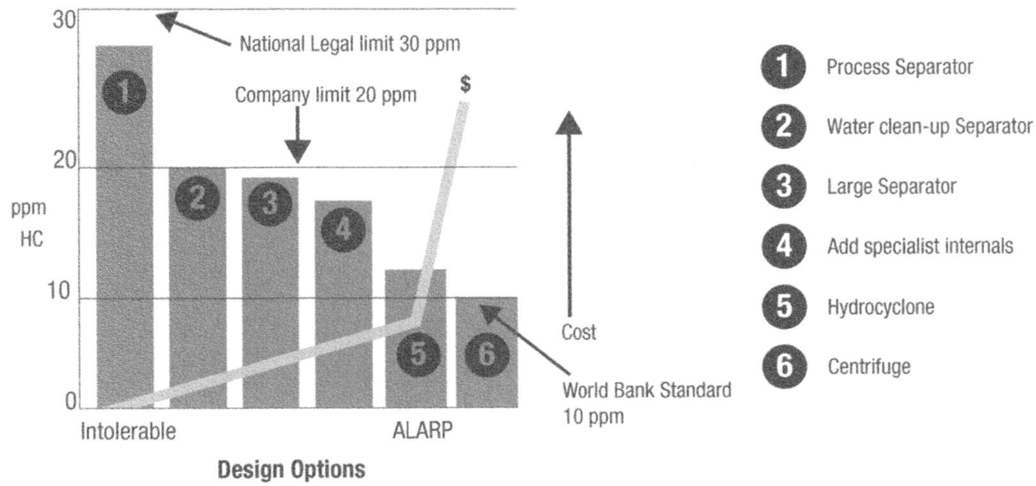

FIGURE 5.11 Selection of ALARP option for a water treatment technology (Killick, 2014). *Source: Adapted from Killick, C. (2014).* Course notes for CHEE4002 - Impact and risk in the process industries. *The University of Queensland.*

and that the best way to determine it is to measure cost and risk reduction impacts delivered for incremental levels of control.

The selection of an ALARP option is further illustrated in Fig. 5.11. In this case, it is illustrating how the risk varies with cost for six different water treatment technology options to treat a hydrocarbon effluent. Options 3–6 can all satisfy the specified company emission limit of 20 ppm. However, the hydrocyclone (option 5) is the ALARP option, because the incremental improvement (risk reduction) offered by the centrifuge (option 6) requires a disproportionate increase in the cost. This example illustrates that you need to go beyond the ALARP option in order to know that you have identified the ALARP option. In this case, that means it was necessary to design and cost the centrifuge in order to determine that the significant increase in cost meant that it was not justified, and thus the hydrocyclone was the ALARP option. This example also illustrates that the company emissions targets may be stricter than the national regulated limits. This is not unusual for large multinational companies.

What would have happened if the project had been required to adopt World Bank standards?

As mentioned, ALARP is a legal term. It is most commonly used when assessing safety and/or fatality risks. However, the principles underlying tolerability and ALARP are applicable to environmental, social, financial and technology risks. They are also linked to risk perception and risk appetite, so they are important concepts to understand.

The outcome of the risk assessment process is a list of risks that are acceptable and a list of risks that are unacceptable and need to be controlled. It is common practice for process companies to record and monitor this list in a **Risk Register**. Depending on the nature of the risk assessment activities, these risks can be articulated as risks, threats, opportunities and/or unwanted event scenarios. In determining the best way to treat a risk, it is often useful to identify the unwanted events that can lead to the unacceptable risks. See Fig. 5.12 for an example of a risk register.

Description of Hazard	Risk scenario	Consequences	Maximum Potential Impact	Estimate of Likehood	Overall Risk Rank	Assessment/recommendation including comments on risk treatment options (i.e. in terms of ISD and DiD)	Responsible person and bowtie ref	What is current effectiveness of the controls?
Petrol which is explosive and flammable	Overfilling of tank	Uncontained pools of liquid find ignition source and catch on fire and/or vapourise to form cloud that explodes	5 Catastrophic	Likely	Very High	Full risk treatment analysis that applies, ISD, DiD and LOPA and outputs a bowtie, control specification and control verification information	Site Executive: *LOC bowtie* info	*Click here to enter text.*
Liquid petrol can become an explosive vapour when heated	Vapour leaks from top of tank	Vapour cloud forms that finds ignition source and explodes killing people and destroying assets.	5 Catastrophic	Moderate	Very High	Full risk treatment analysis that applies, ISD, DiD and LOPA and outputs a bowtie, control specification and control verification information	Site Executive: *LOC bowtie* info	*Click here to enter text.*
Vapours from liquid petrol overpressurise equipment	Vapour leaks from equipment	Vapour cloud forms that finds ignition source and explodes killing people and destroying assets.	5 Catastrophic	Unlikely	Very High	Full risk treatment analysis that applies, ISD, DiD and LOPA and outputs a bowtie, control specification and control verification information	Site Executive: *LOC bowtie* info	*Click here to enter text.*
Equipment containing petrol is impacted by vehicle/mobile plant	Liquid leaks from the system	Uncontained pools of liquid find ignition source and catch on fire injuring people and damaging assets	4. Major	Unlikely	Significant	Full risk treatment analysis that applies, ISD, DiD and LOPA and outputs a bowtie, control specification and control verification information	Site Executive: *LOC bowtie* info	*Click here to enter text.*
Component corrosion/wear in the equipment containing petrol	Liquid leaks from the system	Uncontained pools of liquid find ignition source and catch on fire injuring people and damaging assets	4. Major	Likely	Very High	Full risk treatment analysis that applies, ISD, DiD and LOPA and outputs a bowtie, control specification and control verification information	Site Executive: *LOC bowtie* info	*Click here to enter text.*
Incorrect selection or connection of tank.	Petrol pumped tank containing other hydrocarbon	Contaminated project that cannot be shipped and will need to be reprocessed	2. Minor	Moderate	Medium	Risk treatment analysis that applies ISD to prevent pumping into incorrect tank and to DiD to detect contamination	Engineering Manager: Bowtie info n/a	*Click here to enter text.*
Incorrect monitoring of tank.	Overfilling of tank and/or petrol leaking from incomplete connections	Uncontained pools of liquid find ignition source and catch on fire and/or vapourise to form cloud that explodes	5 Catastrophic	Moderate	Very High	Full risk treatment analysis that applies, ISD, DiD and LOPA and outputs a bowtie, control specification and control verification information	Site Executive: *LOC bowtie* info	*Click here to enter text.*

FIGURE 5.12 Example risk register.

5.4 Risk treatment and management

5.4.1 Overview of risk treatment

Having identified, analysed and evaluated an entity's risks, and compiled the risk assessment information into the risk register, you now need to decide what to do. How will you treat the risks that have been assessed as unacceptable, or not to ALARP standards? This step is called **risk treatment**. The risk treatment options are summarised in Fig. 5.13.

In treating risks, you first look to eliminate or substitute the hazard. You then eliminate exposure occurrences and threats. Of course, the nature of high hazard industries means that we will ultimately be working with hazardous materials, and threats that can release the hazard do exist. In this case, we must implement controls that reduce the likelihood of the occurrence of unwanted events, and mitigate the consequences of unwanted events. This is the focus of this section.

In this book, we extend the ISO31000 framework presented in the Standard, and propose that risk treatment comprises, *treatment identification, control analysis,* and *implementation assurance* as shown in Fig 5.1. The selection, optimisation and management of **controls** is the critical part of risk treatment. Once the controls are implemented, appropriately managed and their performance is being measured, then **residual risk** can be determined by the effectiveness of the controls. If the right suite of effective controls are

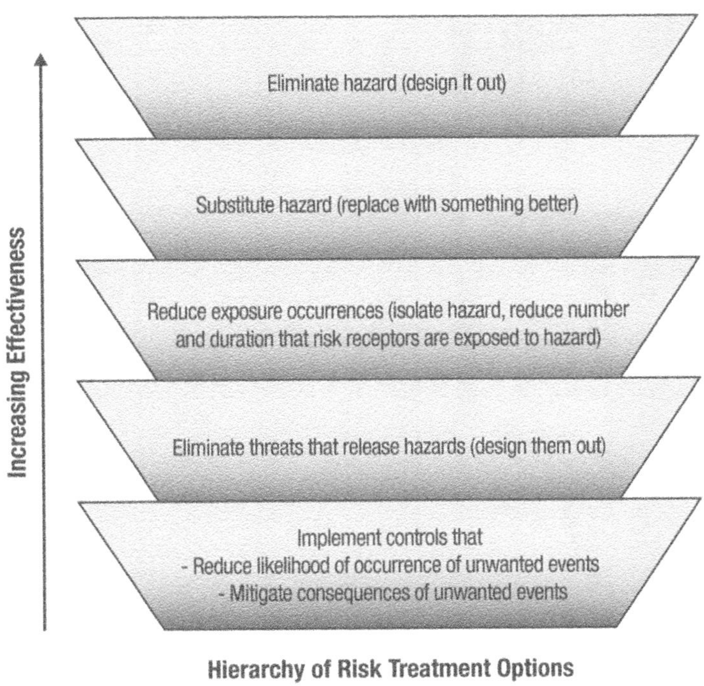

FIGURE 5.13 Risk treatment options for addressing unwanted events (Hassall et al., 2015). *Source: From Hassall, M. E., Joy, J., Doran, C., & Punch, M. (2015).* Selection and optimisation of risk controls (ACARP report C23007). *Retrieved from Australia: http://www.acarp.com.au/abstracts.aspx?repId = C23007.*

implemented and they are well maintained and managed then the level of risk can be lowered to an acceptable or ALARP level as per the example shown in Fig. 5.10.

Having designed and implemented your controls, the **residual risk** is determined by the effectiveness of the controls. If the right suite of effective controls are implemented and they are well maintained and managed then the level of risk can be lowered to an acceptable or ALARP level as shown in Fig. 5.14.

Controlling risk is done with human actions, engineered devices or a combination of the two, that are implemented to reduce the likelihood and severity of unwanted events and increase the likelihood and magnitude of better than expected outcomes. We will focus on selecting and assessing controls using the bowtie approach. Bowtie analysis is introduced in the next section, followed by detailed information to support each stage of the bowtie analysis.

High Risk

Low Risk

FIGURE 5.14 Residual risk is determined by the effectiveness of the controls.

5.4.2 Bowtie analysis

Bowtie analysis is a powerful way to visually represent the suite of controls and associated management systems required to manage a risk. Bowtie analysis, as shown in Fig. 5.15 comprises two parts:

1. The bowtie diagram which depicts:
   a. The **hazard** and the **unwanted event** which is shown in the centre of the diagram (i.e. as the knot of the bowtie).
   b. The **threats** that could release the hazard to cause the unwanted event which are shown on the left side of the diagram.
   c. The **consequences** that could result should the unwanted event, occur which are shown on the right of the diagram. These consequences should reflect the objectives that matter (i.e. those shown on the risk matrix).
   d. The **prevention controls** (also called safeguards) needed to prevent or reduce the likelihood of the threats causing the unwanted event.
   e. The **mitigation controls** (also called recovery barriers) needed to mitigate the unwanted event so that the consequences are minimised or completely averted.
2. A Control Assurance Management System (CAMS) which describes the activities required to monitor, maintain and improve control performance. These activities include:
   f. Operational support activities required to ensure effective control performance (e.g. provision of cognitive aids such as signs, procedures, plans, codes of practice).

	Control Assurance Management System (CAMS)														
	Prevention controls								Mitigation controls						
	Aa	Ab	Ac	Ba	Bb	Ca	Cb	Cc	1.1	1.2	1.2	2.1	2.2	3.1	3.2
Operational activities															
Maintenance activities								*Insert description of activities here*							
Engineering activities															
Management activities															

FIGURE 5.15 Bowtie diagram with associated CAMS table.

g. Maintenance activities required to sustain effective control performance (e.g. training of personnel, maintenance and calibration of technology).
h. Engineering and management activities required to monitor, verify and audit controls to assess their effectiveness and identify any deficiencies or improvement opportunities that should be addressed.

Some (advanced) bowties will also show erosion factors which are factors that can cause a control's performance to decline over time and/or to fail thereby undermining the effectiveness of the controls. An example of an advanced bowtie is shown in Fig. 5.16 which illustrates the linking of the bowties controls with the CAMS table. Doing erosion factor analysis can help ensure the CAMS has all the elements needed to prevent control failures. Sometimes CAMS are represented as control support maps (ConSupMap) using the Accimap process (Rasmussen & Svedung, 2000) as a template as shown in Fig. 5.17. When developing ConSupMaps, the controls from the bowtie are listed in the second bottom row, and important control related reference documents are listed on the bottom row. The control devices and control actions performed by people are listed in the rows above the control row. The frontline operational activities needed to ensure controls are healthy (i.e. maintence of devices and training of people in how to perform control actions and control monitoring) are listed in the row above the control devices and actions. In the rows above management verification, audits and other governance activities then regulatory activities.

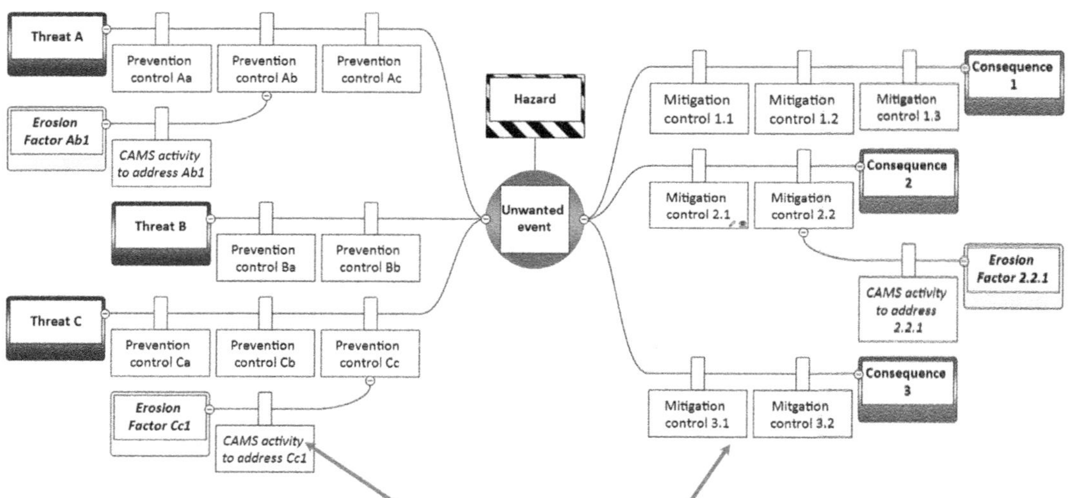

	Control Assurance Management System (CAMS)														
	Prevention controls								Mitigation controls						
	Aa	Ab	Ac	Ba	Bb	Ca	Cb	Cc	1.1	1.2	1.2	2.1	2.2	3.1	3.2
Operational activities															
Maintenance activities							*Insert description of activities here Including ones that address erosion factors*								
Engineering activities															
Management activities															

FIGURE 5.16 Advanced bowtie with associated CAMS table.

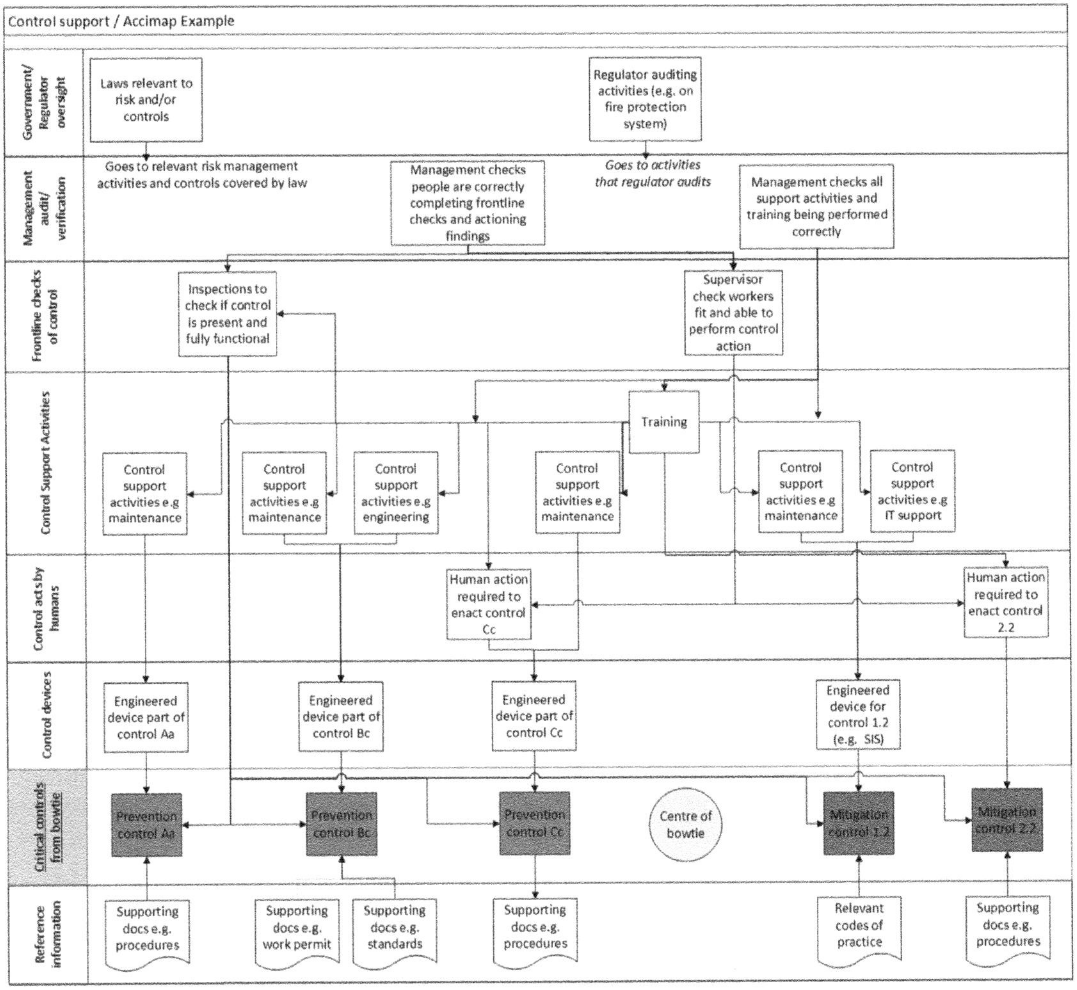

FIGURE 5.17 Control support map (ConSupMap) diagram which illustrates CAMS information.

The ConSupMap can then be annotated to illustrate the information flows and the critical roles which ensure the effective management of the controls. ConSupMaps are revisited in Chapter 6 and demonstrated in Appendix B.

Bowtie analyses may be performed to address threats that lead to unwanted events, or opportunities that lead to a beneficial outcome. However, at present in the process industries, bowtie analyses are conducted to select, assess and communicate the controls needed to address hazards and threats related to unwanted events (i.e. loss reduction). This type of bowtie analysis is a structured activity that comprises the following steps:

1. **Describe the unwanted event to be analysed:** This should reference the risk register and include those risks identified and assessed as being unacceptable.

2. **Determine the scope of analysis:** This is a process of describing what is included and excluded from the analysis in terms of the organisation and the operation. It should include people, technology, operational processes, activities and artefacts, spatial area, and time horizons.
3. **Identify the hazard(s) and the range of threats that can release the hazard to cause the unwanted event:** These threats can exist or emerge from the working environment/context, equipment and technology, people, organisational interaction issues and from external sources including adverse weather or acts of sabotage.
4. **Identify the range of possible consequences that might result if the unwanted event occurs:** The consequences should relate to the system's objectives (e.g., relate to the consequence categories as listed in the matrix). For processing plants this could include adverse impacts to peoples' health and safety, asset integrity and performance, finances, environment, community relations and company reputation.
5. **Identify the suite of controls needed to prevent each threat from causing the unwanted event and to mitigate the unwanted event to avoid the undesirable consequences:** The controls should be positioned on the threat and consequence lines in the order that represents when they act in the event timeline. Then an assessment should be done to determine if sufficient controls have been identified. This assessment should be based on the effectiveness and independence of each control and the overall suite of controls for the environment within which they are to be implemented. The assessment should check to determine if ALARP has been achieved.
6. **Identify failure modes for important controls:** Control failure modes are the things that can cause the control to fail or cause the control performance to erode over time, so it fails to work as required when required.
7. **Determine the CAMS elements that are needed to ensure the controls will function fully and reliably (i.e. work as required when required):** CAMS elements are activities such as monitoring, maintenance, testing, calibration, training and auditing. They can also include artefacts such as signage, procedures, demarcation, codes of practice and checklists. They are the elements required to check and test that control devices are maintained in full working condition and the elements that help to ensure that human control actions are performed correctly and reliably.

The following sections will look at each of these steps in detail.

5.4.3 Unwanted event characterisation

Characterising the unwanted event involves identifying the unwanted event, determining the scope and identifying the threats and consequences.

5.4.4 Unwanted event identification

An understanding of what the unwanted event is should come from the risk register. For the process industries, most catastrophic unwanted events relate to the release or loss of containment of a hazardous substance and/or loss of control of a hazardous energy source. Further unwanted events might be identified from brainstorming exercises, benchmarking

TABLE 5.2 Examples of unacceptable risks and unwanted events from industry.

Unacceptable risk	Example of unwanted events
Human fatalities or permanent disabling injuries or illnesses	Loss of containment of pressurised hydrocarbon, unsafe operation of road transport vehicle, ignition of flammable liquids/gases, persons exposed to edge/dropoff >1.5 m
Permanent and disabling health issues	Diesel emissions above specified limits, undetected depression in employees
Community opposition	Loss/suspension of license to operate, failure/delay to get permits approved, community blockage of access ways, large investor sell-offs
Environmental pollution	Air emissions or noise over license limits, uncontrolled release of toxin, tailings leak
Asset damage causing business disruption	Well-blowout, inundation of flood waters, loss of structural integrity of storage tank/processing plant/other key infrastructure, loss of key supplier
Lost opportunities	Failure to exploit new markets, failure to meet/exceed local supplier targets, failure to adopt beneficial technology, failure to keep key experts

exercises, reviews of past incidents experienced in the industry and expert foresighting exercises. Some examples of unacceptable risks and unwanted events from industry are listed in Table 5.2.

In the bowtie, the description of the unwanted event should ideally describe the point at which the system has gone from being 'in control' to being 'out of control'. This rationale can be applied to all types of unwanted events (e.g., unwanted events with adverse safety, environmental, financial, production losses, community outcomes).

In some cases, it is clear what the description of the unwanted event should be. For example, a fuel leak from a bulk storage tank (loss of fuel containment) could become the description of an unwanted event, rather than the subsequent fire, explosion or pollution which should be considered as a consequence. In this example, it can be helpful to articulate the consequence of interest. For example, fuel leak from bulk storage tank with the potential to catch on fire or explode. However, in other cases it may not be clear what the description should be. In these instances, discussion and discretion will be required to determine the most appropriate description for the unwanted event. To help with this process, some examples of good and poor descriptions of unwanted events are shown in Table 5.3. Using consistent descriptions for the same unwanted event across different sites and divisions within a company will help with comprehension, comparisons and benchmarking of bowtie information. Variability in the descriptions used for the knot will result in variability in the bowties and can compromise the identification of effective controls.

5.4.5 Determine the scope of the analysis

The scope of the analysis needs to be determined and documented to ensure those involved in the analysis, and those who review or reference the analysis at some date in

TABLE 5.3 Descriptions of unwanted events that form the knot at the centre of the bowtie.

Descriptions of unwanted events for the bowtie knot	
Good examples	**Poor examples**
Uncontrolled release of gas from a storage vessel	Gas explosion (This is a consequence)
Overheating of processing plant equipment (with potential to catch on fire)	Conveyor fire (This is a consequence)
Loss of control of motorised vehicle or mobile plant	Human error (Too general, if it relates to incorrect action this needs to be better defined)
Person exposed to live electricity	Incomplete isolation (This is a control failure)
Airborne toxins above acceptable workplace limits	Persons overcome by toxic emissions (This is a consequence)
Person exposed to fall from heights greater than 1.5 m	Fall arrest not used correctly (This is a failed control)
Misfire of explosives	Contaminated explosives (This could be a cause)
Loss of structural integrity of dam	Dam wall failure (This is a consequence)

the future, understand what was included and excluded from consideration. As such, scoping the analysis should involve describing the purpose of the analysis, the audience of the outputs, the personnel involved and the boundaries of the analysis in terms of what is included and excluded.

Understanding the purpose and the audience of the analysis is important to ensure that output produced achieves the purpose and is aligned with the audience's needs. For example, bowtie analyses performed for official safety case documentation may contain barriers that consist of all types of controls and control artefacts. In contrast, bowties constructed to communicate critical controls will highlight the devices and actions that must be implemented and operational because they are relied upon to prevent or mitigate events that could lead to catastrophic outcomes.

Defining the boundaries of the analysis should clearly describe what was included in the considerations and what was excluded from the analysis and the reasons why. A useful framework for describing the scope is the PLEATSO framework (also described in section 5.2):

- P – People
- L – Locations
- E – Equipment (or plant)
- A – Activities and any associated artefacts
- T – Timeframe or the time into the future
- S – Scenarios associated with the unwanted event. Typically these have been identified in risk assessment activities
- O – Other assumptions which should inlcde listing assumptions about the hazardouse nature of substances, and hazard processing parameters (e.g., high pressures, high temperatures etc).

The PLEATSO framework can help to ensure that the scope boundaries consider organisational areas/functions, operational processes/function, spatial area and time horizons. When defining the scope, it can help to reference aerial photographs, organisational hierarchy diagrams, process flow diagrams, engineer drawings and past internal and relevant external incidents. ***It is important to note that for risk treatment analyses the scope should only focus on the specific unwanted event(s) being analysed.***

The scope should also identify the people involved in the analysis. Ideally, bowtie analyses should be performed by a diverse team of people, involving people who understand the bowtie process, the unwanted event, and those who are responsible for actioning, monitoring and maintaining the controls. Other people such as subject matter experts and/or those that understand lead practice may also add value as they have the potential to bring a different perspective to the analysis.

5.4.6 Identify the threats that can cause the unwanted event

Typically, a range of threats could cause any given unwanted event. The threats are those things that could release the hazard in a manner that could result in the unwanted event. It is important that the complete range of threats that could cause the unwanted event are identified, even those that may be perceived as uncontrollable. This can be done by brainstorming, eliciting input from experts and from reviewing relevant incident information. Generic examples of threats are shown in Table 5.4. Specific examples of threats for unwanted vehicle interactions on a work site are shown in Fig. 5.18.

It is important to emphasise that threats should not be failed controls. This is an easy mistake to make for novices, but it is a fundamental error that results in convoluted, poor-quality bowties which can *'weaken the value of the method and hence the potential to reduce major accidents'* (Pitblado & Weijand, 2014, p. 1).

TABLE 5.4 Examples of threats that can lead to unwanted events.

Descriptions of threats that go on left of the bowtie	
Good Examples	**Poor examples**
Work environment issues (e.g. infrastructure not constructed for conditions)	Inadequate inspections (This refers to control checks)
Equipment/technology failure (e.g. wrong equipment, equipment not in serviceable condition, equipment used incorrectly/overloaded)	Failure to respond to alarms Overpressurisation of line caused by safety shutdown system (Both refer to ineffective controls)
Work interaction issues (e.g. congested work activity, interacting transport vehicles etc.)	Worker error/complacency (Poorly defined generalised statements)
Adverse weather conditions (floods, cyclones, earthquakes, bush fires, lightning etc.)	Building collapse (This is a consequence)

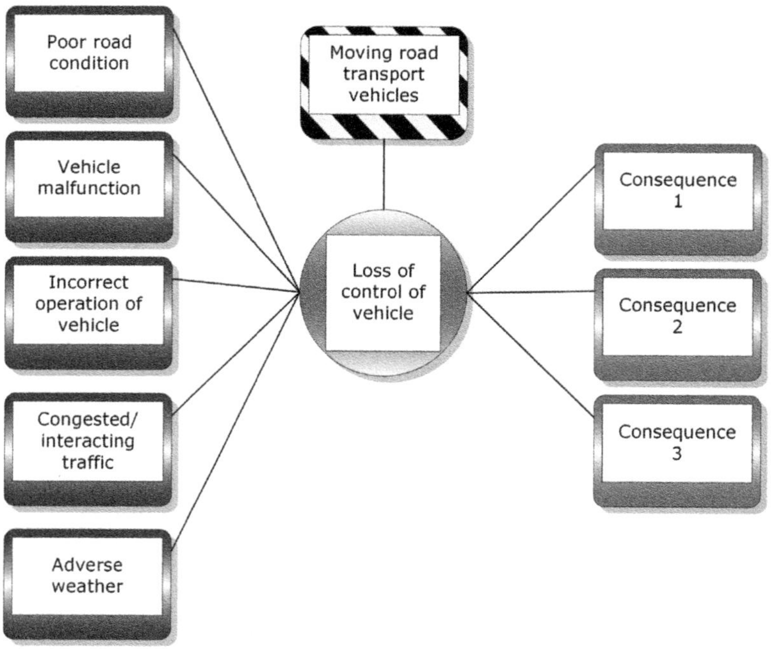

FIGURE 5.18 Examples of threats for an unwanted vehicle interaction.

5.4.7 Identify the possible consequences that could result from the unwanted event

Unwanted events can lead to a variety of consequences that can adversely impact things that matter to an organisation including people, plant, environment, finances, reputation and legal position. The impact of the consequences can range from being negligible to being catastrophic as shown on the risk matrix.

With the bowtie analysis, it is important to identify in detail the full range of possible consequences that might result from an unwanted event before mitigating controls are implemented. The description of the consequence should be specific enough to ensure the most effective suite of controls are selected to address the consequence. For example, instead of stating that a consequence is a 'potential fatality' it is better to be more specific and explicit and state that 'it is a potential fatality from explosion blast pressure', or 'potential fatality from breathing toxic air', or 'potential fatality from smoke asphyxiation'. Consequences from secondary events (domino effects) should also be considered. Examples of secondary events include failure to escape, vehicle crash causing fire, or falling from height into water and drowning. A range of possible consequences are shown in Table 5.5.

Examples of specific consequences that may result from an unwanted vehicle interaction on a work site are shown in Fig. 5.19.

TABLE 5.5 Examples of consequences that can result when an unwanted event occurs.

Descriptions of consequences that go on right of bowtie	
Good examples	**Poor examples**
Human fatalities, injury, acute and chronic health impairments from hydrocarbon vapour cloud explosion and fire	Harm to employees (Inadequate because it doesn't define harm and if potential impact on contractors/public is ignored)
Fire/explosion damage to assets including own and others plant, equipment, infrastructure	Damage to fire protection (This is the erosion of a control)
Environmental damage or contamination of air, land and water both on and off site due to chemical leak in a flood	Climate change (Should be more specific, e.g. increase in carbon dioxide emissions)
Production losses including reduction in product volumes and/or quality	Inadequate plant design (This is a threat)
Reputational damage including damage within company and with communities, regulators and other stakeholders	Adverse news and social media articles (Inadequate because adverse articles may not translate to reputational harm)
Explosion debris cause offsite traffic/pedestrian accidents	Meteor crashing onto site (This is not really a credible consequence)

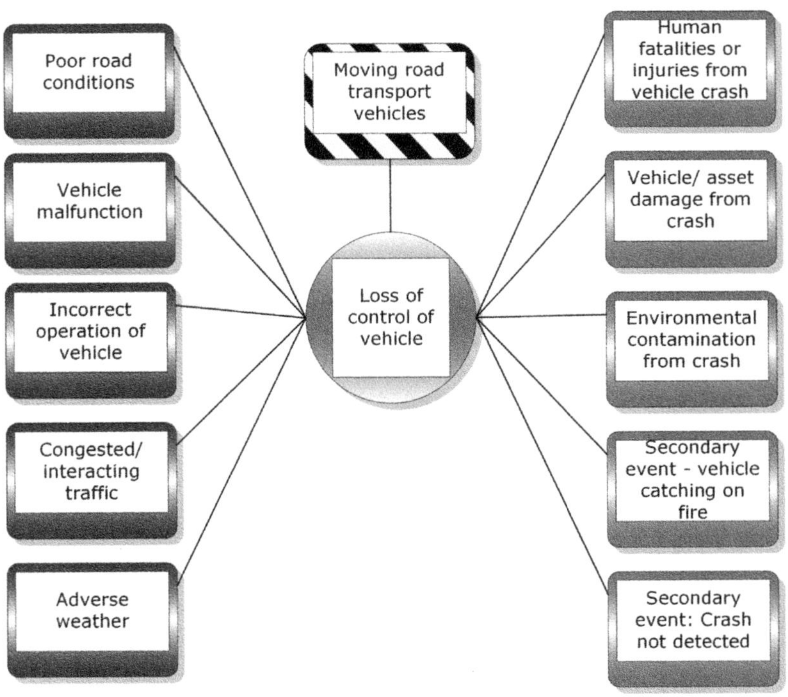

FIGURE 5.19 Example of consequences for an unwanted vehicle interaction.

5.4.8 Control analysis

Risk controls are the interventions taken to manage the risk to an acceptable level. The ISO 31000 standard defines risk control as a 'measure that maintains or modifies risk' (ISO 31000:2018). This definition is quite abstract and many things could be interpreted as controls. In this book, we are advocating a specific, and more workable, definition of controls. The proposed definition is:

> A control is an object and/or human action that of itself will arrest or mitigate an unwanted event sequence and whose performance is specifiable, measurable and auditable. *(Hassall et al., 2015).*

This definition was derived to address operational risks, and therefore focuses on what control actions and devices are needed by frontline personnel to effectively manage risks at the operational interface as shown in Fig. 5.20. The work done by supervisors and others should focus on supporting the frontline workers and assuring the controls (e.g. executing control assurance management or CAMS activities). A decision tree has been developed to help people apply this definition to identify controls and to differentiate controls from CAMS elements as shown in Fig. 5.21.

The identification of risk controls can be done by brainstorming, focus groups, benchmarking and by getting expert advice. Specifying risk controls that meet the proposed definition of 'control' can be challenging. One common mistake is to only specify part of a control. For example, 'gas leak detection' may be specified as a mitigating control for an unwanted event such as 'loss of containment due to a small leak' in a gasoline pipeline. However, as can be seen from the decision chart in Fig. 5.21, leak detection does not, of itself, directly mitigate an unwanted outcome. So what is the control? Well, gas leak detection is important, but it is only part of the control. In this case, the control may be 'automatic shut-off in response to leak detection'. This control would mitigate an unwanted event, and it is specifiable, measurable and auditable. It would also, of itself, directly prevent or mitigate an unwanted event or outcome.

Another example of a control that meets the criteria set out in the decision tree (Fig. 5.23) is 'Workers will be continuously and securely attached while aloft on a communications tower'. The performance is specifiable, that is workers need to wear fall arrest harnesses attached to nominated anchor points with approved lanyards when they are off the ground. It is also measurable through human observation or with cameras. It is also auditable as demonstrated by Lekutis (2009), who found that the risk control focused approach used in Canada and UK delivered the lowest fatality rates for communication tower workers. This control would, of itself, mitigate the potential of a fatality, and its performance is specifiable, measurable and auditable.

If you are unsure whether something is a control or not, Hassall et al. (2015) suggest it is worth asking the following questions:

- If you were the person in harm's way, would the element be something that is going to help prevent you, the plant and the environment from being harmed? Can you check to see if the element is/will work as required when required? If the answer is yes to both these questions then the element is probably a control. If the answer

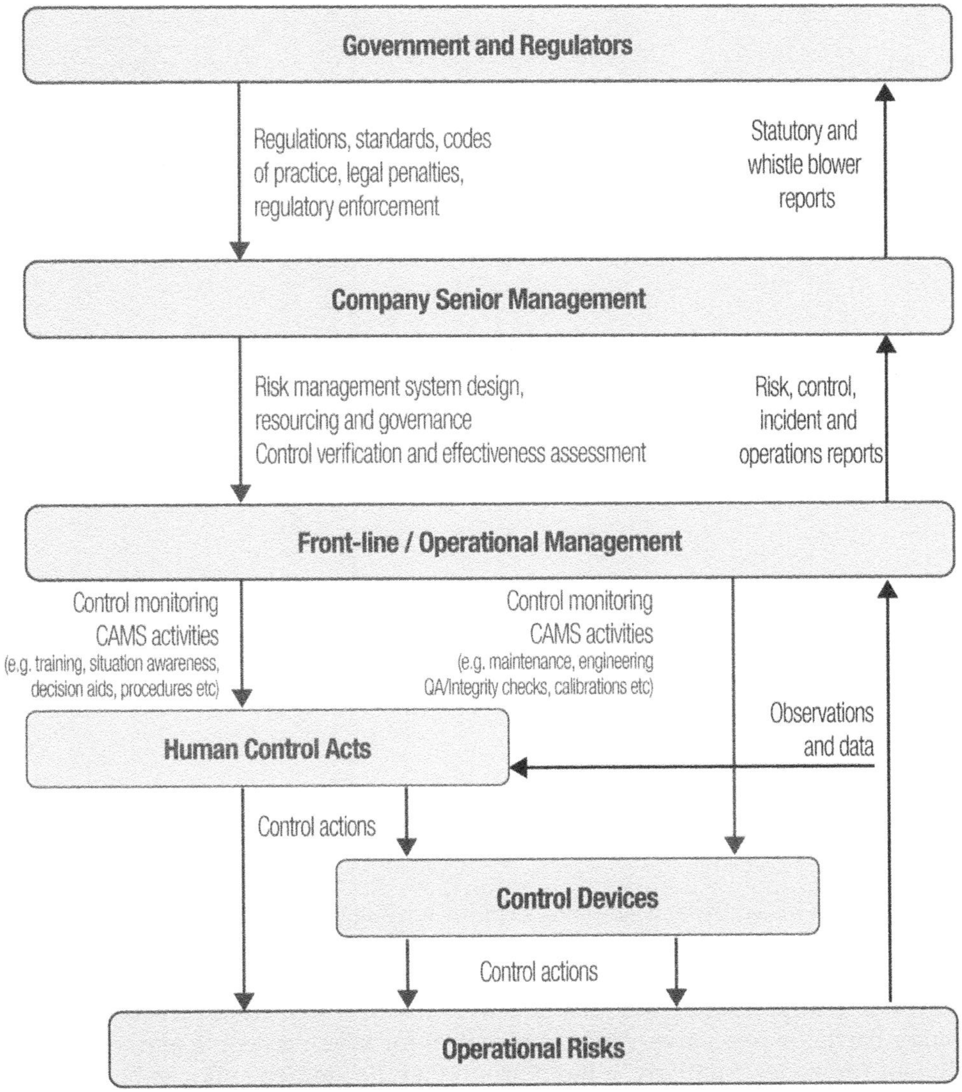

FIGURE 5.20 Risk control activities by organisational level.

is no to one of these questions, then the element may be a control failure prevention element or a control assurance management plan element.

- Is the element critical to preventing or mitigating an unwanted event? If so, it is probably best placed as a control.

Another tip to aid in specifying active devices and/or socio-technical controls is that controls should have *detect*, *decide* and *act* components (American Institute of Chemical

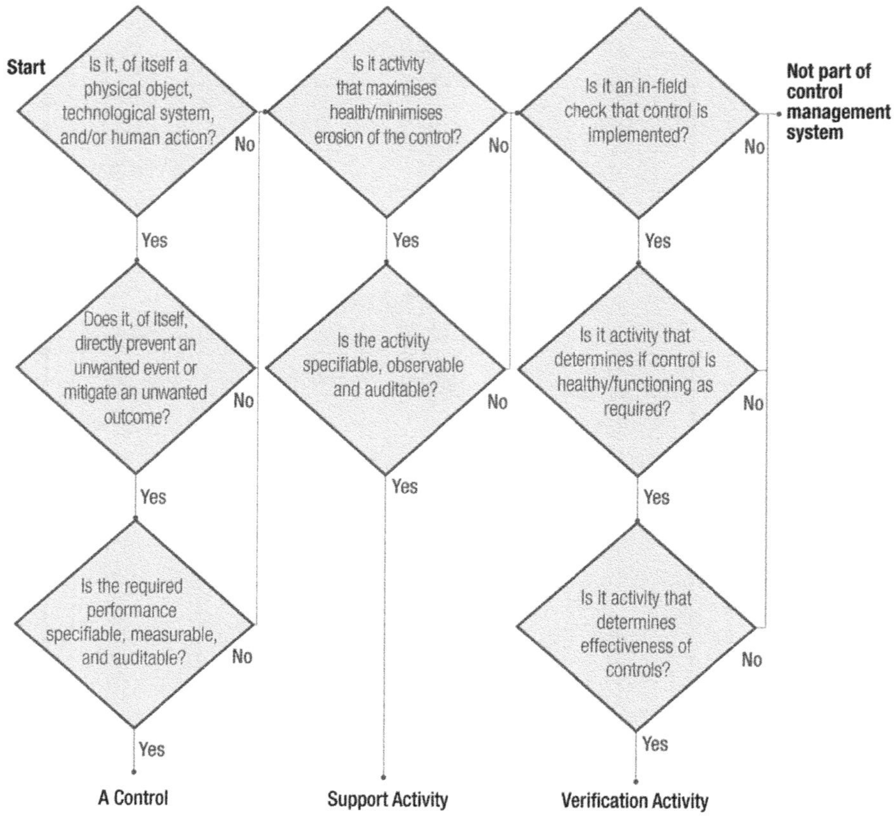

FIGURE 5.21 Defining a control (Hassall & Harris, 2017).

Engineers, 2018). This approach helps to explain why 'gas leak detection' is not a control, but that 'automatic shut-off in response to leak detection' is a control. The former only detects, but the latter detects, decides and acts.

However, it is important to note that effective controls can also be passive. For example, a stationary barrier is often a very effective control for keeping people and/or vehicles out of the line-of-fire zones. Similarly, a bund around a bulk storage tank can be very effective in preventing leaks from contaminating the environment. In all cases, the control should, of itself, directly prevent or mitigate an unwanted event or outcome.

When selecting controls, a check should be made to ensure that there are an adequate number of independent controls that address all the threats and consequences. For a control to be independent, it needs to be:

- Independent of the unwanted event (i.e. still able to function should an unwanted event occur). Controls that are not independent of the unwanted event cannot be counted as being able to reduce the risk at all.
- Independent of other controls (i.e. not be reliant on other controls working for it to work and have no common supports such as human actions, and/or supply of energy).

Two controls that are not independent counted both be counted as being able to reduce the risk.

Understanding the independence of controls is important as it determines what suite of controls will actually mitigate a risk and what selection of controls can be undermined by single/common erosion factors. When questions arise about the independence of a control this should prompt an engineering-based review of whether:

- There are ways to make the control independent of the unwanted event and any common supports (e.g. power supply, same person needed to respond to two controls at same time).
- The named control is really 'of itself' the thing that prevents or mitigates the risk – often this results in rewording controls so they capture the group of activities that reflect an independent control.

Ideally, the aim of bowtie analysis is to identify a suite of independent controls that intervene at all stages of the event sequence, from early to late, on both sides of the bowtie. This is illustrated in Fig. 5.22 and explained in Table 5.6. An example of a layered bowtie is also provided in Fig. 5.23. It is equally important that the analysis specifies the CAMS required to ensure the controls are implemented and monitored and maintained so that they remain effective.

Once the suite of controls are selected for a given risk, the controls should then be documented in the risk register as shown in Fig. 5.12. It can also be done in a more formal way through bowtie analysis as explained in the next subsection.

It is important to identify the failure modes for controls so that these may be addressed with the control management system. Control failure modes analysis should identify any factors that could cause a control to fail or could cause a control's performance to erode over time. As discussed previously, failure modes and corresponding CAMS elements can be visually shown using advanced bowties. As a minimum, identifying control failure modes should be considered for controls that have a significant impact arresting a threat

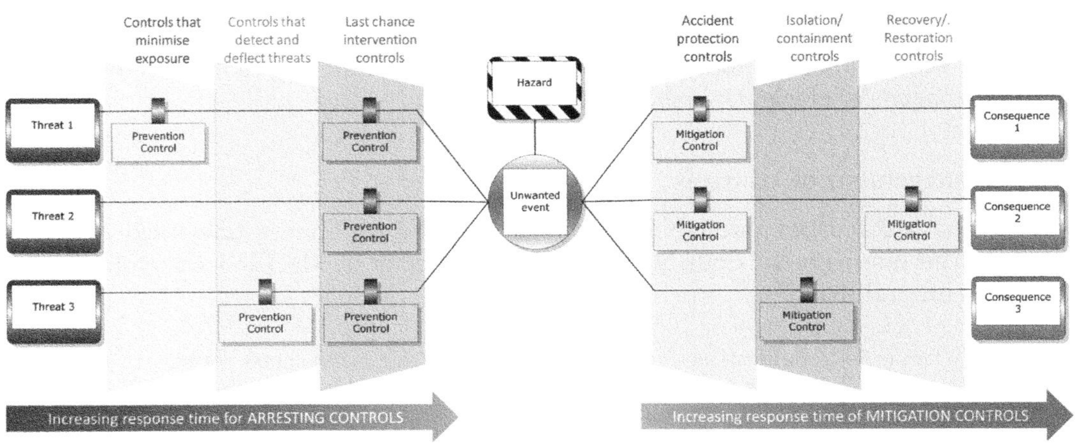

FIGURE 5.22 Layers and types of control by function and response time.

TABLE 5.6 Types of controls.

Control function		Intervention time	Examples of 'unwanted vehicle incident' controls for the threat: *Poor intersection design*
ARRESTING CONTROLS	Controls that minimise/eliminate exposure to a threat	Years/ months/days	Design out intersections, reroute vehicles to reduce number of vehicles that need to go through intersection
	Controls that detect and deflect threats	Minutes/ seconds	Traffic lights at intersection
	Last chance intervention controls	Seconds/ milliseconds	Human defensive driving, 'Collision Prediction' system
UNWANTED EVENT			
MITIGATING CONTROLS	Protection controls that provide instantaneous protection against harm	Milliseconds/ seconds	Airbag and seatbelt
	Isolation/containment controls that minimise damage to people, assets, environment production and reputation	Seconds/ minutes/ hours	Vehicle shutdown and fire suppression systems; treatment of injured people; site 'All Stop' system, accident scene barricading and management; fuel spill containment; event communication process
	Recovery/restoration controls which ensure system is returned to safe options	Minutes/ hours/days	Injured person return-to-work processes; damaged asset return-to-work processes; accident scene restoration processes

or a number of threats, or have a significant impact mitigating the severity of a consequence or a number of consequences. The FMEA approach (Section 5.3) can be used to perform control failure mode analyses. Examples of some control failure modes factors and failure prevention elements are listed in Table 5.7. An advanced bowtie for an unwanted vehicle interaction, illustrating some control failure modes and corresponding control management elements, is shown in Fig. 5.24.

5.4.9 Management of controls

As we have discussed throughout this book, we still experience unwanted events in industry. The International Council of Mining and Metals (ICMM) looked at the fatality related events and found:

> The top factors for ... incidents are people not properly identifying risks, controls not being in place, or the controls not being effectively implemented or maintained. *(ICMM, 2015)*.

This statement highlights that just doing the analysis to determine effective risk treatments is not sufficient – those treatments need to be effectively implemented and

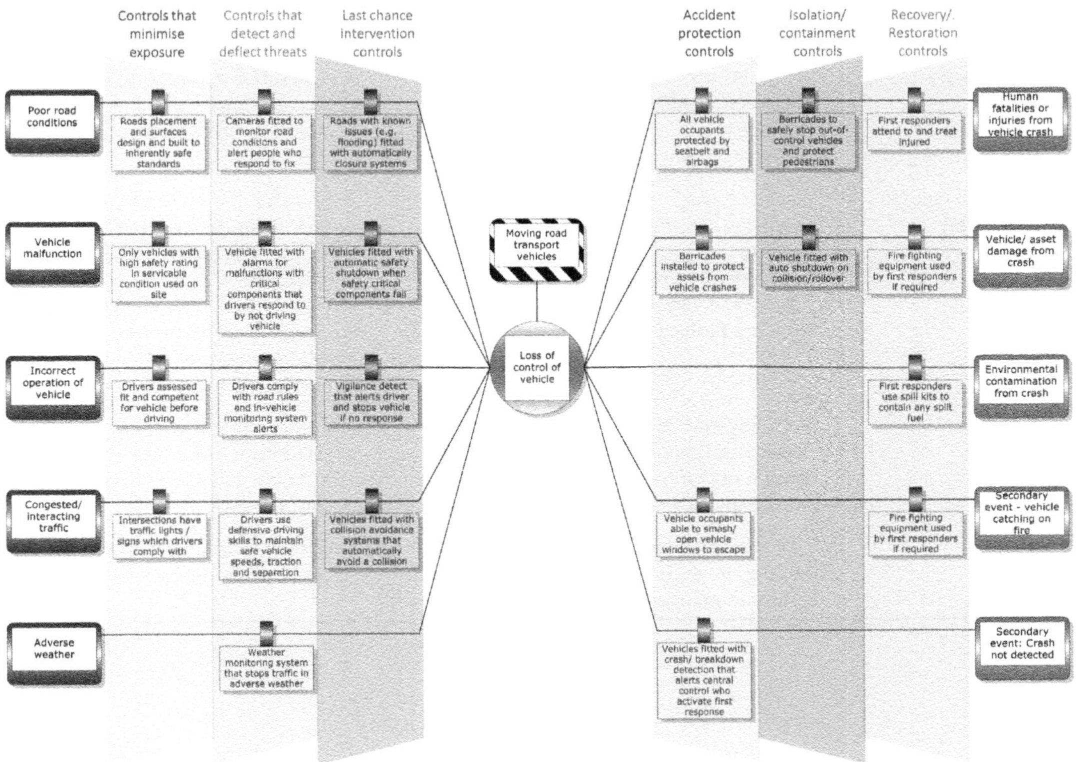

FIGURE 5.23 Layered bowtie for vehicle interactions.

maintained over time. The ICMM produced guidelines that recommend extra verification on those controls that are critical for preventing significant unwanted events. Verification activities are management's check that the people involved know what to do, that controls are implemented and that control actions and CAMS items are being done in a timely manner and to a high degree of quality (i.e. they are not been overlooked or rushed in terms of implementation and maintenance and that the checking process has not become a 'tick and flick' paper exercise). Verification activities should also be checking and seeking feedback from the people involved about the reliability and effectiveness of controls in order to answer the question – are we doing the best we can do?

The Bhopal accident (1986) is a horrific and pertinent illustration of the consequences of controls not being effectively implemented or maintained. In that case, despite several controls being designed and implemented that had the potential to prevent and/or mitigate the loss of containment of the MIC, they were ineffective at the time of the accident. These included a missing blind plate, a vent gas scrubber and a flare tower that were out of service, and an undersized water spray system (Slater, 2014; National Geographic., 2013).

Fig. 5.24 shows how monitoring and verification activities can be allocated in an organisation. Table 5.8 is an example of the assurance and verification activity details for a specific control.

TABLE 5.7 Examples of control erosion factors and erosion prevention elements by control type.

Type of control	Examples of control erosion factors / failure modes	Examples of failure prevention elements
Physical (passive) controls	• Wear and tear • Corrosion/weathering • Damage Incorrect placement/set up of temporary objects	• Condition monitoring and maintenance programs • Damage protection and response processes
Technological (active) system controls	• Wear and tear • Corrosion/weathering • Damage • Component failure • Failure of services (e.g. electricity, instrument air) • Software faults, cyber threats	• Condition monitoring and maintenance programs • Damage protection and response processes • Fail to safe software systems • Cyber security protections • Management of change process including audit
Human action	• Normalisation of deviance • Desensitisation (e.g. to alarms, signs) • Erosion of competencies and skills • Availability factors (e.g. workload, distractions)	• Induction and training programs • Mentoring programs • Competency-based assessments and reviews
Other	• Poor situation/environmental conditions (e.g. lighting, noise, housekeeping) • Failure to manage changes	• Work environment standards and monitoring programs

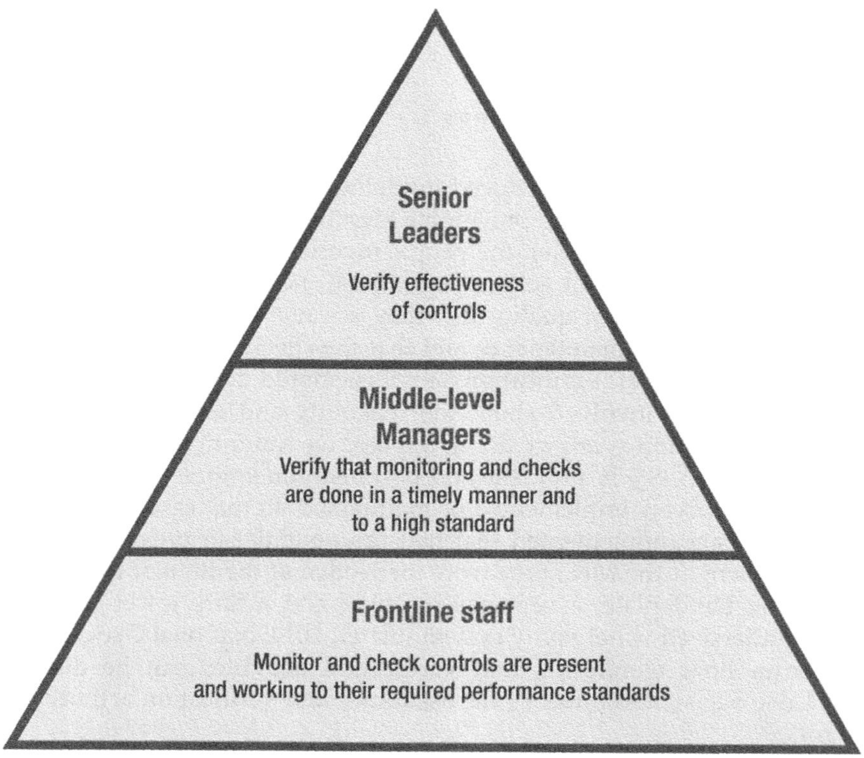

FIGURE 5.24 Control monitoring and review activities assigned to organisational levels.

TABLE 5.8 Example of control specification, monitoring and verification information.

1. What is the unwanted event?		
Loss of control of vehicle on site.		
2. Name of the control		
Intersections have traffic lights/signs which drivers comply with.		
3. What are the specific objectives of the control related to the unwanted event? Include a description of good/ideal control performance.		
Objective: To ensure prevent collisions at intersections by ensuring drivers follow signals/rules stating which vehicle(s) have 'right of way' and which vehicle(s) have to yield/giveway. **Good/ideal performance:** All road intersections have functional traffic lights or stop or giveway signs installed to be visible 100m from intersection. Intersections also be marked with clear painted line showing drivers where intersection is and where to stop vehicle		
4. Support activity		
Maintenance	**Training**	**Other**
Intersection traffic lights and street lighting maintained so remain fully functional Intersection signs and lines are cleaned and maintained so remain in good condition	Competency based training every 12 months for: - Drivers - Maintainers - Civil engineer	Civil engineer conducts annual review of intersection design, condition and visibility against relevant standards
5. Control monitoring requirements		
Daily inspection	**Answer**	**Comments and actions**
Night shift supervisor: Are all intersection lights and signs present, working and visible from 100m?	☐ Yes ☐ No ☐ Not applicable	
All supervisors: Review traffic reports (e.g. IVMS or camera footage) - Are drivers/vehicles complying to intersection rules?	☐ Yes ☐ No ☐ Not applicable	
All supervisors: Can workers explain what intersections rules are and is their training up-to-date?	☐ Yes ☐ No ☐ Not applicable	
Manager Verification questions	**Answer**	**Comments and actions**
Do inspection walk-through with supervisors - Are supervisors able to demonstrate correct daily inspection and reporting requirements.	☐ Yes ☐ No ☐ Not applicable	
Do intersection inspection with civil engineer – Is he/she able to describe how intersection compliance check is done and reported?	☐ Yes ☐ No ☐ Not applicable	
Is intersection maintenance up-to-date with no outstanding actions?	☐ Yes ☐ No ☐ Not applicable	
Check traffic reports across site to determine effectiveness of control. Is control effectiveness acceptable?	☐ Yes ☐ No ☐ Not applicable	Control effectiveness =

5.5 Summary

Identifying, evaluating and treating risk is crucial for ensuring that an organisation achieves its objectives. It is often the overlooked or underestimated risk that causes companies losses from incidents and/or lost opportunities. This chapter has introduced the theory and methodology for performing risk assessments and risk treatment activities.

Risk assessment is a structured process with three distinct steps – risk identification, risk analysis and risk evaluation. The tools are not conceptually complex, and can be quite straightforward to use. The best way to fully understand the process of risk assessment is to do it. However, within process industries, it is critical that this process is implemented rigorously, with suitably experienced and qualified people. Novices beware! There is no good substitute for experience and relevant knowledge when performing quality risk assessments.

Risk treatment is required when an inherent risk is deemed unacceptable or intolerable. Risk treatment involves identifying the unwanted event, analysing controls to select the ones required to prevent and mitigate the unwanted event, then determining the management activities required to ensure that the controls are implemented, monitored, maintained and effective at addressing the unacceptable risk.

In this book, we strongly advocate the use of the bowtie technique to assist with risk treatment and management. It is a powerful way to visually represent hazards, unwanted events, threats, consequences, and the prevention and mitigation controls. In combination with a CAM, this provides a structured, rigorous and effective way for specifying and managing critical controls.

Review questions

1. What is the first step in the risk management process?
2. What are the three activities that comprise risk assessment? List them in the order that they should be conducted.
3. What are the three activities the comprise risk treatment in the extended version of the risk management process?
4. State one example of a hazard identification and analysis technique.
5. State one example of a process deviation analysis technique.
6. State one example of a plant failure analysis technique.
7. A key lesson learned from using risk matrices is that high consequence, low likelihood events can often get overlooked. Comment on this statement and explain under what circumstances it is not necessary to consider the likelihood that a consequence might occur.
8. What is ALARP? Give an illustrative example?
9. List the hierarchy of risk treatment controls in order of increasing effectiveness.
10. What is the name of the controls on the left side of the bowtie?
11. What is the name of the controls on the right side of the bowtie?
12. What is the difference between a basic and advanced bowtie?

13. What is the role of a CAMS?
14. Which of the following is an example of a good description of an unwanted event? If it is not, explain why.
    a. Gas explosion
    b. Incomplete isolation
    c. Uncontrolled release of gas from a storage vessel
    d. Dam wall failure
    e. Airborne toxins above acceptable workplace limits
15. Which of the following is an example of a good description of a threat? If it is not, explain why.
    a. Equipment not in serviceable condition due to corrosion
    b. Worker error
    c. Building collapse
    d. Failure of pressure relief valve
    e. Inadequate inspections
16. Which of the following is an example of a good description of a consequence? If it is not, explain why.
    a. Harm to employees
    b. Malfunctioning control system
    c. Uncontrolled release of gas from a storage vessel
    d. Human fatalities from hydrocarbon vapour cloud explosion
    e. Water contamination due to chemical run-off
17. What is the definition of a control? Provide an example of:
    - A poorly specified control. What is wrong with it?
    - A well-specified control.
18. List the three types of arresting controls in order of their intervention time (decreasing).
19. List the three types of mitigating controls in order of their intervention time (increasing).
20. State three examples of control erosion factors.

References

American Institute of Chemical Engineers. Center for Chemical Process Safety. (2018). *Bow ties in risk management: A concept book for process safety*. New York, NY: John Wiley & Sons, Inc, Hoboken, NJ.

Hassall, M.E., & Harris, J. (2017). *Risk controls knowledge: Determining leading practice from case study analysis (ACARP report C25036)*. Australia. Retrieved from https://www.acarp.com.au/abstracts.aspx?repId = C25036.

Hassall, M.E., Joy, J., Doran, C., & Punch, M. (2015). *Selection and optimisation of risk controls (ACARP report C23007)*. Retrieved from: http://www.acarp.com.au/abstracts.aspx?repId = C23007; Australia.

ICMM. (2015). *Health and safety critical control management good practice guide*. Published by the International Council on Mining and Metals (ICMM), London, UK. Retrieved from https://www.icmm.com/en-gb/publications/health-and-safety/health-and-safety-critical-control-management-good-practice-guide; London, UK.

IOGP. (2011). *Human factors engineering in projects*. Report No. 454. International Association of Oil & Gas Producers (IOGP).

Killick, C. (2014). *Course Notes for CHEE4002 - Impact and risk in the process industries*. The University of Queensland.

Lekutis, C. (Producer). (2009, August 14, 2009). *Canada and the UK enjoy enviable safety records - Scheule and margin pressures seen as possible reasons for higher death coiunt of techs in the US.* Retrieved from http://wirelessestimator.com/content/articles?pagename = International%20Tower%20Climber%20Deaths.

National Geographic. (2013). Seconds From Disaster: Bhopal, Series 4, Ep 1 of 6. National Geographic. Broadcast Date: 30 Nov 2013; Duration 45 min., 8 sec. Source: Informit. Melbourne.

NOPSEMA (Ed.). (2015). *Guidance note - ALARP (N-04300-GN0166 Revision 6).* Available from https://www.nopsema.gov.au/assets/Guidance-notes/A138249.pdf. Australia.

Pitblado, R., & Weijand, P. (2014). Barrier diagram (Bow Tie) quality issues for operating managers. *Process Safety Progress, 33*(4), 355–361. Available from https://doi.org/10.1002/prs.11666.

Rasmussen, J., & Svedung, I. (2000). *Proactive Risk Management in a Dynamic Society.* Sweden: Risk & Environmental Department, Swedish Rescue Services Agency.

Slater, D. (2014). *Bhopal: A tragedy of unintended consequences. The Chemical Engineer. IChemE.* (December), 32–36.

Solicitors Regulation Authority (Ed.). (2014). SRA regulatory risk framework. Available from http://www.sra.org.uk/risk/risk-framework.page.

Snyder, B. (2017). 7 insights from legendary investor Warren Buffett. Available from https://www.cnbc.com/2017/05/01/7-insights-from-legendary-investor-warren-buffett.html#:~:text = %E2%80%9CRisk%20comes%20from%20not%20knowing,nicely%20with%20the%20next%20quote. (2017) Accessed 12.02.23.

UK Health and Safety Executive. (n.d.). *Guidance on ALARP Decisions in COMAH.* Retrieved November, 2021, from https://www.hse.gov.uk/foi/internalops/hid_circs/permissioning/spc_perm_37/.

CHAPTER

6

Monitor and review risks

OUTLINE

6.1 Introduction 126

6.2 Why perform event investigations? 128

6.3 Purpose and theory behind investigations 130

6.4 Incident investigation techniques and application considerations 132

6.4.1 User beware 134

6.4.2 Timeline 134

6.4.3 5 Whys analysis 135

6.4.4 Fishbone 138

6.4.5 Human Factors Analysis and Classification System 142

6.4.6 Bowtie analysis 145

6.4.7 Accident Analysis Mapping 148

6.4.8 Strategies Analysis for Enhancing Resilience 150

6.5 Integration of learning back into the business 154

6.6 Summary 154

Review questions 155

References 157

Fundamentals of Risk Management for Process Industry Engineers
DOI: https://doi.org/10.1016/B978-0-12-820320-0.00010-6

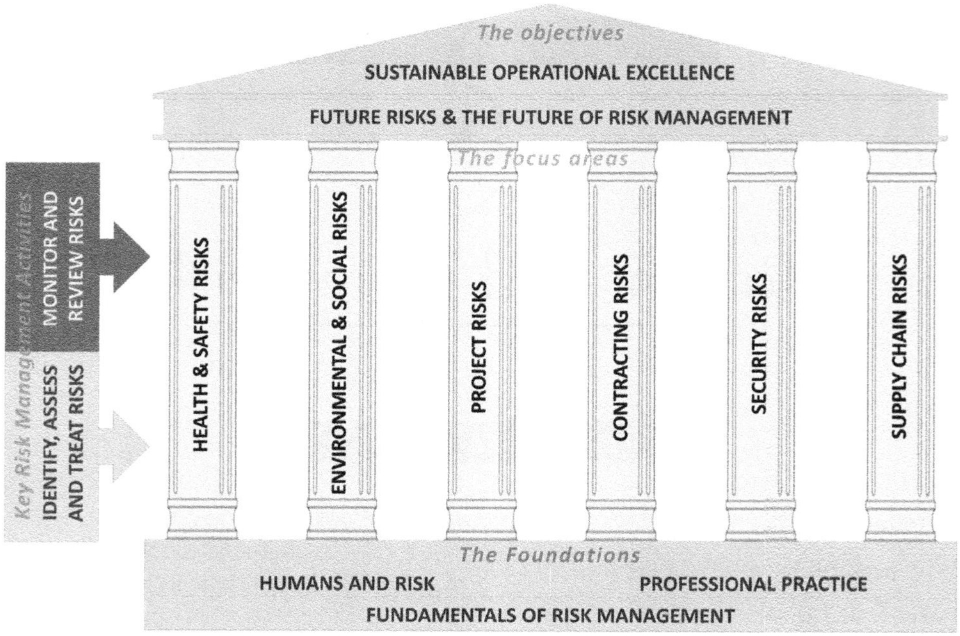

'One of the great mistakes is to judge policies and programs by their intentions rather than their results' Milton Friedman (Phelan, 2018).

6.1 Introduction

In the last chapter we learned how to identify, assess and treat risks in order to prevent incidents. In this chapter, we will focus on monitoring and reviewing risks, as shown on the right side of the ISO31000 Risk Management Process (Fig. 5.1). In the process industries, monitoring and reviewing risks entails:

- Monitoring current and emerging risks (as discussed in Chapters 1 and 5) to keep risk register up to date.
- Monitoring the status, adequacy and effectiveness of controls implemented to treat risk (as discussed in Chapter 5).
- Investigating successful and unwanted events to capture learnings in order to prevent the recurrence of accidents and to promote the recurrence to beneficial events.
- Monitoring and reviewing to continually improve the overall risk management process.

Monitoring and reviewing risks is done to continually check for new or changing risks as well as to check that the required risk treatments are implemented and effective. This should be done prospectively by focusing on the proactive risk management processes. It also should be done retrospectively to capture lessons learned from incidents as shown in Fig. 6.1.

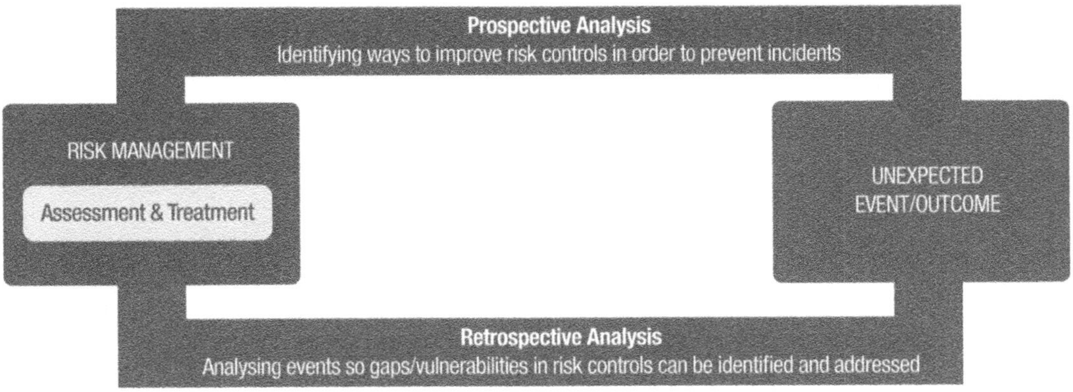

FIGURE 6.1 Prospective and retrospective risk monitoring and review analysis.

In addition, it is also important that there is oversight on the prospective and retrospective processes to ensure they are occurring in a timely manner, being performed to the right degree of rigour, and that the insights and learnings are used to inform the overall approaches to managing risks. Typically, in the process and other industries this oversight is conducted using three lines of defence. The three lines of defence model, shown in Fig. 6.2 highlight that good oversight or monitoring and reviewing of risk management processes requires:

- A first line of defence where operation management takes ownership, responsibility and accountability for assessing and controlling risks, using the required frameworks and approaches to deliver the agreed level of risk.
- A second line of defence where senior management provides the required framework and resources to effectively manage risk and monitors compliance and provides feedback on performance.
- A third line of defence which comprises an independent in-depth review or audit of the performance of the risk management framework and the first two levels of defence.

As mentioned above, the monitoring and reviewing of risks should include investigating events that have produced adverse or beneficial outcomes. Event investigations allow us to identify and capture lessons that should be learned in order to prevent future accidents and promote future beneficial outcomes. Event investigations are conducted in a structured and systematic way in order to ensure all important learnings are identified along with recommendations that inform decision makers on how to improve risk management activities. In the process industries, it is important to take a whole of systems perspective when investigating events to ensure all learnings and recommendations are captured, as highlighted with the following quote:

> *Many accident investigations make the same mistake in defining causes. They identify the widget that broke or malfunctioned, then locate the person most closely connected with the technical failure: the engineer who miscalculated an analysis, the operator who missed signals or pulled the wrong switches, the supervisor who failed to listen, or the manager who made bad decisions.* ***When causal chains are limited to technical flaws and individual failures, the ensuing***

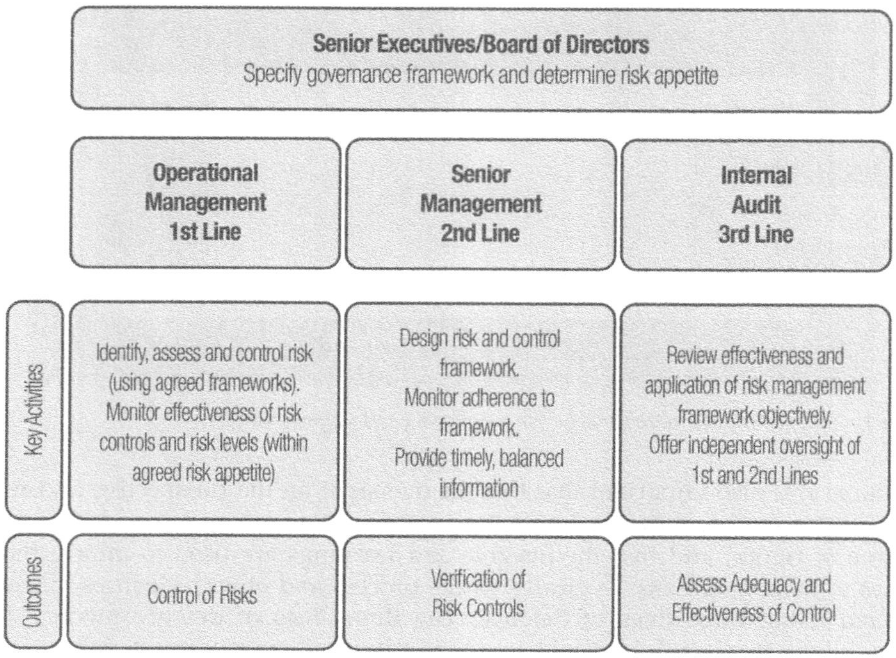

FIGURE 6.2 Three lines of defence model for monitoring and reviewing risk management activities within a company.

> ***responses aimed at preventing a similar event in the future are equally limited:*** *they aim to fix the technical problem and replace or retrain the individual responsible. Such corrections lead to a misguided and potentially disastrous belief that the underlying problem has been solved* (Columbia Accident Investigation Board, 2003, p. 177).

The Columbia Accident Investigation Board emphasised that NASA's ***organisational culture and structure*** were key factors underpinning the Columbia accident. Due to the importance of investigations in the process industries, the remainder of this chapter will focus on the theory and approaches associated with incident investigations.

At the end of this chapter you should be able to apply a suite of tools to perform an incident investigation, and to demonstrate insight by making relevant recommendations in order to prevent reoccurrence of an incident. You should also be able to recount several key process incidents, summarise key lessons learned, and highlight key recommendations to prevent reoccurrence of the incident.

6.2 Why perform event investigations?

Most operating in the process industries experience a range of event outcomes as shown in Fig. 6.3. The occurrence of novel and reoccurrence of unexpected adverse and successful events is an indication that there are shortcomings in risk management processes and process plant operational performance.

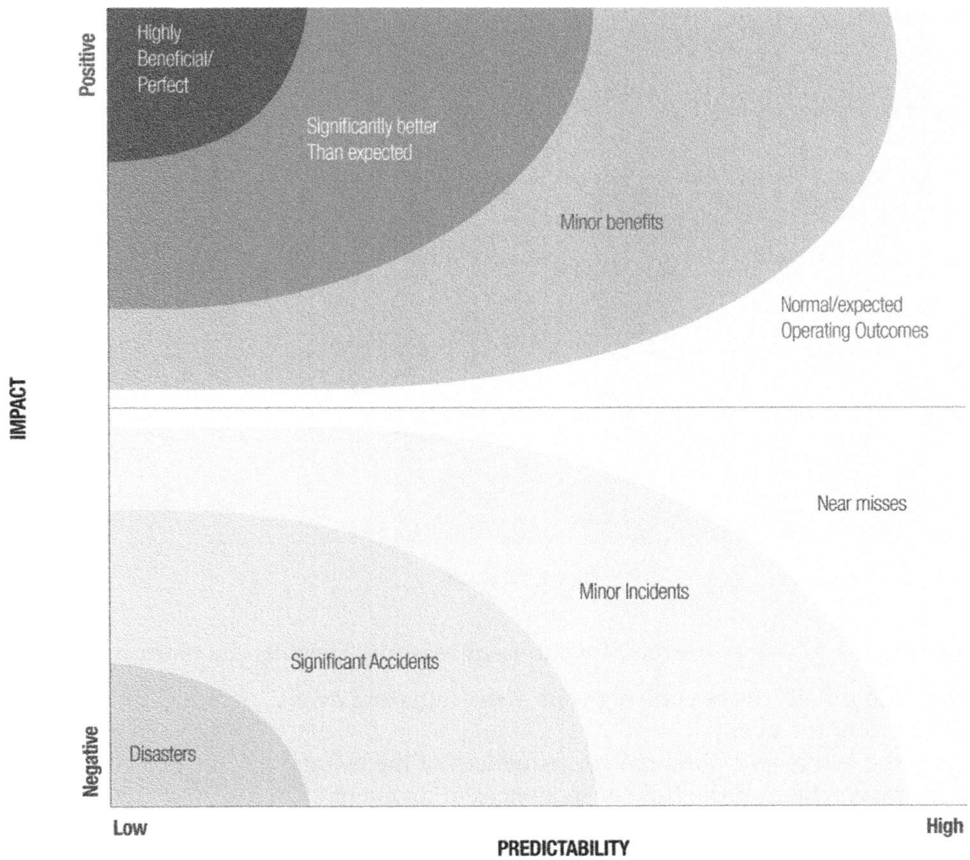

FIGURE 6.3 Range of event outcomes experienced in industry (Hollnagel, 2011). *Source: Adapted from Hollnagel, E. (2011). Prologue: The scope of resilience engineering. In E. Hollnagel, J. Pariès, D. D. Woods, & J. Wreathall (Eds.), Resilience engineering in practice: A guidebook. Surrey UK: Ashgate.*

The reason why we choose to investigate events is because we want to learn in order to improve. Investigating both adverse and successful events can provide insights that help improve and strengthen risk management processes. So the events that should be investigated are the ones we can and want to learn useful lessons from. But sometimes we need to investigate to comply with the law as shown in Fig. 6.4.

In process industries, most events investigated are adverse events or incidents because they have significant actual or potential impacts on health, safety, environment, and operational performance. The adverse events in the process industries are predominantly reoccurring events. Reoccurring events are those events that are similar to previous incidents (e.g. Fishwick, 2014; Gill, 2013; Waite, 2013). Research has shown that adverse events often result from a failure to identify risks properly, and a failure to implement and maintain known controls for known hazards (ICMM, 2013; Noetic Solutions, 2014). Investigations, if done well, should identify and address shortcomings in risk identification and control.

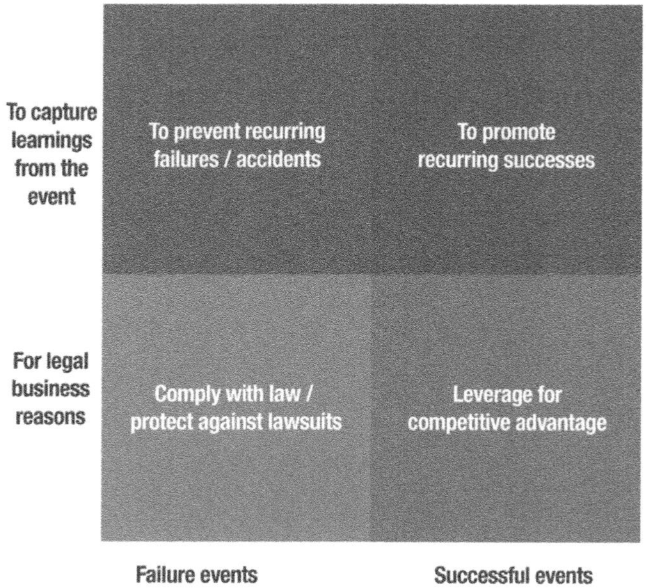

FIGURE 6.4 Reasons for investigating different types of events.

Investigations can be performed for different reasons. They can be performed to:

- Identify the initial loss of control event – the initiating event.
- Find causes of the event.
- Identify the actual and potential consequences of the event.
- Collect information on the full consequences of an event.
- Provide insights into the effectiveness of involved controls.
- Collect information to use for legislated reporting requirements.
- Collect information for legal reasons including for use in legal proceedings.
- Collect and disseminate learnings from the event (that may include findings from some or all of the above points).

6.3 Purpose and theory behind investigations

The purpose of investigating events is to learn about what happened and to gain insights into how to prevent negative and promote positive outcomes in the future. Specifically, event investigations seek to achieve at least one of the following:

- To determine the circumstances that led to unwanted outcomes in order to identify ways to prevent reoccurrences.
- To determine the circumstances that led to positive outcomes in order to identify ways to promote reoccurrences.

Before conducting incident investigations, it is important to determine:

- Whether we should do an investigation?

- Why we are doing the investigation?
- What are we investigating?
- How are we going to go about doing the investigation?

Determining whether to do an investigation is the first decision to be made. According to van Kampen and Drupsteen (2017), choosing whether to investigate or not should include consideration of the following:

1. The actual severity or potential severity of the incident (i.e. such as fatalities, extensive property damage, a large loss of containment).
2. The local legal/regulatory or company procedural requirement to do so.
3. The occurrence of recurring incidents of a similar nature either within the same company or sector.
4. 'Near miss' incident where the occurrence of a potentially severe accident was just barely avoided.

In addition, van Kampen and Drupsteen (2017) state that investigating the 'near miss' or minor outcome incidents which could have been severe accidents allows companies to:

> *'identify and control unforeseen hazards or inadequate control measures before they cause a more serious incident. In this way regular incident investigation can help to improve safety performance continuously/.*

Determining ***why*** an investigation is being performed is important because it can help guide decisions associated with the scope and focus of the investigation, as well as the tools to be used.

Determining ***what*** to investigate involves making decisions about the scope of the investigation. Events often emerge from a series of preceding activities, and decisions that can stretch a long way back in time and extend beyond the affected part of the organisation to other departments and organisational levels. The impact of an event can also stretch forward in time and in spatial area.

Determining the ***scope*** is about setting the time, physical area, organisational and external limits of the investigation. It is about clearly specifying what is in the scope of investigation. For example, an incident investigation might be limited to the events that occurred in the week leading up to the accident and exclude activities and decisions that occurred outside that timeframe. In addition, the incident investigation might be confined to looking at the actions of those directly involved, rather than all potentially contributing actions including those of leaders and regulators. Incident investigations might also be limited to looking at the immediate causes and acute consequences of an event and not include investigating the latent causes of the event and potential chronic consequences that could result from the event. Finally the investigation might include only things that fall within the spatial area controlled by the company and thus exclude things outside that area such as issues associated with supply chains, the company's reputation, environment and/or community health.

Determining ***how*** to investigate involves determining what incident investigation process and tools will be used, and what resources will be involved in the investigation. This step clarifies if the investigation is going to: (1) follow a formal or informal process; (2) involve predefined and specified techniques and tools; and (3) be conducted by an individual, team or prescribed set of people. The importance or significance of the event can dictate how formal and how in-depth the investigation process is going to be. The complexity of the event to be analysed can also be used to guide decisions and the processes, tools and resources required.

Steps to consider when conducting an incident investigation process are:

STEP 1: Describe the context, which involves articulating:
The event to be investigated and the outcome of that event
The purpose of the investigation
The scope of investigation (what's in and what's out of scope)

STEP 2: Investigate the accident to determine ***what*** happened.
The output of Step 2 usually is a timeline and/or a narrative that outlines what happened, where it happened, when it happened and who was involved. This step is often done iteratively with determining the scope of the investigation.

STEP 3: Analyse the accident to determine ***why*** it happened.
The aim of Step 3 is to identify the factors that initiated, escalated (either positively or negatively) and/or thwarted the event (i.e. to prevent a more positive or negative outcome). Investigation techniques are used to help investigators identify technical, human and system contributions to accidents. These are discussed in more detail in the next section.

STEP 4: Recommend ways to ***prevent or promote*** similar unwanted or wanted events.
The aim of Step 4 is to summarise the findings of the incident investigation into actionable recommendations that are **s**pecific, **m**easurable, **a**ssignable, **r**ealistic/relevant and **t**ime-bounded (SMART).

To ensure quality incident investigations and analyses, Ryan (2015, p.827) produced a list of eight requirements that need to be met which are quoted below:

1. *Have a clear scope for analysis (e.g. whether it should focus at the level of the work and the technological system, or more broadly at influences from government and regulators)*
2. *Be influenced by a model or group of models*
3. *Provided a detailed description of the accident, including a visual representation of the accident sequence if appropriate*
4. *Search for and reveal underlying causes*
5. *Contribute to understanding of prevention (e.g. safety barriers)*
6. *Help in generating recommendations*
7. *Give consideration to practical aspects, such as level of education and training that is needed to use the method*
8. *Be valid and reliable*

6.4 Incident investigation techniques and application considerations

When selecting incident investigation tools or techniques it is important to ensure the incident investigation will deliver the desired outcome. Different tools and techniques will help analysts identify and analyse different facets of an event. Table 6.1 provides a summary of commonly used and/or contemporary incident investigation techniques. It is good practice to employ a range of different techniques, as each may provide a different, but complementary, perspective. For example:

TABLE 6.1 Summary of incident investigation techniques.

Technique	Cause	Prevention control	Initiating event	Mitigation control	Consequence	Human/ Organisational Factors	Example References
Timeline (sequence of events)	Unclear/ Sometimes	Unclear/ Sometimes	Yes	Unclear/ sometimes	Yes	No	
5 Whys	Yes	No	Yes	Not	Yes	Yes	Serrat (2017)
Incident Cause Analysis Method (ICAM)	Yes	Unclear/ Sometimes	Yes	Unclear/ sometimes	Yes	Yes	De Landre et al. (2007)
Root Cause Analysis	Yes	No	Yes	No	No	Unclear/ Sometimes	Harich & Rosas (2020)
Swiss Cheese Accident Model	Yes	Yes	Yes	Unclear/ Sometimes	Yes	Yes	Reason (1997); Reason et al. (2006)
Fault Tree Analysis	Yes	Yes/ Sometimes	Yes	No	No	Unclear/ Sometimes	Pilot (2002)
Event Tree Analysis	No	No	Yes	Yes	Yes	No	
Bowtie Analysis	Yes	Yes	Yes	Yes	Yes	Yes/ Sometimes	Hassall et al (2015); Pitblado and Weijand (2014); Saud et al. (2014)
Human Factors Analysis Classification System (HFACs)	Unclear/ Sometimes	No	Unclear/ Sometimes	No	No	Yes	Shappell and Wiegmann (2000) and Wiegmann and Shappell (2001)
Accident Analysis Mapping (Accimap)	Yes	Unclear/ Sometimes	Yes	Unclear/ Sometimes	Yes	Yes	Svedung and Rasmussen (2002); Underwood and Waterson (2014)
Systems Theoretic Process Analysis (STPA)	Yes	Yes	Yes	No	Unclear/ Sometimes	No	Leveson (2004, 2011); Leveson and Thomas (2018)
Functional Resonance Analysis Method (FRAM)	Yes	No	Yes	No	Yes	Yes	Hollnagel (2012) and Macchi et al. (2009)
Strategies Analysis for Enhancing Resilience (SAfER)	Yes	Yes	Yes	Yes	Yes	Yes	Hassall et al. (2014)

(adapted from Dodshon and Hassall, 2017).

- The *scope* of the investigation, and determination of *what happened*, can be derived using **timeline** analysis. A descriptive narrative can also be useful.
- *The root or primary cause(s)* of the event may be determined by using **5 whys** analysis, and **Root Cause Analysis** (RCA).

- *The range of causes or contributing factors* (e.g. technical, human, organisations systems and environmental) can be considered using the **fishbone technique**. Other examples include **Essential Factors Analysis**.
- The *human contribution* to the event across different organisational levels can be described and analysed using the Human Factors Analysis and Classification System **(HFACS)** technique.
- The role that *organisation systems* (e.g. communication, contractor management, training systems) played in the event can be elucidated using the Accident Analysis Mapping **(Accimap)** technique.
- The *role that risk controls and risk control management plans played* in the event can be evaluated using the **bowtie analysis**.
- Design interventions *that help humans successfully manage similar situations* in the future can be facilitated by use of the Strategies Analysis for Enhancing Resilience **(SAfER)** technique.

In the remainder of this chapter we will describe the more prevalent and newly emerging techniques: timeline (sequence of events), 5 whys, causal analysis (fishbone), Accimap, HFACS (human error analysis), fault and event tree analysis, bowtie analysis, and SAfER technique. These techniques are demonstrated in Appendix B. TapRoot and ICAM are not discussed as these are commercially available software-based incident investigation frameworks that incorporate the philosophies of the other techniques covered in this chapter. Systems theoretic accident model and processes (STAMP and its STPA derivation) and the function resonance analysis method (FRAM) are approaches that typically require expertise to administer so they are only introduced here.

6.4.1 User beware

Analogous to the guidance we provided in Chapter 5, we would like to provide a word of caution before discussing incident investigation techniques in detail. This continues a key theme that we are expounding throughout the book – namely that the effectiveness of the use of any of the techniques presented will heavily rely on the background, competence and knowledge of the person or team using and interpreting the techniques.

When conducting incident investigations, it is important to understand that personal and group biases can affect the quality of the analysis and recommendations. We discussed bias earlier in Chapter 3. Many of these biases are applicable in event investigations. There is also the phenomena of *'what you look for is what you find'* as highlighted by Lundberg et al. (2009).

6.4.2 Timeline

Timeline (also known as *event timeline*) is used in incident investigations to state **what** happened and **when**. It is a statement of facts that may include:

- the initial events that led to the loss of control incident.
- the subsequent events that escalated the loss of control incident to the actual consequence.
- the events associated with the response and recovery of the situation.

The event timeline *'is the workhorse in an event investigation because it provides a systematic tool to separate events in time to allow events that may be critical to determining appropriate causal factors to be seen and acted upon'* (U.S. Department of Energy, 2012, pp. 2–44). Thus timelines are a way of articulating the scope of the investigation because they highlight the start point, end point and events considered.

Timelines can be represented diagrammatically or in tabular or descriptive form. Fig. 6.5 shows a diagrammatic timeline representation of the Macondo well blowout which led to the explosion and fire that resulted in the sinking of the Deepwater Horizon drilling rig in 2010. The timeline clearly illustrates the sequence of key events from the Deepwater Horizon drilling rig arriving at the site in January 2010, through to the explosion and fire on April 20, then the post-event sequence through to the wellbeing sealed on September 19. Table 6.2 presents a timeline for the toxic chemical release that occurred at the DuPont La Porte Chemical Facility in Texas USA on November 15, 2014. It shows the timeline of events from unit startup on November 12 through to the emergency response then stopping the chemical leak on November 15, 2014. In this particular case, there is no information provided post-event. This is not always the case, and is context specific, depending on the purpose of the investigation. For detailed reviews of the Macondo/Deepwater Horizon and DuPont incidents, see the full investigation reports published by the U.S. Chemical Safety and Hazard Investigation Board (U.S. Chemical Safety and Hazard Investigation Board., 2016, 2019) respectively.

The degree of detail placed in timelines can vary. The important features are that it clearly:

- Defines the start and end points of the investigation.
- Describes the key events to help readers understand what happened and when.
- Focuses on the events 'that matter' because they contributed to the initial loss of control, the escalation of the event, and the recovery of the situation. To test whether an event matters, a good question to ask is – if this event didn't occur, would the accident have been prevented, or would the consequences of it have been significantly less?
- Does not include events that did not happen – for example, the valve was not closed; the tank was not drained.
- Does not include analysis or supposition.
- Represents the information in a readable and easy-to-understand format.

6.4.3 5 Whys analysis

The 5 whys incident investigation technique involves taking the key events from the timeline and asking why did they happen? The aim of the technique is to force people into thinking about the root causes of an event. The 5 whys technique is illustrated in Fig. 6.6. The timeline of events is at the top, and the why's are asked for each event to determine root causes.

The technique is a quick technique to use that helps people to think more deeply about the causes of an event. However, without a good timeline and without good guidance, the technique tends to produce different results when different investigators use it. It is one of the techniques where 'what you look for is what you find' (Lundberg et al., 2009).

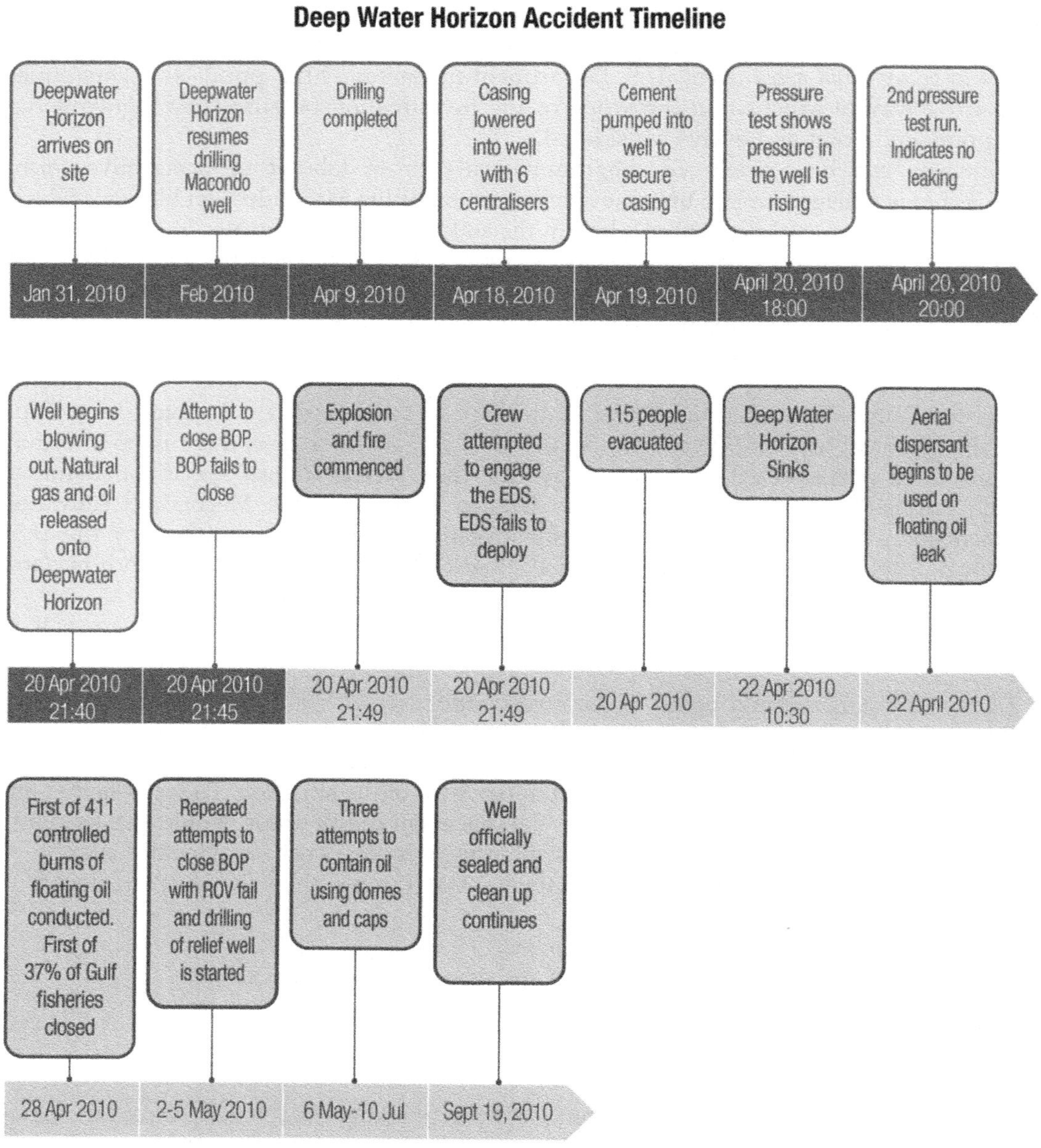

FIGURE 6.5 Timeline for the Macondo well blowout (Graham et al., 2011; U.S. Chemical Safety and Hazard Investigation Board., 2016). *Source: From Graham, B., et al. (2011). Deep water: The Gulf oil disaster and the future of offshore drilling: Report to the President. USA, National Commission on the BP Deepwater Horizon Oil Spill and Offshore Drilling. and From U.S. Chemical Safety and Hazard Investigation Board. (2016). Investigation report - Executive summary - Drilling rig explosion and fire at the Macondo Well - Report No. 2010-10-I-OS. U.S. Chemical Safety and Hazard Investigation Board.*

TABLE 6.2 Timeline for Toxic Chemical Release at the DuPont La Porte Chemical Facility 2014 (U.S. Chemical Safety and Hazard Investigation Board., 2019).

Date	Time	Events
12-Nov-14	14:00-15:00	Operators attempt to restart Lannate® process after a unit shutdown. Flow is not established
Co	15:00	Operators begin troubleshooting to clear plugging.
14-Nov-14	10:00-03:02	Methyl mercaptan released. Detectors inside and outside the manufacturing building activate alarms 32 separate times
15-Nov-14	02:51-02:54	High pressure in equipment inside the manufacturing building activates control system alarms. Liquid methyl mercaptan flowing into the waste gas vent header piping due to troubleshooting efforts to clear plugging
	02:58-03:03	Shift Supervisor and Operator 1 separately go to the manufacturing building to try to reduce the equipment pressure
	03:01-03:13	Two drain valves on the waste gas vent header are manually opened. Highly toxic and flammable liquid methyl mercaptan flows out
	03:24-03:26	Three methyl mercaptan detectors inside the manufacturing building activates alarms at the control panel
	03:30	Operator 1 makes an urgent distress call over the radio. Control room operator requests more information but receives no response
		Operator 2 and Operator 3 run from the control room into the manufacturing building to respond to the distress call from Operator 1. They do not wear any respiratory protection
		Operator 4 sees Operator 2 and Operator 3 running into the manufacturing building. Operator 4 follows them. They are not wearing respiratory protection
	03:30-03:40	Operator 5 is exposed to methyl mercaptan in the manufacturing building. They escape from the building and lie down on grass to recover
	03:35	The Board Operator attempts to contact the Shift Supervisor, Operator 1 and Operator 2 over the radio. They do not respond
		Operator 6 takes 3 escape respirators into the manufacturing building
	03:40	The manufacturing building fume release alarm sounds. It is manually activated
		Operator 6 encounters Operator 4 in a stairwell and puts an escape respirator on them. Operator 4 recovers and escapes the manufacturing building
	03:50	The Board Operator calls for the plant emergency response team (ERT). The ERT are not informed there is a toxic chemical release
	03:57	The Board Operator calls the security guard and requests the guard to call 9-1-1
	03:58	Emergency responders from the ERT arrive at scene. They do not have appropriate PPE to enter a building with a hazardous chemical release
	04:05	Site Incident commander calls for the ERT to come to the scene with self-contained breathing apparatus (SCBA)

(Continued)

TABLE 6.2 (Continued)

Date	Time	Events
	04:10	Truck containing the SCBA fails to start and cannot come to the scene
	04:20	Operator 3 regains consciousness and exits the manufacturing building
	04:25-04:45	First ERT team members enter the manufacturing building. The Shift Supervisor, Operator 2 and Operator 6 are found unconscious. [It is later determined they all died from toxic exposure]
	05:08	First group of external firefighters arrive on-site
	05:15	ERT responder closes an open drain valve from which a gas is escaping
	06:02	Methyl mercaptan storage tank pump is turned off
	06:10-09:30	ERT make three more entries into the manufacturing building to find Operator 1. They cannot find Operator 1
	10:07	The methyl mercaptan tank emergency isolation valve is activated to stop methyl mercaptan from flowing to the process
	11:15-11:55	ERT recover Operator 1 from the manufacturing building. They are later determined to be deceased
	11:40	Waste gas vent header drain valve closed to stop methyl mercaptan release

In addition, questions arise such as 'what is a root cause?' and 'how far back do you go before you claim you have root cause?'. These are good questions. For the root causes to be insightful, you are looking for a range and variety of root causes. You can test if you have found a root cause by continuing to ask why. Further answers should not be providing any further insight, and may become more generic and less actionable. In this case, look at previous whys to identify a meaningful root cause. The root cause is often a failed control, and it is best if it is actionable (i.e. can you identify a clear action to resolve it?). As Fig. 6.5 highlights, root causes may be identified after any number of whys, and for some events there is potentially more than one root cause. Table 6.3 summarises the strengths and weaknesses of the 5 whys analysis.

Fig. 6.7 presents a condensed example of a 5 whys analysis for the Texas City Refinery explosion that occurred in 2005 (U.S. Chemical Safety and Hazard Investigation Board., 2007). Whilst simplified, this example is still very informative, focuses on critical events leading up to the incident, and identifies seven distinctly different contributing factors. In this case, the factors range from faulty equipment, design flaws, unsafe acts, resourcing failures and inadequate management oversight or verification.

6.4.4 Fishbone

The fishbone diagram, also known as the Ishikawa diagram, can be used by incident investigators to diagrammatically represent the range of causes and contributing factors (or secondary causes) to an event. An example of a fishbone diagram is shown in Fig. 6.8. The fishbone diagram requires analyses to identify the classes of causes and factors that

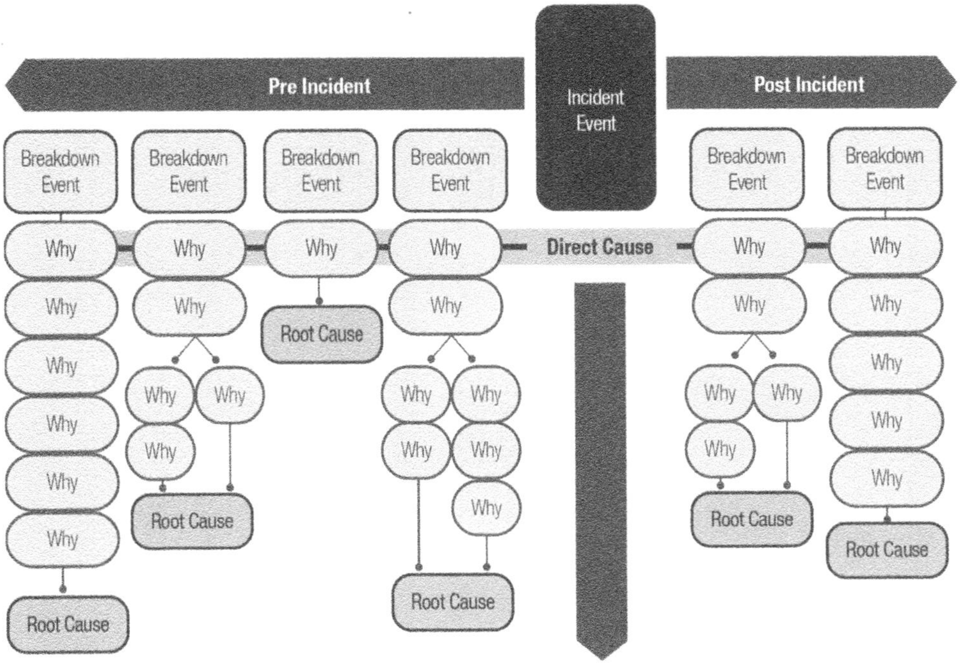

FIGURE 6.6 Illustration of 5 whys analysis (Killick, 2014). *Source: Killick, C. (2014).* Five whys. *Slide presented by Clive Killick to MINE4200 class, July 2014.*

TABLE 6.3 Strengths and weaknesses of 5 whys.

Strengths	Weaknesses
Easy and quick to use	Can be superficial
Can understand multiple aspects of an incident	Limited by the users understanding of the incident
Does not require deep technical understanding	Easy to miss (overlook) large areas of enquiry
Straightforward to communicate	

need to be considered. These can include equipment, process, people, material, environment and management. Other classes of causes that might be relevant to the process industries could include procedures or methods, data and communications. An example of an annotated fishbone diagram is shown in Fig. 6.9. The fishbone diagram is another technique that is limited to what the investigators focus on, and therefore different investigators will produce different diagrams for the same event. Another major limitation of the fishbone technique is that it focuses on the identification of things that could be done to prevent the incident, and the factors that caused the incident, and excludes anything after the event (e.g. escalation and recovery issues).

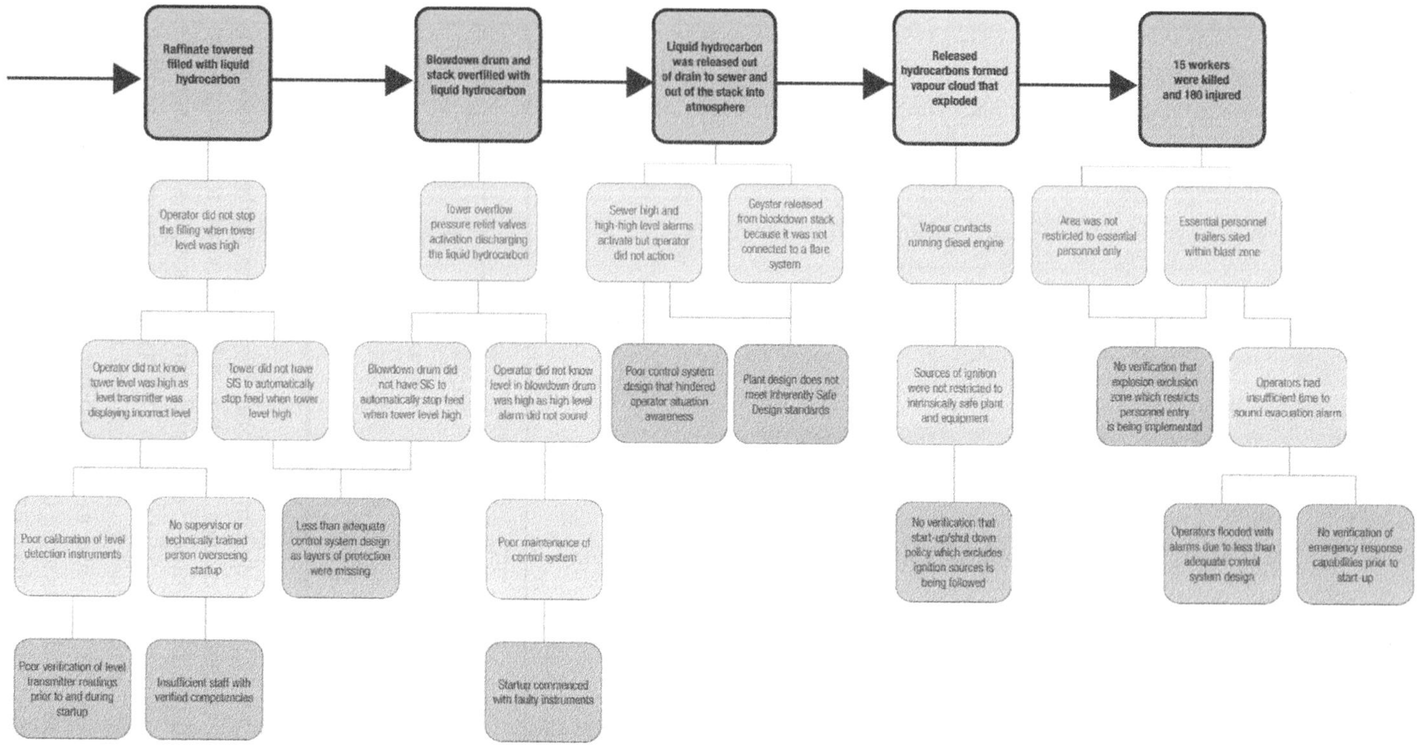

FIGURE 6.7 Summary of 5 whys analysis for Texas City Refinery explosion (U.S. Chemical Safety and Hazard Investigation Board., 2007). *Source: Adapted from U.S. Chemical Safety and Hazard Investigation Board. (2007).* Investigation report-Refinery explosion and fire - BP, Texas City, March 23, 2005. *U.S. Chemical Safety and Hazard Investigation Board.*

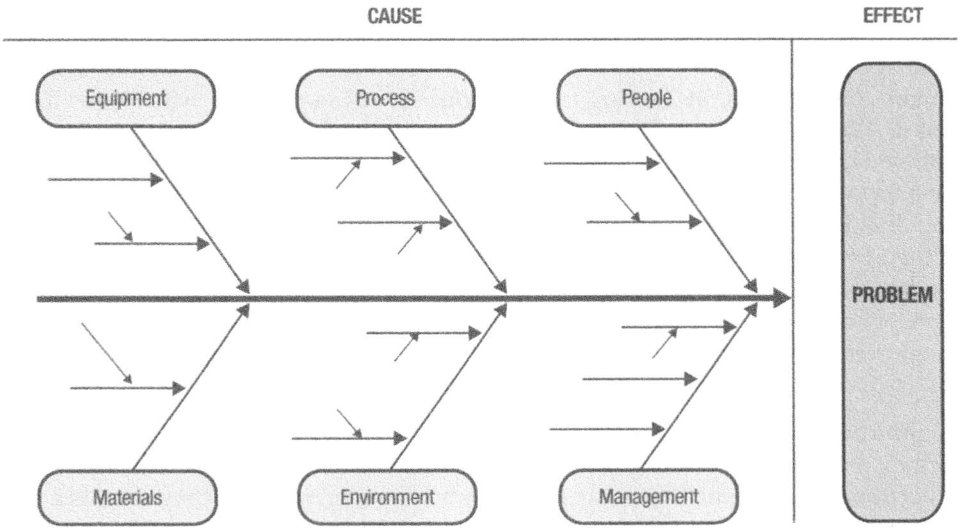

FIGURE 6.8 Blank fishbone diagram (Wikipedia, 2021). *Source: Adapted from Wikipedia. (2021).* Ishikawa diagram. *Wikimedia Foundation, Inc. https://en.wikipedia.org/wiki/Ishikawa_diagram.*

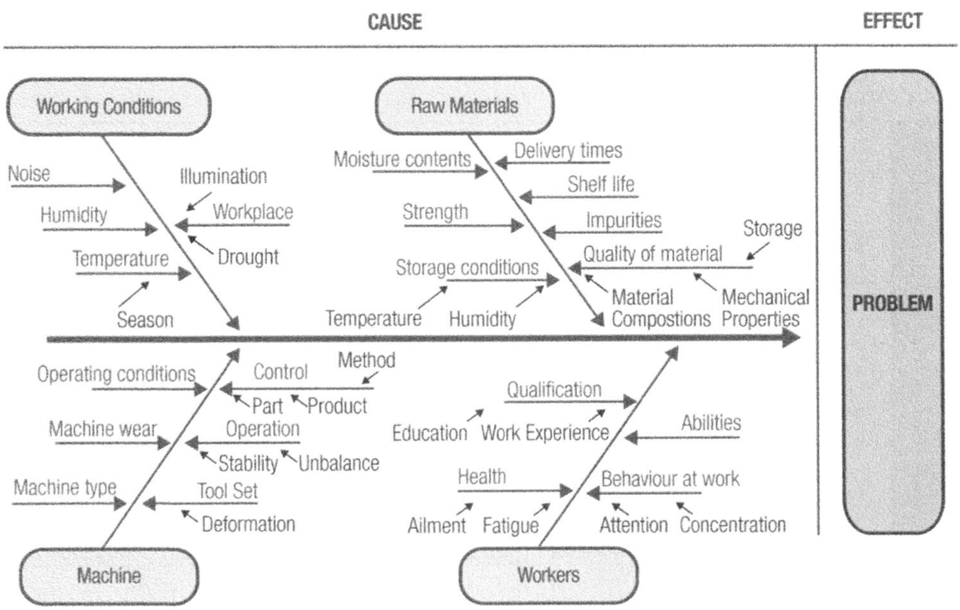

FIGURE 6.9 Example of fishbone diagram (ConceptDraw, 2023). *Source: Adapted from ConceptDraw. (2023). Fishbone Diagram. https://www.conceptdraw.com/solution-park/business-fishbone-diagram [Accessed 15 Feb 2023].*

6.4.5 Human factors analysis and classification system

The HFACS is a technique designed to help incident investigators identify the human errors and human contributions, at various organisational levels, associated with a given accident (Shappell & Wiegmann, 2000). The HFACS framework is shown in Table 6.4. Examples for each of these HFACS items derived from research are summarised in Table 6.5.

Table 6.6 presents an HFACS analysis for the Texas City Refinery explosion. As is usual for major incidents, there are many organisational and management factors that contributed to the event.

As these examples show, HFACS focuses on intentional and unintentional acts of human error with some exploration of a select range of underlying factors that may have influenced people in a way that could result in an error that may have been a factor in an incident. However, HFACS does not help analysts decipher:

- the significance of the identified issues as contributors to the event being analysed (i.e. were they a significant or insignificant contributor).
- whether the issues identified always lead to incidents or whether they can also contribute to successful outcomes.

Therefore HFACS by itself does not help analysts identify, nor prioritise, recommendations for preventing the recurrence of incidents. A summary of the strengths and weaknesses of HFACS is provided in Table 6.7.

TABLE 6.4 The Human Factors Analysis and Classification System framework.

Organisation factors	Organisational culture/climate	Operational process	Resource management				
	– –	– –	– –				
Supervisory factors	Inadequate supervision	Planned inappropriate operations	Failed to correct problem	Supervisory violation			
	– –	– –	– –	– –			
Preconditions for unsafe acts	Situational factors		Conditions of operators			Personnel factors	
	Physical environment	Tools/technology	Mental states	Physiological states	Physical/mental limitations	Communication coordination planning	Fitness for duty
	– –	– –	– –	– –	– –	– –	– –
Unsafe acts	Errors			Violations			
	Decision error	Skill-based error	Perceptual error	Routine violation	Exceptional violation		
	– –	– –	– –	– –	– –		

Adapted from (Shappell and Wiegmann, 2000).

TABLE 6.5 Human Factors Analysis and Classification System item examples from the airline, healthcare and defence industries.

HFACS Category	Example Prompts
ORGANISATIONAL INFLUENCES	
Organisational culture	• No blame culture for reporting noncompliance • Organisational overconfidence or underconfident • Unclear structure, authorities, accountabilities
Operational process	• Inadequate/incorrect/out-of-date/unclear policies/standards/procedures • Prioritisation of production over safety by allocation of funds, incentives, measurement, etc. • Inadequate governance of risks
Resource management	• Inadequate/insufficient/incorrect selection, training, promotion, and firing of personnel • Poor design/selection, maintenance/use/retention of plant/equipment resources • Inappropriate funding, e.g. excessive cost cutting, lack of funding
SUPERVISORY FACTORS	
Inadequate supervision	• Failure to provide adequate guidance/instruction • Failure to provide adequate oversight/mentoring/role modelling • Inadequate provision of proper training • Failure to assign task to proficient people • Failure to supervise/ensure adequacy of risk assessment • Failure to monitor performance
Planned inappropriate operations	• Incorrect data used in planning • Plan not appropriately risk assessed • Plan not in accordance with policy/standards/regulations, etc. • Inappropriate allocation/resourcing of tasks • Inadequate briefing of personnel on the plan and/or its risk
Failure to correct problem	• Failure to identify and correct risky/unsafe practices • Failure to identify and correct errors in procedures and policies • Failure to report incidents, unsafe practices and conditions • Failure to investigate incidents, unsafe practices and conditions • Failure to track the implementation of corrective measures
Supervisory violation	• Failure to comply with or enforce policies/standards/regulations/procedures • Permitting/overlooking/instructing people to do unauthorised workarounds/shortcuts • Authorising unnecessary hazards and/or hazardous operations • Authorising unqualified people to do tasks
PRECONDITIONS FOR UNSAFE ACTS	
SITUATIONAL FACTORS	
Physical environment	• Environmental/climatic conditions • Presents of obstructions and hazardous objects • Impeding levels of noise, light, vibration, visibility • Airborne contaminants • Slip, trip obstacles
Tools/technology	• Poorly designed/installed or faulty plant/equipment/tools/PPE • Lack of availability of appropriate plant/equipment/tools/PPE • Lack of or faulty of deficient safety protection devices, warning/alarm systems, control system software/hardware and/or communication systems

(*Continued*)

TABLE 6.5 (Continued)

CONDITION OF OPERATORS	
Operators' mental states	• Distraction/inattention • Loss of or incorrect situation awareness • Mental illness/fatigue • Over stressed/cognitive overload/perceived pressure to complete task fast • Confirmation bias/cognitive tunnelling • Overconfident /underconfident • Misplaced or lack of motivation/complacency • Wilful disregard or unconscious of potential risks
Operators' physiological states	• Physical fatigue/illness/injury/impairment/incapacitation • Heat stress/sun stroke/dehydration • Cold exposure/hypothermia • Impairment due to diet, nutrition, substance intake • Loss of consciousness
Operators' physical/ mental limitations	• Body size/weight/movement limitations • Physical strength limitations • Coordination/reaction time limitations • Sensory limitations (e.g. sight, hearing, smell, touch) • Memory retention limitations for procedures, tasks, emergency response. • Mismatch between task and intelligence/aptitude/physical abilities
PERSONNEL FACTORS	
Communication, coordination and planning	• No/inadequate/poorly timed/inaccurate/confusing/conflicting communication – leader to team or within team or between different leaders and teams • No/inadequate/planning for tasks • No/inadequate/poorly timed/confusing/conflicting delegation of tasks to appropriate people/teams • Failure to coordinate activities within crew/team or between crews/teams • Verification of communication, coordination and planning activities not done
Fitness for duty	• Less than adequate fitness • Lack of adequate rest • Inadequate clothing, footwear and PPE • Inadequate licenses/training/qualifications
UNSAFE ACTS	
ERRORS	
Decision errors	• Inadequate experience/exceeded ability/competence • Inadequate risk assessment • Inadequate or incorrect procedure selected • Failure to prioritise tasks correctly • Failure to plan tasks correctly • Inadequate/incorrect information/procedures provided • Ignored or misinterpreted information including cautions/warnings • Selected wrong/defective tools, instruments • Fault incorrect information/procedures provided • Poor judgement/decision during normal operations • Incorrect choice of action during normal operations • Misdiagnosed or wrong judgement for urgent/emergency situations • Wrong response to urgent/emergency situation

(*Continued*)

TABLE 6.5 (Continued)

Skill-based errors	• Incorrect execution of procedure includes omitting or not completing steps or doing them in wrong order • Poor operating technique • Rushed/delayed necessary action • Poor or ill-timed execution of action including over/under/early/late/incorrect/incomplete inspections, responses to alarms, etc. • Inadvertent/unintended operation or execution of control action when not required • Incorrect application of PPE
Perceptual errors	• Misinterpreted/misread instrument and control system readings • Misperceived equipment (identified wrong tank, pump, conveyor, etc.) • Misperceived change in environment/context/plant/task • Distorted perception of time, distance, speed, etc.
VIOLATIONS	
Routine	• Inadequate inspections of equipment, controls and control support systems • Regular use of unauthorised workarounds/shortcuts to execute activities • Commonly violates procedures/policies/orders (e.g., Fails to use PPE, does not isolate equipment as required, runs operations with disabled safety controls and warning systems) • Routinely does work not qualified/authorised to perform
Exceptional	• An inadequate inspection of equipment, controls and control support systems • Used an unauthorised workaround/shortcut to execute an activity • Violated a procedure/policy/order • Performed work not qualified/authorised to perform • Breached or disabled safety systems (guards, alarms, interlocks, etc.) • Inadequate inspection or unauthorised operation during start-up

Adapted from Department of Defense, 2015; Diller et al., 2013; Shappell & Wiegmann, 2000; Wang et al., 2020.

6.4.6 Bowtie analysis

As mentioned in Chapter 5, bowtie analysis is a visual representation of the controls and control assurance management system elements needed to prevent and mitigate an unwanted incident. Bowties can also be used in incident investigations to highlight the controls that were absent, present but ineffective, and present and effective as shown in Fig. 6.10. Using bowtie analysis in incident investigations guides analysts to identify threats or causes, effective and ineffective controls, as well as actual and potential consequences. Once the absent and ineffective controls have been identified, the analyst can then explore the control assurance management system to determine why the control was absent or ineffective.

The process for using bowtie analysis in incident investigation is as follows:

1. Construct the bowtie for the unwanted event (see Chapter 5 for details).
2. Develop control assurance management system elements for each control on the bowtie (also described in Chapter 5).
3. Determine which controls on the bowtie were missing, which controls were present but ineffective or failed, which controls were present and effective, and which control status was unknown. This information can be shown visually on the bowtie as illustrated in Fig. 6.10.

TABLE 6.6 Human Factors Analysis and Classification System analysis for Texas City Refinery explosion.

Organisation Factors	Organisational culture/climate	Operational Process	Resource Management
	- Previous incidents/near misses not investigated - Emphasis on occupational safety over process safety - Lack of awareness of major safety risks	- Lack of human-fatigue prevention policy - Poor quality procedures	- Insufficient staff to handle board operator workload during startup - Equipment does not meet industry standards

Supervisory Factors	Inadequate Supervision	Planned Inappropriate Operations	Failed to Correct Problem	Supervisory Violation
	- Ineffective supervisory oversight and technical assistance during startup - One shift supervisor off site the other preoccupied with another unit - Lack of hazard and abnormal situation training	- Startup commenced with known instrumentation defects - Improper staffing	- Level instrumentation not repaired prior to startup -	- Faulty instrument status not communicated to operators - Shift supervisor left site during shift

Preconditions for Unsafe Acts	Situational Factors		Conditions of Operators			Personnel Factors	
	Physical Environment	Tools/ Technology	Mental States	Physiological States	Physical/ Mental Limitations	Communication Coordination Planning	Fitness for Duty
	- Poorly, design control system and interface	- Status of tower not communicated across shifts	- Mental fatigue - Poor situation awareness	- Physical fatigue -	- Critical information not known -	- Poor quality shift handover	- -

Unsafe Acts	Errors			Violations	
	Decision Error	Skill-based Error	Perceptual Error	Routine Violation	Exceptional Violation
	- -	- -	- Control system did not allow detection of overfill	- Common practices of deviating from procedures - Overfilling tower was common during startup	- -

Adapted from (Lilburne, 2017; U.S. Chemical Safety and Hazard Investigation Board., 2007).

TABLE 6.7 Strengths and weaknesses of Human Factors Analysis and Classification System.

Strengths	Weaknesses
Focused on the human contributions	Specialised technique that needs specific expertise to be fully effective
Allows understanding of situations that led to incorrect human actions	Limited by the amount of detail known about the situation, people and organisation
Identification of key factors that are not directly linked with the specific incident	Singular focus on the human contribution

4. For the missing, ineffective and failed controls, use the control assurance management system information to investigate what control support activities failed to ensure that the control was implemented, monitored and maintained to the required standard.

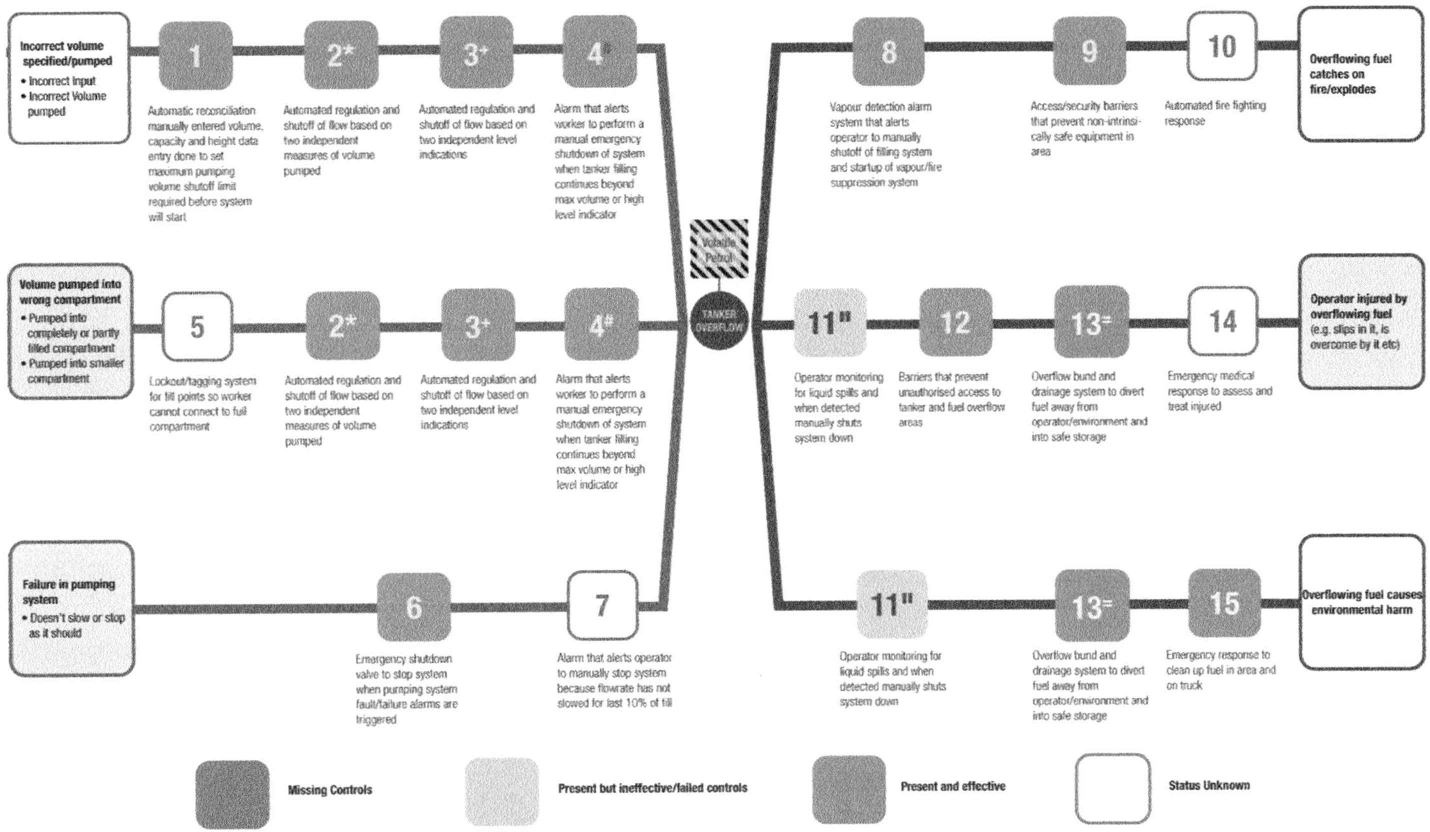

FIGURE 6.10 Bowtie highlighting presence and effectiveness of risk controls during a fuel tanker overfilling incident (Hassall, 2017).

5. From the insights gained on the effectiveness of the controls and control assurance management system elements, make recommendations to improve the effectiveness of the prevention and mitigation controls.

6.4.7 Accident analysis mapping

Accimap analysis of incidents involves mapping the critical event and its contributing factors at various organisational levels. Accimaps can be used to highlight issues with decisions, communications and information flows between different organisational levels and between different people within an organisation. Therefore Accimaps can be used to extend the analysis beyond the direct event chain to uncover useful insights and recommendations regarding organisational, latent factors or preconditions that permitted the initiation, escalation or failure to detect the event. Information on generic Accimaps can be found in Rasmussen and Svedung (2000). In the process industries, the Accimap process has been adapted to focus on the controls and control management framework needed to prevent accidents.

The Accimap process can also be used in conjunction with the bowtie analysis to map failures in control assurance management systems to produce ConSupMaps as described in Chapter 5. A ConSupMap can be developed that shows the control assurance management system elements required to implement, monitor and maintain the controls (as shown on the bowtie). This ConSupMap can then be referenced in an event investigation to highlight what elements were present and effective, present but ineffective or missing. Fig. 6.11 shows an ConSupMap drafted to investigate a fuel tanker overfilling incident. It is clear which controls and control assurance elements were missing or ineffective.

The steps for performing a retrospective Accimap analysis for a process industry incident are:

1. Identify the key controls comprising the event sequence, and place them along the base of the ConSupMap. Key events can also be added into the sequence.
2. Identify key human activities required to enact the control and also those required to monitor if the control is implemented and effective and put them directly above the event sequence. For complex controls, another row can be added between key worker activities and controls rows that highlights the key technological aspects of the control (Note – this row is not shown in the example).
3. Identify important control support activities that are required to ensure (a) controls are maintained and remain fully functional over time and (b) workers are competently able to perform activities require to enact and/or monitor controls.
4. If there are important documents that should be referenced (e.g. hot work permits, isolation procedures, safework practices), another row can be added below controls to identify these documents (Note – this row is not shown in the example but is shown in Appendix B).
5. Determine the managerial verification and audit activities that are required to ensure controls are implemented, are fully functional and workers are performing their activities as required when required.

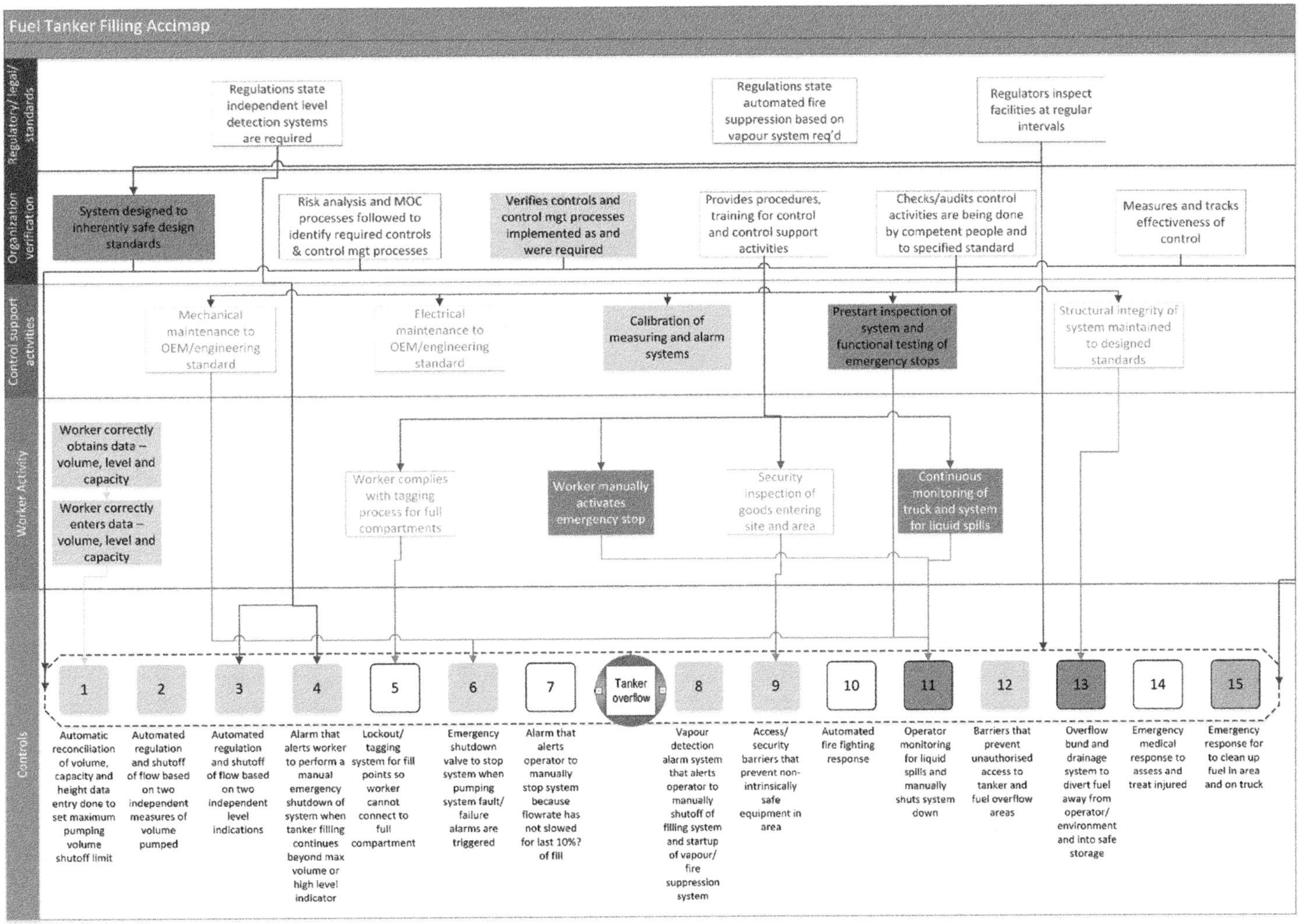

FIGURE 6.11 Accimap showing missing control assurance elements (in grey) for road tanker overfilling incident (Hassall, 2017).

6. Identify any important regulation, legal or standards-related checks that should be conducted and add these to the top row of the diagram.
7. Identify the flows between elements on the diagram (i.e. with arrows). Some of these flows can be labelled to highlight whether they related to data, observations, verbal or digital communications, etc.
8. Analyse and refine the diagram so readers can clearly identify the 'things that matter'. For retrospective analysis this can include highlighting:
   a. What was missing/absent?
   b. What was present and ineffective?
   c. What was present and effective?

The strengths and weakness of this type of Accimap analysis are listed in Table 6.8.

6.4.8 Strategies Analysis for Enhancing Resilience

In dynamic work situations, people often have a range of options for the strategies (decisions and actions) they can take. The selection of a strategy is informed by a person's assessment of the situation as wells as their, perception of risk, time pressure and the difficulty of the task as discussed in Chapter 3 (Section 3.5).

In process industry contexts, some strategies that may be preferred or successful in one operation scenario (e.g. normal conditions) may be unsuccessful or undesirable in another context (e.g. abnormal conditions). For example, an expert using an intuitive approach might be preferred in normal operating conditions but not preferred when the situation is novel and abnormal. Similarly, the use of an avoidance strategy (not doing or deferring the task) might be appropriate if the situation is unsafe to proceed, but inappropriate if a critical alarm is activated and needs attention to avoid a catastrophe. If the critical event being analysed is associated with a dynamic work situation, then the SAfER approach can be used to gain insights that would lead to meaningful recommendations about how to prevent future reoccurrences of an adverse outcome, or to promote future reoccurrences of a beneficial outcome.

The SAfER approach has been designed to help investigators identify the important cues that facilitate good situation assessment, and the range of strategies that might be used to control operations and the circumstances in which these strategies will lead to successful and unsuccessful outcomes (Hassall et al., 2014, 2016). SAfER provides a set of generic strategies (shown in Table 3.2) that investigators can use as prompts to determine which strategies might be adopted in the situation they are investigating, and also the system attributes (or performance shaping factors) that could lead a person to select or not select a given strategy.

TABLE 6.8 Strengths and weaknesses of Accimap.

Strengths	Weaknesses
Allows deep and detailed analysis	Complex – it is difficult to construct an Accimap
Looks vertically inside and outside of the organisation	Requires a lot of information
Not limited to the event sequence	Risk of bias in the analysis (e.g. WYLFIWYF)

TABLE 6.9 Example of a simplified strategies analysis for enhancing resilience analysis for filling a road tanker with fuel.

Critical situation assessment factors	List the factors that need to be monitored to ensure safe operation	Recommendations for design interventions to make the critical factors more salient
Plant/process factors	• Status of pipes, connections, valves, pumps, etc. involved in fuel transfer. • Status of instruments required to monitor fuel transfer. • Contents and flowrate within pipes. • Contents and level (absolute and rate of change) in storage and tanker (e.g. reconciled mass balance for fuel flowing out of storage into tanker). • Current versus projected fill time.	• Interface showing tanker fill level, and actual and projected fill times during filling or mass balance status for tanker and system (e.g., mass out of system = mass into tanker, or system and/or tanker are losing/gaining mass). • Instrument functional/fault alarms. • Colour coding for different types of fuel stored and flowing.
People factors	• Real-time location and vigilance of tanker filling operator during filling process. • Real-time location and status of others in vicinity.	• Interface to include vision from camera monitoring of tanker and key parts of terminal. • Radio contact with field operator.
Context factors	• Current and forecast weather. • Presence of other fuel types and tankers. • Other [unusual] work/activity in vicinity.	• Lightning warning system. • Localised work tracking and communicaiton system.

Generic strategy	What decision/actions might be associated with this generic strategy	Should design promote, prevent or tolerate strategy?	Recommendations for design improvements for both safe and unsafe operations
Avoidance = Not done, defer, or forget to do	For normal operations: • Operator doesn't start loading or stops loading because road tanker, filling station plant or people not assessed as safe and/or ready.	Promote.	1. Ensure operator has E-stops that operator can activate to stop filling.
	For abnormal operations: • Operator does not address loss of containment (e.g. leakages/overflows) because it isn't detected.	Tolerate.	2. Vapour/liquid monitoring systems to detect leaks. 3. Mass balance showing expected vs actual flows with alarms interlinked to automatic shutdown if losses or gains are detected.
Intuitive = automatic response, done without explicitly or deliberately using thought processes	For normal operations: • Operator starts loading assuming everything is ok (tanker, plant and instrumental componentry are all functional, connections from fill station to road tanker are correctly made, fuel type and quality are to spec, etc.).	Tolerate.	4. Automating system checks interlocked with loading pumps so system can't start or stops when safety critical equipment is not functional.
	For abnormal operations: • Operator assumes only one loss of control/containment event is occurring rather than multiple.	Tolerate.	As per 2. above.

TABLE 6.9 (Continued)

Generic strategy	What decision/actions might be associated with this generic strategy	Should design promote, prevent or tolerate strategy?	Recommendations for design improvements for both safe and unsafe operations
Arbitrary-choice = guessed, scrambled haphazard or panicked response	For normal operations: • Operator guesses which fuel to put in road tanker and which piping and tank on truck to use.	Prevent.	5. Tanker and filling station has colour coding of tanks and pipes with unique connectors to prevent wrong type of fuel (ULP, diesel, etc.) going into wrong tank. 6. Tanker has liquid heights shown on display and side of tank.
	For abnormal operations: • Operator guesses size of spill and how best to respond.	Prevent.	7. System has ability to do real-time reconciliations and display loss of containment volumes. 8. System automatically alarms, shuts flow and starts deluge system if loss of containment is detected by reconcilation system or on area monitoring cameras.
Imitation strategies = copy how others do it or copy what has worked in the past	For normal operations: • Operator copies how previously unloaded ULP (but could be different tanker, different volume, etc.).	Tolerate.	9. Automatic shut off system if road tanker level gets too high. 10. Further work required to determine best intervention(s) to allow underfilling strategy to be used without causing adverse outcome.
	For abnormal operations: • Operator copies previously used emergency response which may not be correct for fuel type/volume.	Tolerate.	11. Conduct emergency response drills.
Cue-based strategies = select Chosen Option using the Observed Info/Cues and Predict Consequences results	For normal operations: • Operator closely monitors loading process on screens and by watching tanker and surrounds.	Promote.	12. Employ forcing function technology to ensure loading only progresses when operator is monitoring (eye tracking, acknowledge buttons, etc.).
	For abnormal operations: • Operator looking for and acts on 'weak signals' of abnormal operations (chronic unease).	Promote.	13. Camera and interface systems allow operator to do 'deep dive' interrogations and projections.
Compliance-based strategies = following procedures as they are written/practiced	For normal operations: • Operator follows SOP.	Promote.	14. Embed SOP within interface/ monitoring system as a checklist process so detailed procedural reading not required (integrate with 12.).
	For abnormal operations: • Operator follows Emergency response plan.	Promote.	

Generic strategy	What decision/actions might be associated with this generic strategy	Should design promote, prevent or tolerate strategy?	Recommendations for design improvements for both safe and unsafe operations
			15. Create ERP checklist (similar to aviation) to help operator expediently activate and monitor emergency response [do in conjunction with 11].
Analytical Reasoning strategies = using analytical thinking to reason out the best way to perform task	For normal operations: • Operator goes back to first principles and checks and double checks everything before starting the loading process (could significantly delay loading).	Tolerate.	As per 10. above.
	For abnormal operations: • Operator thinks about and develops own emergency response.	Prevent.	As per 11 and 15 above.

To perform a simplified SAfER analysis, an investigator should:

1. Identify the important cues or critical situation assessment factors that need to be monitored to correctly diagnose the state of the operations – whether the operations are normal/abnormal or safe/unsafe.
2. Identify the strategies that operators might adopt in normal/abnormal or safe/unsafe situations.
3. Determine whether the design should promote the strategy because it will produce a successful outcome, prevent the strategy because it will produce an unsuccessful outcome, or tolerate the strategy because it cannot be prevented and therefore needs to be tolerated in a manner that will not lead to an adverse outcome.
4. Identify recommendations that will:
   a. make the important cues that indicate safe versus unsafe operations more salient
   b. allow the design to promote, prevent or tolerate each strategy in both normal/abnormal and safe/unsafe situations

An example of a SAfER analysis conducted for filling a fuel tanker is shown in Table 6.9. For more information on performing detailed SAfER analysis refer to Hassall et al. (2014, 2016).

6.5 Integration of learning back into the business

Once an investigation has been conducted and recommendations made, these recommendations need to be communicated back to the business and then *actioned*. The integration of learnings back into the business is an important way to prevent reoccurring unwanted events, and to promote the reoccurrence of wanted events. This integration should involve updating the risk assessment information (e.g. risk register, bowtie and CAMs) and ensuring that the updated information is disseminated, understood, and implemented with the recommended improvements being maintained into the future.

Some of the factors that can impede the integration of learning back into the business include:

- No identification of SMART (**s**pecific, **m**easurable, **a**ssignable, **r**ealistic/relevant, **t**ime-bounded) actions or no sense of urgency to implement actions due to the belief that the event was unique.
- Actions are not effective because they are poorly identified, no one was given the accountability to implement, and/or there was insufficient time to implement properly (Drupsteen & Hasle, 2014).

Good and pertinent critiques on organisational learning are provided by Lundberg et al. (2009) and Kletz (2009).

6.6 Summary

Investigating both adverse and successful events can provide insights that help improve and strengthen risk management processes and process plant operations. Investigations can be performed for different reasons. They can be performed to:

- Find the causes of an event

- Collect information on the full consequences of an event
- Provide insights into the effectiveness of involved controls
- Collect and disseminate learnings from an event
- Collect information to use for legislated reporting requirements or for use in legal proceedings

Before performing an investigation, it is important to determine, ***why***, ***what*** and ***how***.

We have introduced a structured methodology for performing an investigation, and we have introduced several techniques that can be used as part of the methodology. Incident investigations must be conducted by competent people in a structured and systematic way in order to ensure that all important learnings are identified, and that appropriate recommendations are made to improve risk management activities. In the process industries, it is critical to adopt a whole of system perspective when investigating incidents to ensure that all learnings and recommendations are captured. By employing multiple, diverse and complementary techniques, a range of causes or contributing factors can be identified, including technical, human, organisational and environmental factors.

Event investigations help practitioners identify lessons to be learned, but to prevent reoccurrence of adverse events, the 'lessons learned' also need to be embedded back into the organisation. This should involve updating the risk assessment information (e.g. risk register, bowtie and CAMs) and ensuring that the updated information is disseminated, understood, and implemented with the recommended improvements being monitored and maintained into the future.

Review questions

1. What is the difference between prospective and retrospective risk monitoring and review? Provide an example for each.
2. What does it mean to *monitor and review risks* in the process industries? What specific tasks is this referring to?
3. The Columbia Accident and Investigation Board (Columbia Accident Investigation Board, 2003) report included the following statement:

   > *"When causal chains are limited to technical flaws and individual failures, the ensuing responses aimed at preventing a similar event in the future are equally limited".*

   What did they mean by this, and thus what key factors underpinning the Columbia accident are they referring to?
4. What are the three lines of defence referred to in the 'Three lines of defence' model?
5. What are the reasons for investigating events in industry?
6. Major incidents in high hazard process industries mostly occur because they are novel events. Comment.
7. What are the four steps for conducting an incident investigation?
8. Compare and contrast these incident investigation techniques: timeline, 5 whys, fishbone and HFACS.

9. The bowtie was introduced in Chapter 5 as a very useful tool to help with risk assessment and risk treatment. Why is the bowtie also useful for incident investigations?
10. State two common impediments that prevent the effective operationalisation of key learnings from incident investigations.
11. Which of the following are key lessons learned from the investigation into the Columbia space shuttle disaster?
    a. Identification of the immediate root causes of the accident is critical
    b. Identification of causes should focus on technical flaws and individual failures
    c. Organisational culture and structure can be key causal factors of major incidents
    d. Organisational culture is a vague concept, and as such, is not relevant to the identification of real causes of accidents
12. Which of the following statements are correct?
    a. Event investigations are conducted to specifically analyse unsuccessful events (failures)
    b. It is a legal requirement that all events must be formally investigated
    c. Most incidents are associated with a failure to implement well-known controls for well-known risks
    d. The first step in conducting an incident investigation is to determine what happened
13. Which of the following tools are helpful in analysing the role of organisational factors in an event?
    a. HFACS
    b. Fishbone
    c. 5 Whys
    d. Accimap
14. Which of the following is not a level of human failure as stated in the Swiss Cheese model of Reason (1990)?
    a. Human error
    b. Unsafe acts
    c. Unsafe supervision
    d. Organisational influences
15. Select a contemporary process industry incident and conduct the following analyses:
    a. Timeline
    b. 5 Whys analysis
    c. Bowtie analysis of the controls
    d. Accimap analysis of the controls and control assurance management system

Based on this analysis, draft a list of recommendations (with SMART actions) to avoid reoccurrence of the incident.

The U.S. Chemical Safety and Hazard Investigation Board provides an excellent repository of incident investigation reports (http://www.csb.gov). These reports provide an excellent basis for performing the above analyses. Some contemporary incidents that you may like to consider investigating are:

- Bethune Point Wastewater Plant Explosion (2007) – focus on conducting hot works

- Imperial Sugar Company Dust Explosion and Fire (2009) – focus on risks of dusts and dust explosions
- Xcel Energy Company Hydroelectric Tunnel Fire (2010) – focus on contractors and working in confined spaces
- Non-Condensable Gas System Explosion at PCA DeRidder Paper Mill (2017) – focus on conducting hot works and management of change

References

Columbia Accident Investigation Board. (2003). *Columbia accident investigation board report volume 1.* https://history.nasa.gov/columbia/Troxell/Columbia%20Web%20Site/CAIB/CAIB%20Website/CAIB%20Report/Volume%201/introduction.pdf.

ConceptDraw. (2023). Fishbone Diagram. Available from https://www.conceptdraw.com/solution-park/business-fishbone-diagram. Accessed 15.02.23.

De Landre, J., Irving, M., Hodges, I., & Weston, B. (2007). *Learning from accidents and incidents.* Multimodal safety management and human factors: Crossing the borders of medical, aviation, road and rail industries (pp. 131–142). Aldershot, Hampshire: Ashgate Publishing.

Department of Defense. (2015). *Human factors analysis and classification systetms (DOD HFACS) version 7.0.* https://www.uscg.mil/hr/cg113/docs/pdf/DoD_HFACS7.0.pdf and from http://www.public.navy.mil/NAVSAFECEN/Documents/5102/DOD_HFACS_v7.0_Guide.pdf.

Diller, T., Helmrich, G., Dunning, S., Cox, S., Buchanan, A., & Shappell, S. (2013). The human factors analysis classification system (HFACS) applied to health care. *American Journal of Medical Quality, 29*(3), 181–190. Available from https://doi.org/10.1177/1062860613491623.

Dodshon, P., & Hassall, M. E. (2017). Practitioners' perspectives on incident investigations. *Safety Science, 93,* 187–198. Available from https://doi.org/10.1016/j.ssci.2016.12.005.

Drupsteen, L., & Hasle, P. (2014). Why do organizations not learn from incidents? Bottlenecks, causes and conditions for a failure to effectively learn. *Accident Analysis & Prevention, 72,* 351–358. Available from https://doi.org/10.1016/j.aap.2014.07.027.

Fishwick, T. (2014). Recurring accidents: Slips, trips and falls. *The Chemical Engineer,* 28–34.

Gill, G. (2013). Recurring accidents: Inadequate isolations. *The Chemical Engineer* (870), 52–56.

Graham, B., Reilly, W. K., Beinecke, F., Boesch, D. G., Garcia, T. D., Murray, C. A., & Ulmer, F. (2011). *Deep water: The Gulf oil disaster and the future of offshore drilling: Report to the President.* National Commission on the BP Deepwater Horizon Oil Spill and Offshore Drilling 2011.

Harich, J., & Rosas, M. K. (2020). Process-driven Problem Solving with Root Cause Analysis: Adapting powerful business tools to fit the sustainability problem. *Earth System Governance conference November 6 to 9, 2019 - Oaxaca, Mexico.*

Hassall, M. E., Sanderson, P. M., & Cameron, I. T. (2014). The development and testing of SAfER: A resilience-based human factors method. *Journal of Cognitive Engineering and Decision Making, 8*(2), 162–186. Available from https://doi.org/10.1177/1555343414527287.

Hassall, M.E., Joy, J., Doran, C., & Punch, M. (2015). Selection and optimisation of risk controls (ACARP report C23007).

Hassall, M. E., Sanderson, P. M., & Cameron, I. T. (2016). Incident analysis: A case study comparison of traditional and SAfER methods. *Journal of Cognitive Engineering and Decision Making, 10*(2), 197–221. Available from https://doi.org/10.1177/1555343416652749.

Hassall, M.E. (2017). *Incident investigation masterclass: Introduction of proposed solution.* Paper presented at the UQ R!SK Incident Incident Investigation Masterclass, Brisbane, QLD.

Hollnagel, E. (2011). Prologue: The scope of resilience engineering. In E. Hollnagel, J. Pariès, D. D. Woods, & J. Wreathall (Eds.), *Resilience engineering in practice: A guidebook.* Surrey UK: Ashgate.

Hollnagel. (2012). *FRAM: The functional resonance analysis method: modelling complex socio-technical systems.* Ashgate Publishing, Ltd.

ICMM (Ed.). (2013). *Requests for proposals – Health and safety risk managing in the mining and metals sector.* London.

Killick, C. (2014). *Five whys.* Slide presented by Clive Killick to MINE4200 class, July 2014.

Kletz, T. A. (2009). *What went wrong? Case histories of process plant disasters and how they could have been avoided.* Burlington, MA: Gulf Professional Pub.

Leveson, N. G. (2004). A new accident model for engineering safer systems. *Safety Science*, *42*(4), 237–270. Available from http://uq.summon.serialssolutions.com/link/0/eLvHCXMwVV3NCsIwDC6C4MXLQPsUHbbdX8_D4UEQREGPSZscB-LeH9MxQY-5BZJ8-SFfkmnAUdKEdA7UgcPo46Fmsuw6geNka8K_WccPmg-FWtG4U_fheOtPZnkGYGI-2WUqC5Fr11BiRvYepDhABN-1ASTYc5vDgSVfRcTgKHj0jW2AA9mqBZ_sXm0hL42P00wuS1qtWSxMOqOuFg202jxCf35erotYfMXyPTOgytekBeRnBzG2PHwAJuA6aA.

Leveson, N. G. (2011). *Engineering a safer world: Systems thinking applied to safety*. Cambridge, Mass: The MIT Press.

Lilburne, C. (2017). *HFACs analysis of Texas City Refinery explosion*. Slide developed by Chris Lilburne for CHEE4002 class, 2017.

Lundberg, J., Rollenhagen, C., & Hollnagel, E. (2009). What-you-look-for-is-what-you-find - The consequences of underlying accident models in eight accident investigation manuals. *Safety Science*, *47*(10), 1297–1311. Available from https://doi.org/10.1016/j.ssci.2009.01.004.

Macchi, L., Hollnagel, E., & Leonhard, J. (2009). *Resilience Engineering approach to safety assessment: An application of FRAM for the MSAW system*. Paper presented at the EUROCONTROL Safety R&D Seminar.

Macchi, L., Hollnagel, E., & Leonhard, J. (2009). Resilience Engineering approach to safety assessment: An application of FRAM for the MSAW system. Paper presented at the EUROCONTROL Safety R&D Seminar.

Noetic Solutions. (2014). *MSAC fatality review 2013-14: Report for NSW safety advisory council.*

Phelan, J. (2018). Milton Friedman: Judge public policies by their results, not their intentions. Available from https://www.americanexperiment.org/milton-friedman-judge-public-policies-by-their-results-not-their-intentions/. Accessed 12.02.23.

Pilot, S. (2002). *What is fault tree analysis?, American Society for Quality*. Milwaukee.

Pitblado, R., & Weijand, P. (2014). Barrier diagram (Bow Tie) quality issues for operating managers. *Process Safety Progress*.

Rasmussen, J., & Svedung, I. (2000). *Proactive risk management in a dynamic society*. Karlstad, Sweden: Risk & Environmental Department, Swedish Rescue Services Agency.

Reason, J. (1997). *Managing the risks of organizational accidents* (Vol. 6). Aldershot: Ashgate.

Reason, J., Hollnagel, E., & Paries, J. (2006). Revisiting the "Swiss cheese" model of accidents. *Journal of Clinical Engineering*, *27*, 110–115.

Ryan, B. (2015). Incident reporting and analysis. In S. Sharples, & J. Wilson (Eds.), *Evaluation of human work* (fourth ed.). Boca Raton: CRC Press.

Saud, Y. E., Israni, K. C., & Goddard, J. (2014). Bow-tie diagrams in downstream hazard identification and risk assessment. *Process Safety Progress*, *33*(1), 26–35.

Serrat. (2017). The Five Whys Technique. In O. Serrat (Ed.), *Knowledge Solutions: Tools, Methods, and Approaches to Drive Organizational Performance* (pp. 307–310). Singapore: Springer.

Shappell, S.A., & Wiegmann, D.A. (2000). *The human factors analysis and classification system - HFACS* (DOT/FAA/AM-00/7).

Shappell, S. A., & Wiegmann, D. A. (2001). Applying reason: The human factors analysis and classification system (HFACS). *Human Factors and Aerospace Safety*.

Svedung, I., & Rasmussen, J. (2002). Graphic representation of accidentscenarios: Mapping system structure and the causation of accidents. *Safety Science*, *40*(5), 397–417.

U.S. Department of Energy. (2012). *DOE Handbook: Accident and operational safety analysis - Volume I: Accident analysis techniques (DOE-HDBK-1208–2012)*. Washington D.C; https://energy.gov/sites/prod/files/2013/09/f2/DOE-HDBK-1208–2012_VOL1_update_1.pdf.

Underwood, P., & Waterson, P. (2014). Systems thinking, the Swiss Cheese Model and accident analysis: A comparative systemic analysis of the Grayrigg train derailment using the ATSB, AcciMap and STAMP models. *Accident Analysis & Prevention*, *68*(0), 75–94. Available from https://doi.org/10.1016/j.aap.2013.07.027.

U.S. Chemical Safety and Hazard Investigation Board. (2007). *Investigation report-Refinery explosion and fire - BP, Texas City, March 23, 2005*. U.S. Chemical Safety and Hazard Investigation Board.

U.S. Chemical Safety and Hazard Investigation Board. (2016). *Investigation report - Executive summary - Drilling rig explosion and fire at the Macondo Well - Report No. 2010-10-I-OS*. U.S. Chemical Safety and Hazard Investigation Board.

U.S. Chemical Safety and Hazard Investigation Board. (2019). *Toxic Chemical Release at the DuPont La Porte Chemical Facility La Porte, Texas - Incident Date: November 15, 2014 - Report Number: 2015-01-I-TX*. U.S. Chemical Safety and Hazard Investigation Board.

van Kampen, J., & Drupsteen, L. (February 21, 2017). *Accident investigation and analysis.*

Waite, P. (2013). Recurring accidents: Overfilling vessels. *The Chemical Engineer* (861), 40–44.

Wang, J., Fan, Y., & Gao, Y. (2020). Revising HFACS for SMEs in the chemical industry: HFACS-CSMEs. *Journal of Loss Prevention in the Process Industries*, *65*, 104138. Available from https://doi.org/10.1016/j.jlp.2020.104138.

Wiegmann, D. A., & Shappell, S. A. (2001). Human error analysis of commercial aviation accidents: Application of the Human Factors Analysis and Classification System (HFACS). *Aviation, Space, and Environmental Medicine*, *72*(11), 1006–1016.

Wikipedia. (2021). *Ishikawa diagram*. Wikimedia Foundation, Inc. https://en.wikipedia.org/wiki/Ishikawa_diagram.

CHAPTER

7

Health and safety risks

OUTLINE

7.1 Introduction 162

7.2 What are health and safety risks? 163

7.3 Managing safety risks 166

7.3.1 Inherently safer design 173

7.3.2 Defence-in-depth 174

7.3.3 Resilience engineering 177

7.4 Safety culture 178

7.5 Process safety effectiveness 180

7.6 Safety management systems 181

7.6.1 Overview of ISO45001 181

7.6.2 Overview of process safety management systems 183

7.7 Case study – process safety consideration for filling bulk storage tanks 183

7.8 Summary 187

Review questions 188

References 188

Fundamentals of Risk Management for Process Industry Engineers
DOI: https://doi.org/10.1016/B978-0-12-820320-0.00004-0

The principle of risk being owned and managed by those who create it was enshrined in UK health and safety law in the 1970s ... and its effectiveness is clear and demonstrable (Hackitt, 2018).

7.1 Introduction

A key objective for all organisations is to sustain and maintain the health and safety of staff, contractors, customers, suppliers and community members. Therefore a key focus area of any risk management programme within industry should be the identification, assessment and treatment of health and safety risks. Managing health and safety risks is an ethical and legal requirement, and it also makes good business sense. In the process industries there are numerous hazardous chemicals and work situations that can adversely impact on humans' health and safety. An introduction to some of these chemicals and work situations is described in Appendix A.

In this chapter we will introduce the four types of health and safety risks, namely occupational health, human wellbeing, personal safety and process safety. We will also introduce the many hazards that exist in process plant, and also discuss the importance that organisational safety culture plays in delivering safe processing facilities. Several seminal case studies will be introduced throughout the chapter. These will highlight the criticality of health and safety risks, and we will highlight the many lessons learned.

7.2 What are health and safety risks?

Health and safety objectives and risks are defined and scoped differently by different companies. At a high level, health and safety typically covers occupational health, worker wellbeing, personal safety and process safety (Fig. 7.1).

Each of these elements are not mutually exclusive, as there is overlap and interactions between the different factors. For example, issues with worker wellbeing can directly impact personal and process safety outcomes. The Germanwings Flight 9525 disaster (2015) is a very sad example of how worker wellbeing issues can seriously impact safety outcomes (BEA, 2016).

Conversely, safety incidents can impact the occupational health and wellbeing of workers and others. For example, the Bhopal disaster highlights the devastating impacts that process safety incidents can have on human health. At the Bhopal Union Carbide plant in 1984, methyl isocyanate (MIC) escaped from the plant due to a process incident. It has been estimated that over 500,000 people were exposed to MIC as the highly toxic gas cloud travelled into the nearby town. There were a horrific number of deaths (several thousand people), injuries (reportedly 200,000 people needed medical treatment), permanent disabling injuries and ongoing issues with ill-health and gas-related disease. The lessons from Bhopal should not be forgotten as highlighted by Slater (2015) who wrote, and we quote:

- *When the right choice is not the most economical choice, remember Bhopal*
- *When you choose not to investigate a chronic failure, remember Bhopal*
- *When designing a solution that manages a hazard instead of eliminating it, remember Bhopal*
- *When operators have concerns with a decision you are about to make, remember Bhopal*

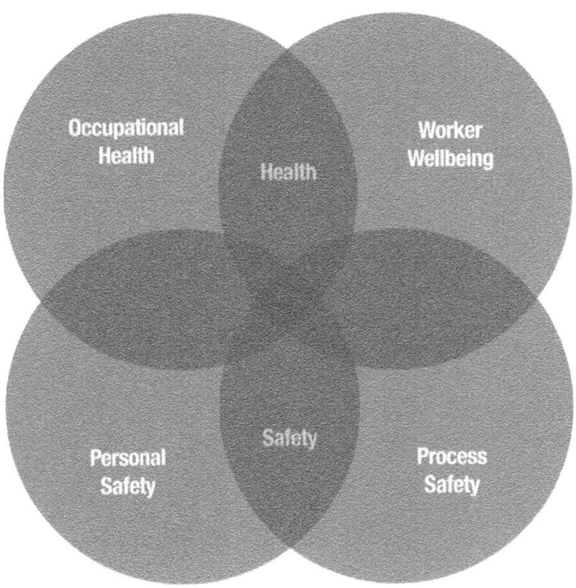

FIGURE 7.1 Classifications of health and safety risks in industry.

- *When you perform a safety audit, remember Bhopal*
- *For the sake of thousands of innocent people affected, we must all remember Bhopal*

Occupational health is defined in a variety of ways across different organisations. In general, occupational health covers all aspects of a workplace that could lead to physical and/or mental ill-health, disease, sickness or impairment. Occupational health covers factors that could lead to musculoskeletal disorders, hearing loss, vibration syndrome, illnesses and diseases such as cancer, respiratory, circulatory and communicable diseases and allergic reactions (Table 7.1). Occupational health also covers factors that could lead to psychological, physiological and social disorders. Examples of these types of disorders include excessive stress, work-induced anxiety or depression, fatigue, obesity and social issues caused by bullying, harassment, conflict, exclusion and disempowerment.

Managing occupational health risks should be done in accordance with ISO31000. However, when identifying consequences and assessing impact of occupational health risks, analysts should consider both the acute, or short-term, immediate impacts as well as the chronic or long-term implications. Recent events such as the re-emergence of black lung in the Australia Coal mining industry, the rapid rise in the emergence of silicosis for workers working with 'manufactured stone' as outlined in Fig. 7.2, and the recurring incidents of legionnaires' disease have highlighted the importance of industry continuing to vigilantly control known health risks.

Safety covers both **personal safety** and **process safety**. After the BP Texas City Refinery disaster (2005), the Baker Commission report identified the need to carefully distinguish between process and personal safety, and that these two types of safety need to be managed differently. It states '*Personal or occupational safety hazards give risk to incidents – such*

TABLE 7.1 Industry-related health risks.

Occupational health				Worker wellbeing		
Musculoskeletal disorders	Hearing loss	Excessive exposure to vibration	Exposure to illness or disease-causing substances (both chemical and biological)	Psychological factors including occupational stress, anxiety and depression	Physiological factors including fatigue, physical activity/inactivity, diet/obesity	Social factors such as bullying, harassment, level of conflict, inclusion/exclusion, empowerment/disempowerment

The world is failing on silicosis

Silicosis is caused by the inhalation of respirable crystalline silica dust which, over time, leads to lung inflammation and fibrosis. The condition is progressive and almost always fatal.

The disease is entirely preventable. Interventions such as wet-cutting, good ventilation practices, and perhaps, most importantly, effective, well fitted supplied-air respirators, can all protect workers. These measures are known to be effective.

Editorial, The Lancet Respiratory Medicine, March 11, 2019.
[DOI: https://doi.org/10.1016/S2213-2600(19)30078-5]

FIGURE 7.2 Silicosis.

as slips, falls and vehicle accidents – that primarily affect one individual worker for each occurrence. Process safety hazards give risk to major accidents involving the release of potentially dangerous materials, the release of large amounts of energy through fires and explosions, or both' (Baker et al., 2007, p. x).

Personal safety usually comprises the slips, trips, strains, sprains, cuts and abrasion type events that typically occur over a short timeframe and usually involve the actual or potential injury to one or a few people and possibly some other minor or localised damage. Examples of personal safety events include falls from heights, electric shocks, entanglement in equipment, contact with or by moving objects and tripping over type events. Personal safety refers to events that are high frequency and low consequence (Fig. 7.3).

In contrast, process safety comprises the loss of control, or the loss of containment, of a hazardous process which can result in system wide damage, or damage that goes beyond the system boundaries that often leads to actual or potential catastrophic outcomes. Consequences can include multiple fatalities or severe injuries, significant environmental harm and extensive asset damage. Process safety events can unfold over hours, days, and even months or years. Examples of process safety events include fires, explosions and release of toxic chemical clouds.

Most high hazard industries have invested time, resources and money into improving safety. This effort has resulted in significant improvements in personal safety outcomes with reductions in lost time injuries and total recordable injuries being reported by many companies and across many jurisdictions (Fig. 7.4).

Unfortunately, the same improvement has not been reported for process safety incidents. It is now recognised that the approaches that worked to reduce personal safety incidents, which focussed on systems, culture and behaviour, have not reduced the occurrence of process safety incidents. Several recent examples of process safety accidents which have occurred over the last decade are presented in Table 7.2.

According to the Baker report *'Process safety incidents can have catastrophic effects and can result in multiple injuries and fatalities, as well as substantial economic, property, and environmental damage'*

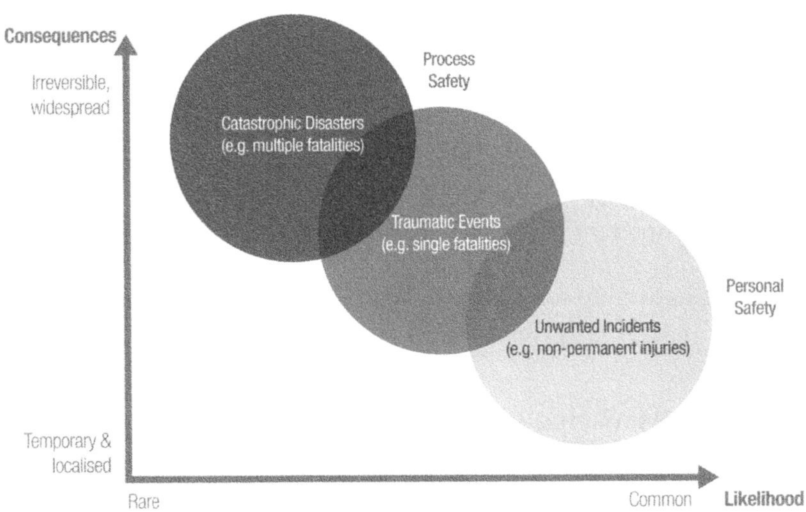

FIGURE 7.3 The relationship between process safety and personal safety.

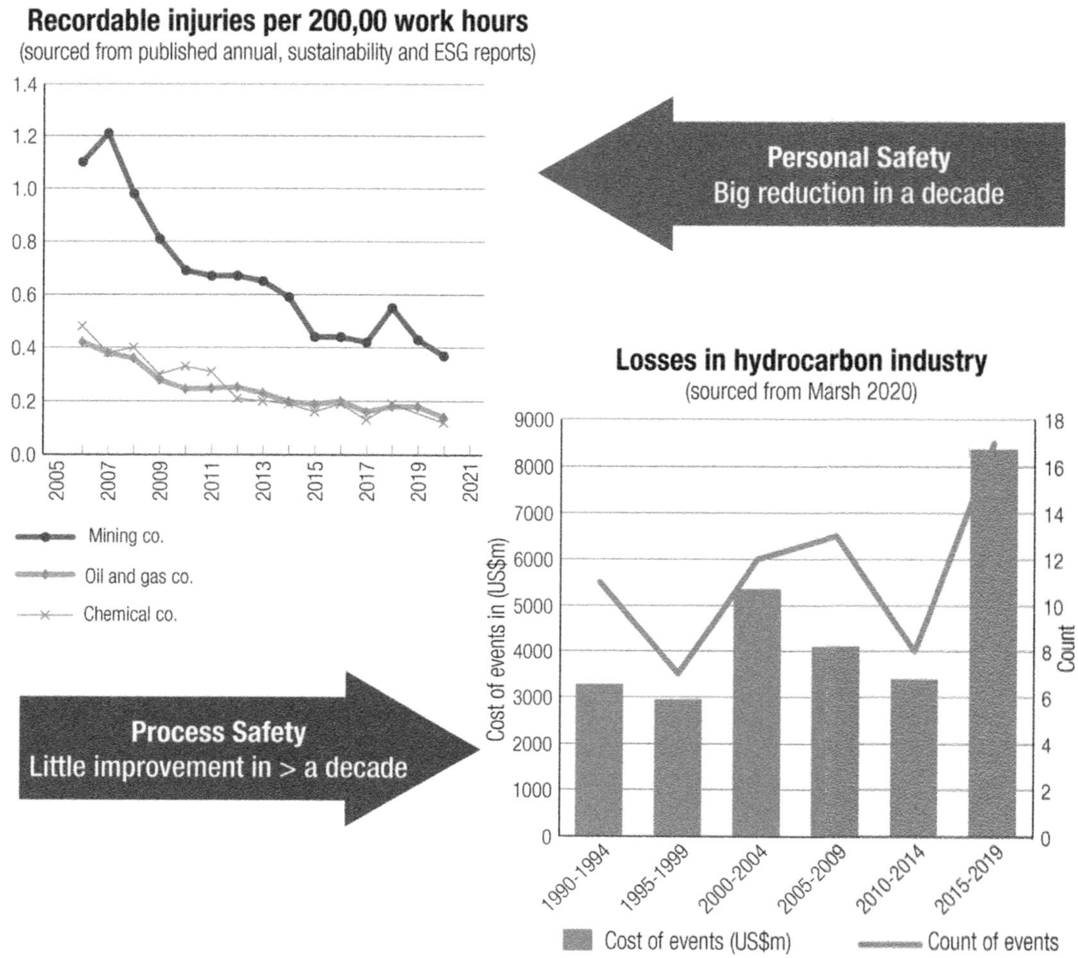

FIGURE 7.4 Personal and process safety trends over time.

(Baker et al., 2007, p. 10). Process safety activities focus on the prevention of leaks, spills, equipment malfunctions, overpressures, excessive temperatures, corrosion, metal fatigue and other similar conditions. It is particularly important because the process industries handle and manufacture many hazardous goods.

Process engineers should take a leadership role in improving process safety performance.

7.3 Managing safety risks

So how should we manage safety risks in the process industries? First, we need to differentiate between the management of personal and process safety activities.

TABLE 7.2 Summary of some recent process safety accidents.

Accident	Estimated impact	Reported causes
2010 Deepwater Horizon oil rig blowout in Gulf of Mexico, United States	A blowout of the Macondo well caused explosion and fire on the Deepwater Horizon rig which killed 11 people, injured 16 more and caused the largest oil spill in U.S. history (Graham et al., 2011). The cost of the disaster to BP alone is estimated to be near to $65 billion (Vaughan, 2018)	'A complex and interlinked series of mechanical failures, human judgments, engineering design, operational implementation and team interfaces came together to allow the initiation and escalation of the accident' (BP, 2010, p. 5). The rig and crew were deprived of the safeguards required to (1) protect against blowout and (2) mitigate the blowout if it should occur (Graham et al., 2011)
2010 Ajka red mud dam failure, Hungary	Ten people died, 286 people required medical treatment and a further 790 people and 1017 ha of land were affected by the mud (Kátai-Urbán & Cséplő, 2010; Turi, Pusztai, & Nyari, 2013)	The failure was caused by poor siting, poor design and poor construction combined with unfavourable environmental conditions and regulatory deficiencies (Turi et al., 2013)
2010 Tesoro oil refinery explosion, United States	The explosion from the catastrophic rupturing of a heat exchanger killed seven people, and cost the company a $39 million in settlement (Stensland, 2014)	The event was caused by high temperature hydrogen attack which weakened the heater exchanger steel resulting in it rupturing (U.S. Chemical Safety & Hazard Investigation Board, 2014)
2011 Fukushima nuclear power plant disaster, Japan	An explosion and radiation leak from the plant resulted in relocation of people, shutdown of the plant and cleanup which cost approximately $130 billion (Elliott, 2012)	A tsunami from the Tōhoku earthquake flooded the nuclear reactors' backup power generators which stopped the cooling system and the reactor overheated and exploded. According to the official report the loss of backup power was caused by 'organisational and regulatory systems that supported faulty rationales for decisions and actions' (The National Diet of Japan, 2012, p. 16)
2011 Explosion at Sendai refinery Japan	$691 million in damage to fluid catalytic cracker but no fatalities as workers were evacuated (Marsh, 2020)	Fire and major explosion occurred after the 2011 Tōhoku earthquake and tsunami (Marsh, 2020)
2011 Chevron Pembroke refinery fire, Wales	Storage tank explosion killed 4 workers and severely injured one other (Health & Safety Executive, 2020). The costs of this incident included $6.4 million in fines (Doyle, 2019).	The explosion occurred when its flammable atmosphere was ignited while the tanked was being cleaned in preparation for maintenance (Health & Safety Executive, 2020).
2011 Explosion in oil sands refinery, Fort McKay, Canada	$451 million costs with five workers injured (Marsh, 2020)	Explosion followed a fire that occurred when hot hydrocarbon was released when a valve was opened (Marsh, 2020)

(Continued)

TABLE 7.2 (Continued)

Accident	Estimated impact	Reported causes
2012 Neptune Sherbrooke krill oil plant explosion, Quebec, Canada	An uncontrolled release of chemicals exploded killing 3 and injuring 18 workers (Neptune Technologies & Bioressources Inc., 2012). The facility was damaged extensively and $48 million was spent rebuilding it from November 2011 to June 2014 (Anonymous, 2014)	The cause of the chemical leak – believed to be acetone – was the 'lack of control of a chemical product, some building aspects not up to code, and deficiencies in training of staff' (Ayala, 2014)
2012 Explosion in propane and butane storage, Falcon State, Venezuela	Pressurised stored propane and butane exploded killing 48 people, injuring 80 more, damaged a further 9 tanks and cost $382 million in property damage (Marsh, 2020). Also destroyed was sections of plant, pumps, pipelines and other facilities with the total cost estimated to be between $1.1 and 1.84 billion (Parraga, 2013)	Propane gas leaked during pumping which formed a hydrocarbon cloud that exploded with the cause of the leak being either poor maintenance or sabotage of the pump (Parraga, 2013)
2013 West fertilizer plant fire, Texas, United States	Fire in the fertilizer storage facility detonated killing 15 people and injuring more than 260 including emergency service personnel with the estimated cost being in excess of $245 million (U.S. Chemical Safety & Hazard Investigation Board, 2016b)	A fire started in seed storage and spread to the room where the ammonium nitrate fertilizer was store – possibly contaminating it and causing it to explode (U.S. Chemical Safety & Hazard Investigation Board, 2016b)
2013 Fire at refinery in La Plata Districk, Ensenada, Argentina	Flash flood caused by heavy rain resulted in an explosion and two fires that cost $280 million in damage and no injuries or fatalities (Marsh, 2020)	Flash flooding caused the refinery to lose power which resulted in a fire in the coke plant that ignited the flammable atmosphere and then exploded (Marsh, 2020; Parks, 2013)
2013 Explosion at Williams Olefins Plant Louisiana, United States	The overpressurising of a boiler caused it to rupture resulting in an explosion and fire which killed 2 workers, injured 167 others and the plant to remain down for 18 months (U.S. Chemical Safety & Hazard Investigation Board, 2016a)	Management of change, pre-start safety reviews and PHAs were not performed adequately and failed to detect/correct fact that boiler could be overpresssured (U.S. Chemical Safety & Hazard Investigation Board, 2016a)
2013 Daelim Industrial storage tank explosion, Yeosu South Korea	Workers were welding on a cleaned high-density polyethylene storage tank when it exploded killing 6 workers and injuring 11 others (조지현, 2013)	The tank was not fully cleaned and it had some residue gas which was ignited by the welding (Hazardex, 2013; 조지현, 2013)
2014 Toxic chemical release at DuPont's LaPorte Texas chemical facility	Four workers died from inhaling the released highly toxic chemical methyl mercaptan (U.S. Chemical Safety & Hazard Investigation Board, 2019)	The design of the methyl mercaptan system lacked the necessary safeguards required to prevent a leak and mitigate fatal outcomes of a leak (U.S. Chemical Safety & Hazard Investigation Board, 2019)

(Continued)

TABLE 7.2 (Continued)

Accident	Estimated impact	Reported causes
2014 Gas explosion Andhra Pradesh India	A gas leak from the GAIL pipeline exploded killing 23 people, injuring 16 more and it also harmed animals, coconut trees and infrastructure (Lakshmi & Kumar, 2015)	Corrosion of the gas pipeline that was not replaced (Lakshmi & Kumar, 2015)
2014 Freedom Industries Chemical release into Charleston, West Virginia drinking water	A chemical spill from Freedom Industries contaminated West Virginia residents drinking water resulting in over 300 people seeking medical treatment and nearly 300,000 people with restricted water use (U.S. Chemical Safety & Hazard Investigation Board, 2017a)	The spill happened when the floor of a chemical storage tank corroded and the chemical seeped into the soil then into the river because the secondary containment failed and the underground culvert deteriorated (U.S. Chemical Safety & Hazard Investigation Board, 2017a)
2014 Kaohsiung City vapour cloud explosion, Taiwan	Leaking liquid from underground propylene pipe formed vapour cloud in the pipe trench which ignited killing 32 people and injuring more than 300 (Yang et al., 2016)	The leak resulted from corrosion in the pipe that went undetected due to lack of inspection and maintenance (Yang et al., 2016)
2015: Port of Tianjin (Binhai China) container storage fire and explosion	The fire and explosion killed 165 people, injured 798 people and damaged over 200 buildings, 7500 containers, nearly 12,500 vehicles and resulted in direct losses of 6.87 billion yuan (Zhang, Jing, & Sun, 2018)	The rough handling of nitrocellulose caused it to spontaneously combust and the resultant fire spread to ammonium nitrate that was stored, without permission, nearby which triggered two major explosions (Fu, Wang, & Yan, 2016)
2015 ExxonMobil Refinery explosion, California, United States	Hydrocarbons backflowed into electrostatic precipitator causing explosion that injured four and caused catalyst particles to be released into community and onsite equipment damage (U.S. Chemical Safety & Hazard Investigation Board, 2017b)	Insufficient safeguards implemented and inadequate measurement of safeguard performance and condition (U.S. Chemical Safety & Hazard Investigation Board, 2017b)
2015 Propylene release and fire, Litvinov, Czech Republic	The loss of containment of propylene which ignited causing fire and explosion damage resulting in losses of more than $500 million	Loss of cooling water to propylene condenser resulted in pressure increase that led to loss of containment, fire and explosion (Herink et al., 2022)
2015 Gunashli Platform fire, Caspian Sea	Fire on the platform killed 30 people, injured others and damaged the platform and ignited a well (Bagirova, 2015)	Storm damaged the high-pressure gas riser and the leaking gas exploded which resulted in the fire but many of the dead died when lifeboat cables broke and it was smashed against the platform (Levy, 2016)

(Continued)

TABLE 7.2 (Continued)

Accident	Estimated impact	Reported causes
2015 Refugio oil spill, California, United States	Extensive environmental damage was incurred when pipeline ruptured and spilt over 100,000 gallons of crude oil near Refugio State Beach which resulted in $22 million in damage claims (DARRP Producer, 2022), $150 million in cleanup costs and $1.5 million in criminal charges (Gorman & McWilliams, 2018)	The rupture was caused by corrosion (Gorman & McWilliams, 2018) and the spill was exacerbated by the fact that the pipeline did not have an automatic valve shut off system (Melley, 2015)
2016 Refinery fire, Sannazzaro, De Burgondi, Italy	The fire caused €217 million in property damage and cleanup costs (Eni, 2017)	The cause of the fire was a mechanical fault in the Eni Slurry Technology unit (Eni, 2017)
2016 Dangyang coal power station explosion, China	An explosion of high-pressure steam pipe in the power station killed 21 and injured 5 others (Wong, 2016)	The plant was under construction and the explosion occurred during a debugging process (Bibby, 2016)
2016 Fireworks explosion Tultepec, Mexico	The explosion at the largest fireworks market in Mexico killed 42 people, injured 60 more and completely destroyed the market which remained closed for 2 years (Mexico News Daily, 2018)	'A lack of sufficient safety measures was the likely cause of the blast' as there were six separate explosions that happened (ABC News, 2016)
2016 Pipeline release Ludwigshafen, Germany	Three people died, 8 people were seriously injured and 22 were slightly injured (BASF, 2016)	An angle grinder was used to cut an adjacent pipeline that contained flammable raffinate (BASF, 2016)
2017 UAE refinery fire	Fire in a newly build residual fluid catalytic cracker resulted in $1 billion of losses due to closure and rebuilding costs (Marsh, 2020)	Cause was a loss of containment of hot, light hydrocarbon during maintenance activities (Marsh, 2020)
2017 Hydrogen leak fire in distillation unit Abidjan, Ivory Coast	Fire in a distillation unit caused $175 million worth of damage (Marsh, 2020)	The fire was caused by a hydrogen leak that was ignited (Marsh, 2020)
2017 Titanium dioxide manufacturing facility fire Pori Finland	The fire caused $1 billion of damage to the facility (Swaby, 2019)	The fire started in the electrostatic precipitators and spread quickly along gas pipelines installed in the roof and made of flammable material (Safety Investigation Authority Finland, 2017)
2017 Arkema chemical fire, Crosby Texas	Fumes from the fire caused 21 people to seek medical attention and 200 residents to be evacuated for 1 week	Extensive flooding from Hurricane Harvey caused the plant to lose power and backup power which resulted in the loss of refrigeration and then standby refrigerated trailers were subsequently flooded causing the organic peroxide to decompose and combust (U.S. Chemical Safety & Hazard Investigation Board, 2018)

(Continued)

TABLE 7.2 (Continued)

Accident	Estimated impact	Reported causes
2018 FCCU slide valve failure Wisconsin, United States	The explosion injured 36 people and forced the evacuation of a large proportion of the nearby town and cost $650 million (Marsh, 2020)	Erosion in the fluid catalytic cracking unit slide valve resulted in air missing with hydrocarbon which then triggered a large explosion (Marsh, 2020)
2018 Naphtha hydrotreater explosion Vohburg Germany	Eight employees were injured, the nearby town evacuated and extensive damage was incurred by the vapour cloud explosion with the total cost of the event being $495 million (Marsh, 2020)	A 1.5 metre long crack in the gasoline desulphurisation reactor led to the loss of containment of hot hydrocarbon that then formed vapour cloud which exploded (Hoorelbeke & Roosendans, 2019)
2018 oil spill Balikpapan, Indonesia	Oil spilt from pipeline at bottom on Balikpapan bay, polluting 20,000 ha of water, it then caught on fire killing 5 and resulted in evacuation of 700,000 people (Grupp, 2018)	The cause was a ship dropping anchor on the pipe causing it to break (Grupp, 2018)
2019 Alkylation unit LOC Philadelphia, United States	Large fire and explosion of leaking hydrofluoric acid resulted in $750 million in costs which bankrupted the operator (Marsh, 2020)	The loss of containment occurred when a pipe ruptured at a thinned elbow that was installed in 1973 (Marsh, 2020)
2019 Limbe, Cameroon distillation unit fire	A fire and explosion near the distillation units caused damage costing $600 million (Marsh, 2020).	Cause of the fire is yet to be published
2019 Explosion and fire at AB Specialty Silicones	Four people were killed, another was injured and asset damage was incurred onsite and by nearby businesses from a flammable gas explosion (Chemical Safety & Hazard Investigation, 2021a)	Hydrogen gas was generated during mixing process which formed hazy vapour cloud that flowed out of tanks and exploded (Chemical Safety & Hazard Investigation, 2021a)
2019 Chemical plant explosion Chenjiagang Chemical Industry Park, China	Major explosion in industrial park that produced pesticides and fertilizers caused $800 million in damages and killed nearly 80 people and injured another 640 (Marsh, 2020).	The cause was illegal custody of hazardous chemical wastes (Hazardex, 2020)
2019 Aghorn Waterflood Station Hydrogen Sulphide Release	Two people fatally injured – a worker and his spouse – when they were overcome by toxic H_2S gas (U.S. Chemical Safety & Hazard Investigation Board, 2021b)	Plunger failed in pump releasing gas, pump was not isolated, the pumphouse ventilation was inadequate and the building H_2S system was not functioning and worker did not wear H_2S detector (U.S. Chemical Safety & Hazard Investigation Board, 2021b)

(Continued)

TABLE 7.2 (Continued)

Accident	Estimated impact	Reported causes
2019 Chemical plant explosion Houston, Texas	Three people were injured and approximately 60,000 people evacuated when a butadiene tank exploded which also resulted in $500 million in damages (U.S. Chemical Safety & Investigation Board, 2022)	Popcorn like substance formed from the polymer and built up inside the process vessels causing a rupture that resulted in a loss of containment of butadiene that formed a vapour cloud which exploded (U.S. Chemical Safety & Investigation Board, 2022)
2019 LNG complex fire and explosions Arzew, Algeria	Two explosions followed by a fire in the LNG complex injured four people and caused $400 million in property damage (Marsh, 2020)	Cause of the fire is yet to be published

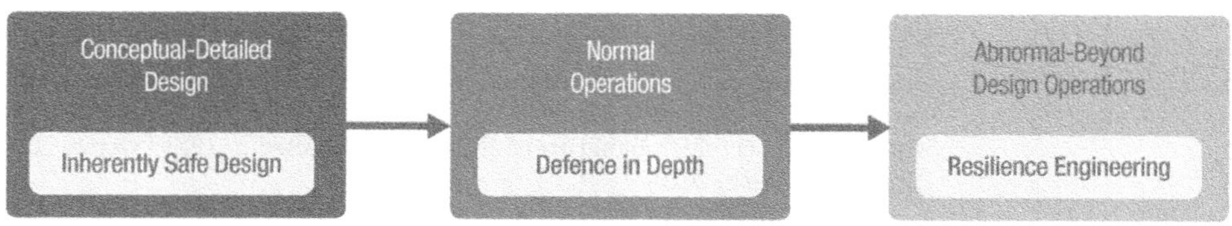

FIGURE 7.5 Approaches to managing process safety.

Managing personal or occupational safety risks should be done in accordance with ISO31000. The effectiveness of risk treatment efforts can be measured by assessing the effectiveness of controls and, because these types of events are frequent, it can also be measured using lagging measures such as the total recordable injury rate.

Managing process safety risks should also be done in accordance with ISO31000. However, because process safety involves the containment and control of hazardous materials within the whole system, and across the whole lifecycle, the management of process safety risks requires a suite of approaches.

If risk treatment is recommended for any hazards or hazardous scenarios identified in the hazard identification stage, then we need to either:

- Eliminate or reduce the risk at the design stage using the principles of inherently safer design, or
- If the hazard cannot be 'designed out', then we need to implement controls using defence-in-depth (DiD) principles.

Inherently safe design allows process safety considerations to be designed into the project, whereas DiD and resilience engineering are approaches to manage process safety in an operating facility (Fig. 7.5). We will consider each of these in more detail in this section.

7.3.1 Inherently safer design

Inherently safer design requires engineers to adopt a critical thinking approach to challenging design assumptions with questions such as *can we eliminate, minimise, or substitute the hazardous material? Can we moderate or simplify the process to reduce severity and improve detection and management of unwanted deviations?* The structured sequence of steps is shown in Fig. 7.6.

Inherently safer design aims to reduce the magnitude and probability of process safety events by:

1. ***Eliminating*** *the hazardous material* – for example the Bhopal disaster could have been avoided if intermediate MIC storage was eliminated, either through immediate reaction or by using a completely different reaction scheme.
2. ***Minimising*** *amount of hazardous material* – for example one of the findings out of the Bhopal disaster was that the large volume of MIC stored on site was not required. A much smaller volume could have been stored and would not have resulted in such widespread harm.
3. ***Substituting*** *hazardous material with a less hazardous one* – for example using a C4 type explosive is less hazardous than a nitroglycerin explosive. Another example is lead-based paints being replaced by water based paints.
4. ***Moderating*** *the strength or severity of consequences* – for example by lowering temperatures, lowering pressures, installing protections with secondary collection systems.
5. ***Simplifying*** *the process* – by designing processes to be less complicated they are less prone to failure.

The inherently safer design approach has been endorsed by many authorities including the U.K. Health and Safety Executive, the U.S. Chemical Safety Board and Safework Australia.

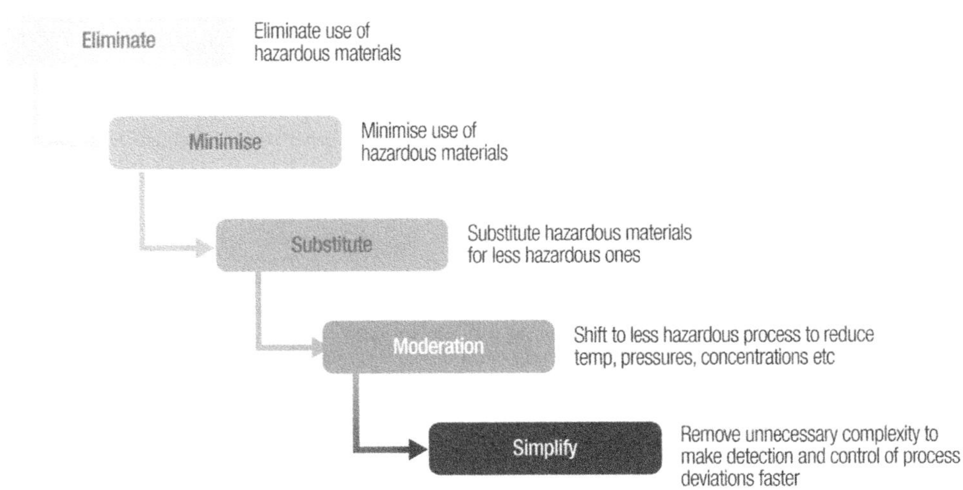

FIGURE 7.6 Inherently safe design principles.

Let us consider the application of inherently safer design principles for a petrol road tanker filling operation. If the process involves filling a petrol road tanker at a fuel tank farm, how might inherently safe design (ISD) beneficially affect the design of this operation? Some example responses are shown in Table 7.3.

7.3.2 Defence-in-depth

If the hazards and threats cannot be eliminated, then adequate layers of controls need to be selected to prevent the unwanted event, and to mitigate the consequences. The DiD approach stipulates the use of multiple independent barriers to stop threats from entering and propagating through a system. Castles provide an excellent illustration of the concept of DiD (Fig. 7.7), whereby strategies included building the castle on top of hills to enable 360-degree views, building high and thick walls, surrounding the castle with moats, and having archers or cannons installed in turrets. For the process industries, the DiD strategy similarly relies on independent layers of barriers to arrest threats and mitigate the effects of unwanted events.

The DiD typically used in the process industries comprises the following layers as shown in Fig. 7.8 and these are described in more detail next.

1. ***Inherently safe design*** (as discussed in the previous section)
2. ***Basic process control system*** entails the equipment that receives inputs from the process and uses them to adjust the operation to keep it within its normal production limits. Examples include:
   a. A heat exchanger temperature control system, whereby the controller adjusts coolant flow rates in order to maintain the output temperature within the desired range.

TABLE 7.3 Inherently safe design (ISD) options for road tanker filling fuel depot.

ISD Questions	Tanker filling example responses
Can we **eliminate** the petrol?	Eliminating the use of petrol is not an option
Can we **minimise** the use of petrol?	Minimising the tank holding volumes and flow rates should be considered to reduce potential spill volumes. However, this assessment should consider other risks associated with having more smaller tanks
Can we **substitute** the petrol with a less hazardous fuel?	Substituting the petrol is not an option
Can we **moderate** the process to less hazardous conditions?	Currently operating at atmospheric temperature and pressure. Could consider refrigeration but this should only be done if risk reduction associated with cooler petrol would outweigh cost and risk increases associated with the refrigeration system
Can we **simplify** the process to remove any unnecessary complexity?	Simplification of the piping, valve location and status, tank level detection and vapour sensing system should be implemented to ensure operators can easily detect the current status, and project the future status of the system, and whether it is in, or trending towards, a safe/unsafe state

FIGURE 7.7 Example of defence in depth used to protect castle residence *Source: Photo courtesy of Chensiyuan, CC BY-SA 4.0, via Wikimedia Commons sourced from https://commons.wikimedia.org/wiki/File:1_carcassonne_aerial_2016.jpg.*

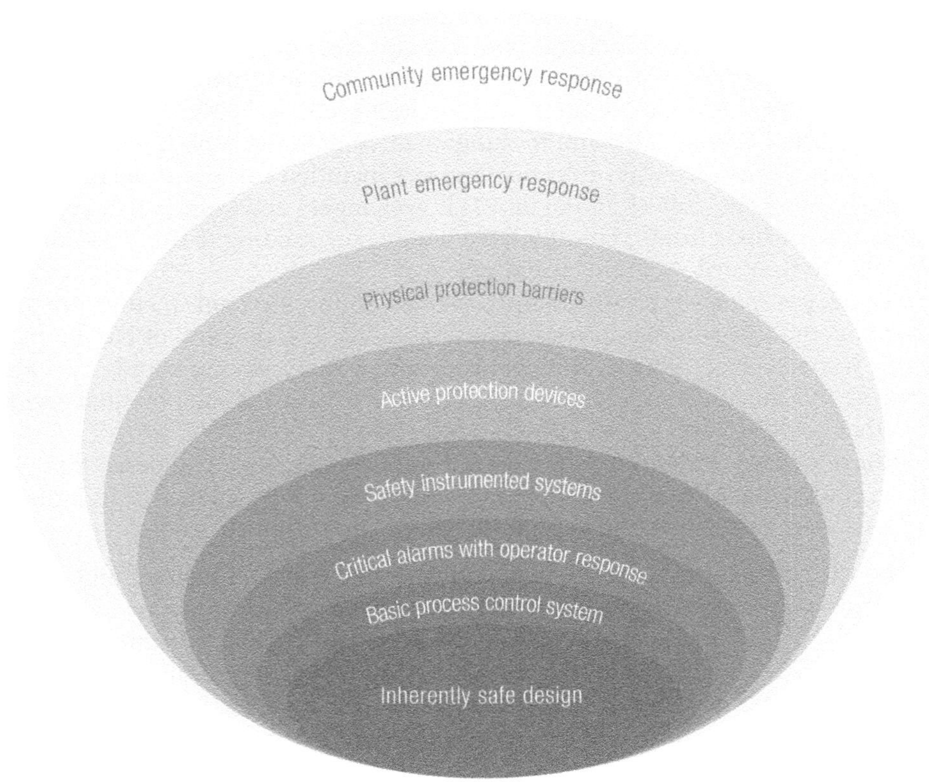

FIGURE 7.8 Defence in depth layers of protection.

   b. A level control system in a vessel, whereby a controller manipulates the outlet flow in order to control the liquid level.

   The aim of the process control system is to sustain normal operating states.

3. ***Critical alarms with operator response*** describes layers of protection that comprise alarms to alert operators that the system is drifting from normal to potentially unsafe operations. The basic process control system has been unable to correct the situation, and so an operator response is required to bring the system back into its normal state. Examples of critical alarms requiring operator response include:
   a. High and/or low outlet temperature alarms on heat exchangers.
   b. High and/or low level alarms on vessels.
4. ***Safety instrumented systems*** are automated control systems which respond to unsafe situations. These systems act independently of operators. Examples of safety instrumented systems include:
   a. Very high (or high high) outlet temperatures on a heat exchanger initiating an action to open a cooling media valve.
   b. Very low (or low low) level measurements on a vessel initiating an action to close the outlet valve(s).

   The aim of safety instrumented systems is to prevent an unsafe operating state escalating into an unwanted event.
5. ***Active protection devices*** and equipment that automatically activate in unsafe or abnormal situations in order to mitigate the consequences of accidents. Examples include pressure burst discs or pressure relief valves, and automatic fire suppression systems. Process vessels often are fitted with pressure relief valves.
6. ***Physical protection barriers*** also aim to mitigate or contain the consequences of accidents. Examples of physical protection barriers include overflow drains, bunds, blast walls/barriers and collection ponds. Heat exchangers and vessels that contain flammable and/or toxic fluids typically have bunds surrounding them. Vessels that can overflow also have overflow drains installed.
7. ***Plant emergency response*** typically comprises fire-fighting, first aid/medical treatment and environmental contamination response capabilities. The purpose of having plant emergency response is to mitigate accident consequences by having people available who can respond quickly and are specifically trained to address the potential accident scenarios that could occur.
8. ***Community emergency response*** is the external public emergency personnel that respond to emergency call numbers (e.g. 112 in Europe, 119 in Japan, 112 in India, 911 calls in North America, 999 in the United Kingdom and 000 in Australia).

The bowtie (discussed in Chapter 5) is a very effective way to illustrate and assess the layers of controls associated with a DiD strategy for a specific unwanted event. The bowtie should clearly show the independent layers of controls that eliminate, minimise exposure and detect and deflect threats. It should also clearly show the layers of controls that protect humans against harms caused by loss of containment or loss of control events, and also layers of controls that help recover and restore the system to safe operations. DiD must work not just to prevent unwanted events but to also mitigate the consequences if an unwanted event occurs.

7.3.3 Resilience engineering

In the process industries, resilience can be defined as the intrinsic ability of a system to adjust its functioning prior to, during or following changes and disturbances, so that it can sustain required operations. Enhancing resilience is about building the capabilities required to respond to emergent situations which are beyond the design of the inherently safe system and the defensive barriers used to prevent and mitigate unwanted accidents.

The three key principles associated with resilience engineering are:

1. Accidents can emerge from novel or unexpected interactions between system components;
2. Humans are best equipped to solve novel problems in real time by identifying and making the timing adaptations needed to restore safe operations; and
3. To allow humans to successfully solve problems in real time, the system needs to be designed to increase the number of things that can go right in addition to preventing the number of things that can go wrong.

To build resilience, an organisation should focus on the attributes shown in Fig. 7.9 which comprises:

- *Surveillance, comprehension* and *actioning* of safety critical parameters to be performed by operations personnel in real time
- *Research* and *preparedness* to identify and address any emerging factors (e.g. change, threats, technologies, etc.) that could impact the safety or the management of safety within the system.

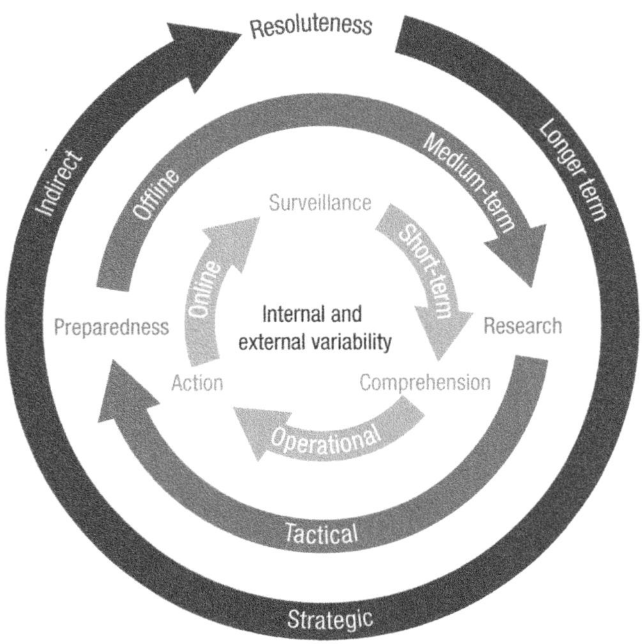

FIGURE 7.9 SCARPaR framework for organisational resilience (Hassall, 2013).

- *Resoluteness* in the leadership in instilling a corporate culture, and implementing the safety management systems that ensures ISD, ALARP standard DiD and which also ensures that Resilience attributes are implemented, sustained and continuously reviewed to identify and action improvement opportunities.

7.4 Safety culture

Safety culture is variously defined and these definitions are often subjected to debate. However, in general, safety cultures refer to the safety-related beliefs, values and attitudes shared by the majority of people within an organisation. It characterises the 'way we do things around here'. To effectively manage health and safety risks, organisations need to adopt the right approaches and need to foster the right culture to sustain operating excellence.

Safety culture has been described as a maturity continuum using three generalised categories which were later expanded to five. These five categories as shown in Fig. 7.10 are:

- Basic: Basic organisations have no safety focus in their culture. There can be an attitude of 'who cares as long as we're not caught' (Hudson, 2007, p. 704).
- Reactive: Reactive organisations recognise that safety is important but activity focusses on responding to accidents after they have occurred (Hudson, 2007).
- Compliant or bureaucratic: These types of organisations ensure they have the systems in place to manage all hazards (Hudson, 2007). However, these organisations can comprise people in departments who 'want to maintain their "turf," insist on their own rules, and generally do things by the book—*their book*' (Westrum, 2014, p. 59).

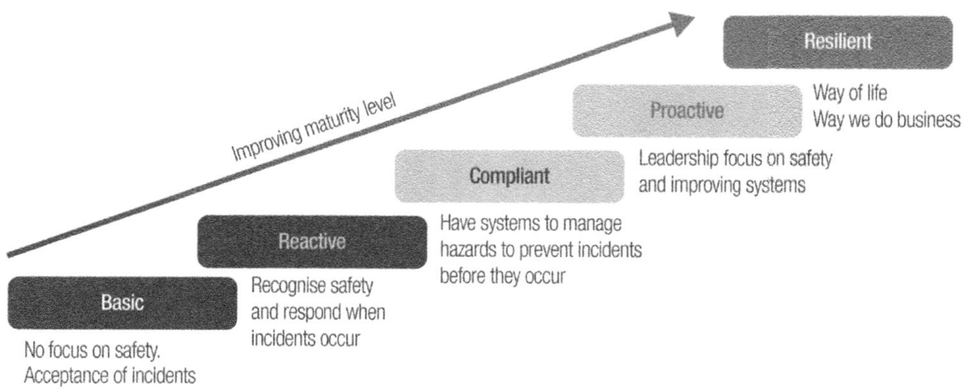

FIGURE 7.10 Safety culture model (Adapted from Hudson, 2007; Stemn, Bofinger, Cliff, & Hassall, 2019; Westrum 2014). *Source: Hudson, P. (2007). Implementing a safety culture in a major multi-national.* Safety Science, *45 (6), 697–722. https://doi.org/10.1016/j.ssci.2007.04.005; Stemn, E., Bofinger, C., Cliff, D., & Hassall, M. E. (2019). Examining the relationship between safety culture maturity and safety performance of the mining industry.* Safety Science, *113, 345–355. https://doi.org/10.1016/j.ssci.2018.12.008; Westrum, R. (2014). The study of information flow: A personal journey.* Safety Science, *67(0), 58–63. https://doi.org/10.1016/j.ssci.2014.01.009.*

- Proactive: Proactive organisations have a safety leadership focussed on continually improving (Hudson, 2007) so that they anticipate and fix safety problems before they arise (Parker et al., 2006).
- Resilient and generative: These organisations embed proactive safety in everything they do, it is not a separate focus. They *'focus on the mission.... Everything is subordinated to good performance, to doing what we are supposed to do'* (Westrum, 2014, p. 59).

Striving for the desired safety culture is a journey that requires leadership. Organisations can move up or down the safety continuum and experience suggests that they tend to move through phases sequentially – they do not jump a phase. Experience also shows that changing safety culture also requires strong leadership.

Reason (1997, p. 195) also identified the following as important subcomponents of an ideal safety culture:

- Having an *informed* culture. *'This means creating a safety information system that collects, analyses and disseminates information from incidents and near-misses as well as from regular proactive checks on the system's vital signs'. In other words, an informed culture is 'one in which those who manage and operate the system have current knowledge about the human, technical, organizational and environmental factors that determine the safety of the system as a whole'.*
- Having a *reporting* culture. *'Any safety information system depends crucially on the willing participation of the workforce, the people in direct contact with the hazards. To achieve this, it is necessary to engineer ... an organizational climate in which people are prepared to report their errors and near-misses'.*
- Having a *just* culture. *A just culture is 'an atmosphere of trust in which people are encouraged, even rewarded, for providing essential safety-related information – but in which they are also clear about where the line must be drawn between acceptable and unacceptable behavior'.*
- Having a *flexible* culture. *'A flexible culture takes a number of forms, but in many cases it involves shifting from the conventional hierarchical mode to a flatter professional structure, where control passes to task experts on the spot, and then reverts back to the traditional bureaucratic mode once the emergency has passed. Such adaptability is an essential feature of the crisis-prepared organization and, as before, depends crucially on respect – in this case, respect for the skills, experience and abilities of the workforce and, most particularly, the first-line supervisors. But respect must be earned, and this requires a major training investment on the part of the organization'.*
- Having a *learning* culture. *A learning culture encompasses 'the willingness and the competence to draw the right conclusions from its safety information system, and the will to implement major reforms when their need is indicated'.*

In addition to the above, one approach to culture that is particularly relevant to process safety is the concept of High Reliability Organisations (HROs). HROs is the term used to describe high hazard organisations that have excellent safety records. 'High reliability organisations have a "group mindfulness," which includes an organisation-wide sense of vulnerability and pessimism about possible failures. Responsibility and accountability for reliability is distributed throughout the organisation. Such organisations aim to increase the quality of attention and alertness across all departments and teams'. (The Health Foundation, 2011, p. 7).

Research has identified that these organisations have cultures characterised by:

- ***Preoccupation with failure***: HROs treat anomalies as symptoms of a problem with the system. The latent organisational weaknesses that contribute to small errors can also contribute to larger problems, so errors are reported immediately so problems can be found and fixed (Weick & Sutcliffe, 2007).
- ***Sensitivity to operations and reluctance to simplify interpretations***: HROs take deliberate steps to comprehensively understand the work environment as well as a specific situation. They are cognisant that the operating environment is very complex, so they continuously monitor the systems' safety and status of controls and remain sensitive to unexpected changed conditions (The Health Foundation, 2011; Weick & Sutcliffe, 2007).
- ***Containment of unexpected events***: HROs develop resilience capabilities so they can detect, contain and recover from unexpected events. They also defer to expertise to solve problems during upset conditions. HROs follow the typical communication hierarchy during routine operations but during abnormal situations, decisions are made by those at the front line and authority migrates to the person who can solve the problem, regardless of their hierarchical rank (The Health Foundation, 2011; Weick & Sutcliffe, 2007).
- ***Mindful leadership and a just culture***: HROs encourage proactive engagement of workers, accountabilities for halting operations on safety grounds, open discussion and reporting of bad news and errors without fear of repercussions (The Health Foundation, 2011; Weick & Sutcliffe, 2007).

From this information, we can see the importance of understanding the organisation culture and the role it plays in assuring safety. It can influence how humans perceive, assess and respond to risks. Therefore understanding the categories and subcomponents of culture is necessary in order to strive for an ideal safety culture and to promote risk optimisation within an organisation.

7.5 Process safety effectiveness

As you have seen, the field of process safety is still developing and improving, and we continue to witness serious process safety incidents at regular occurrence. Effective process safety practice requires many aspects of a plant and the organisation to be managed in a complementary and effective manner.

Because process safety events within a given organisation are rare, it is not appropriate to measure effectiveness based on past incident rates (i.e. lagging indicators such as LTIR). Rather, it is better to measure and report on the effectiveness of the controls. This was so disastrously demonstrated in the Deepwater Horizon accident, where their focus on lagging indicators of personal safety did not help them detect or prevent the disastrous blowout that occurred.

The IChemE Safety Centre Guidance on Lead Process Safety Metrics (IChemE, 2015) provides a suite of lead process safety metrics that cover a range of elements, including *competence, engineering and design, systems and procedures, assurance, human factors* and *culture*. Hopefully you will recognise that these are all critical elements that we have already

TABLE 7.4 Summary examples of lead process safety metrics (IChemE, 2015).

Metric	Example
Conformance with process safety-related role competency requirements	Are we meeting the minimum shift 'manning' requirement for the emergency response team?
Demand on safety critical elements	Monitors the occurrences of safety critical elements – e.g. low level shutdowns being triggered
Barriers failing on demand	Monitors the failure to activate of critical barriers – e.g. the occurrence of a safety valve release may be due to a failure of a pressure trip to activate
Permit to work checks performed to plan	Monitors the number of checks performed on the permit to work system
Number of process safety audits to plan	Monitors whether process safety audits are being conducted – e.g. audit of manual valves
Critical alarms per operator hour	Monitors the total number of annunciated alarms presented per operator

highlighted. Examples of some of these process safety metrics are shown in Table 7.4. Note that these are all leading metrics – that is, measuring the effectiveness of the process safety controls (the failure of which could lead to an unwanted event).

7.6 Safety management systems

To ensure health and safety risks are effectively managed and that the aforementioned activities are performed to a high quality by the right people at the right time requires a management system. Safety management systems outline the framework, processes, resourcing, supporting information and communication processes a company is going to follow to ensure the safety and health of its operations. The International Standard ISO 45001:2018 contains Occupational health and safety management systems guidance for use. An overview of this standard is provided in the following. Guidance of process safety management systems (PSMSs) has been provided by the Centre for Process Safety (American Institute of Chemical Engineers Center for Chemical Process Safety, 2016). This will also be discussed after ISO45001.

7.6.1 Overview of ISO45001

The purpose of a health and safety management system is to provide the framework for managing health and safety risks and opportunities in a manner that prevents injury and ill health and to provide safe and healthy work environment to workers and others impacted by workplace activities. It describes the policies, objectives, processes, roles and responsibilities that should be implemented to achieve health and safety objectives.

Key factors that determine the effectiveness of a safety management system and its ability to achieve its intended outcomes, as stated in ISO45001 (2018, p. vi), include:

1. Top management leadership, commitment, responsibilities and accountability.
2. Top management developing, leading and promoting a culture in the organisation that supports the intended outcomes of the health and safety management system.
3. Communication.
4. Consultation and participation of workers, and where they exist, workers' representatives.
5. Allocation of the necessary resources to maintain it.
6. Health and safety policies which are compatible with the overall strategic objectives and direction of the organisation.
7. Effective process(es) for identifying hazards, controlling health and safety risks and taking advantage of health and safety opportunities.
8. Continual performance evaluation and monitoring of the health and safety management system to improve health and safety performance.
9. Integration of the health and safety management system into the organisation's business processes.
10. Health and safety objectives that align with the health and safety policy and take into account the organisation's hazards, and its health and safety risks and opportunities.
11. Compliance with its legal requirements and other requirements.

ISO45001 outlines guidance for implementing a health and safety management system which includes:

a. **Setting the context:** This includes understanding the organisation and its context as well as the needs and expectations of workers and other interested parties. It also involves determining the scope of the health and safety management system.
b. **Leadership and worker participation:** This includes describing leadership's commitments in delivering health and safe working environments and the documentation of health and safety objectives and commitments in a health and safety policy. The responsibilities, accountabilities and expected consultation and participation of different roles in the organisation are also specified.
c. **Actions to address risks and opportunities:** This section describes the plans used to identify hazards and assess risks and risk treatment actions. It also should outline the legal and other requirements relevant to managing workplace health and safety risks. This section should also contain the plans, roles and responsibility and performance monitoring that will be implemented to ensure the achievement of the objectives.
d. **Supporting activities:** Included in supporting activities is identifying the competencies and awareness required to effectively manage health and safety. It also details the communication and documentation needed to support the health and safety management system as well as the achievement of the organisation's health and safety objectives.
e. **Additional requirements:** The health and safety management system should describe how the organisation is going to manage change, control procurement, manage

outsourcing and deal with emergency in a manner that ensures that it prevents injury and ill health.

f. **Performance evaluation:** The performance of the health and safety management system and the organisation's ability to achieve its health and safety objectives should be monitored, measured, analysed and evaluated against targeted performance on an ongoing basis. In addition, independent audits and management reviews should be periodically conducted. Audits should be conducted to check for compliance. Management reviews should be conducted to check for suitability, adequacy and effectiveness.
g. **Performance improvement:** Opportunities for improvement should be identified and actioned. Incident, nonconformance and other (e.g. benchmarking) investigations should be undertaken to help identify opportunities to improve.

The key factors and implementation guidance discussed above have been developed to focus on ensuring worker's occupational health and safety. Many of the concepts overlap with PSMSs which will be discussed next.

7.6.2 Overview of process safety management systems

Guidance on the implementation of PSMSs have been published by the Centre for Chemical Process Safety (CCPS) (American Institute of Chemical Engineers Center for Chemical Process Safety, 2016) and Occupational Health and Safety Administration (OSHA, 2013). A summary analysis is shown in Table 7.5.

Table 7.5 is structured according to the CCPS 20 elements of risk-based process safety. The 14 elements of OSHA's process safety management programme are then mapped against them to show alignment. Where appropriate, we have also included reference to relevant sections of this book.

7.7 Case study – process safety consideration for filling bulk storage tanks

Assessing safety, particularly process safety, risks in the process industries is a core competency of process engineers. As such, the case study for this chapter is quite extensive. The case study focusses on assessing risks associated with transferring fuel into a bulk storage tank and is presented in Appendix B. We have selected this case study because, whilst filling bulk storage tanks with chemicals is common practice, we continue to see recurring accidents associated with overfilling them. Recent notable examples include the 2011 overflow of gasoline in a Reichstat (Bas-Rhin) refinery (IMPEL - French Ministry for Sustaiable Development, 2013), the 2009 Puerto Rico refinery tank overflow, the 2005 Buncefield terminal tank overflow (Waite, 2013), and the 2005 Texas City Refinery explosion that occurred when a tower and reflux drum were overfilled (U.S. Chemical Safety & Hazard Investigation Board, 2007). The case study includes an example of a safety-focussed incident analysis based on the Buncefield event.

TABLE 7.5 Comparison between CCPS and OSHA PSMS information.

	CCPS risk-based 20 PSM elements (American Institute of Chemical Engineers Center for Chemical Process Safety, 2016; American Institute of Chemical Engineers Center for Chemical Process Safety, 2007)		OSHA PSM required 14 elements (OSHA., 2013)
COMMIT TO PROCESS SAFETY	1. Process safety culture	Develop, implement, sustain, monitor and enhance a sound set of values, practices and behaviours that promote each person having a strong commitment to process safety	Do not apply
	2. Compliance with Standards	Identifying and complying to relevant standards, codes, regulations and laws associated with process safety	Process safety information (PSI) including the codes, standards and practices that were employed and pertain to the selection, operation and management of the process technologies and equipment
	3. Process safety competency	Developing, sustaining and enhancing process safety knowledge to continually improve competencies	Do not apply
	4. Workforce involvement	Actively involving people from all levels of the organisation to support the development, implementation and enhancement of process safety management activities	Employers shall consult with employees and their representatives in developing process hazards analysis and other process safety activities
	5. Stakeholder outreach	Identifying, engaging and sustaining relationships with external stakeholders to share information and promote activities that enhance safe operations	Do not apply
UNDERSTAND HAZARDS AND RISKS	6. Process knowledge management	Developing, documenting and maintaining written technical documents and specifications including engineering drawings and calculations, process design information and specifications for process equipment and other information. This might include chemical hazard information, process technology information and process equipment information	Process safety information (PSI) pertaining to hazards, process technologies, equipment specification information which should include hazardous chemicals information, information pertaining to process technologies and equipment specification information
	7. Hazard identification and risk analysis	Includes all activities involved in the identification of hazards and evaluation of risks to people and the environment. Qualitative techniques such as process hazard analysis (PHA), hazard identification and operability analysis (HAZOP), failure modes effects and criticality analysis (FMECA) and layers of protection analysis (LOPA) along with quantitative techniques such as fault and event tree analyses are used to ask and answer the questions: What can go wrong? How bad could it be (e.g. what are the consequences)? How often might it happen (e.g. what is the likelihood)?	Process safety information (PSI) should include process hazard analysis using methods such as What-If, Checklist, Hazard and Operability Study (HAZOP), Failure Mode and Effects Analysis (FMEA), Fault Tree Analysis or appropriate equipment method. The approach should consider any relevant previous incidents, independence and effectiveness of controls, consequences of failure, location of facility and human factors

(*Continued*)

TABLE 7.5 (Continued)

	CCPS risk-based 20 PSM elements (American Institute of Chemical Engineers Center for Chemical Process Safety, 2016; American Institute of Chemical Engineers Center for Chemical Process Safety, 2007)		**OSHA PSM required 14 elements (OSHA., 2013)**
MANAGE RISKS	8. Operating procedures	Written instructions that list the steps for performing a task and describes how these steps should be performed to ensure safety of people and processes is maintained. Operating procedures should cover the safe operation and maintenance of the facilities and the handling and storage of any hazardous materials	Operating procedures shall be developed and implemented to provide clear instructions for safely conducting activities including startup, normal, emergency and shutdown operations. The operating procedure should have information about the safety systems and their functions
	9. Safe work practices	Describe how to perform nonroutine practices safety. Examples of nonroutine practices include hot work, stored energy (e.g. that needs isolating), opening vessels and confined space entry	The employer shall issue hot work permits for operations involving hot work on or near a hazardous process. The permit shall cover both prevention and protection requirements
	10. Asset integrity and reliability	The inspection, testing, preventative maintenance, repairs and other activities required to ensure important equipment sustains its operational integrity throughout its life.	The employer shall establish and implement written procedures and practices to maintain the mechanical integrity of processing vessels, pump, pipes, relief systems, emergency shutdown systems and control devices
	11. Contractor management	Ensure that there are systems of controls that ensure contracted work is performed in a manner that complies with a company's process and personnel safety requirements	Employers using contractors to perform maintenance, repairs or specialty process-related work shall obtain and evaluate information regarding contractors safety performance prior to selection, inform contractor of known hazards associated with their work, explain to contractors the emergency response processes, develop safe work practices required, presence and exiting of contractors from process, periodically evaluate contractors performance, maintain log of contractor injuries and illness
	12. Training and performance assurance	Provision of practical instruction in requirements to perform job and tasks to specified standards and proficiencies and assuring that they maintain reliable performance on critical tasks	Employers shall provide initial and refresher training to each employee. Training shall cover an overview of process and operating procedures as well as emphasising specific health and safety hazards and emergency response operations

(Continued)

TABLE 7.5 (Continued)

CCPS risk-based 20 PSM elements (American Institute of Chemical Engineers Center for Chemical Process Safety, 2016; American Institute of Chemical Engineers Center for Chemical Process Safety, 2007)		OSHA PSM required 14 elements (OSHA., 2013)
13. Management of change	Recognising change, evaluating hazards and deciding whether change should occur and ensuring necessary risk control and other measures are implemented so change does not increase risk or unknowingly introduce new hazards	The employer shall establish and implement procedures to manage any changes to the process chemicals, equipment, controls, procedures and facilities. The procedures should cover basis for change, modifications made, implications to safety and authorisation requirements
14. Operational readiness	Ensuring startup processes are performed safely over the life of the plant	Pre-start safety reviews: Prior to starting up new or significantly modified facilities, the employer needs to ensure that the construction and equipment complies with relevant design standards, safety, operating, maintenance and emergency procedures are in place and adequate, a process hazard analysis has been performed and recommendations addressed, and employees have been adequately trained
15. Conduct of operations	Developing and sustaining observable standards for conducting operational activities that ensures workers behave and are held accountable for delivering reliable performance that meets the standards	Do not apply
16. Emergency management	– Planning for potential emergencies – Providing resources and training needed to execute the plan. – Practicing and improving the plan	Employers shall establish an emergency response action plan in accordance with relevant regulations and standards
17. Incident investigation	Reporting, formally investigating, and tracking incident trends to learn what was the root causes, how they could be addressed, and to share the learning with other facilities to reduce frequency and/or consequences of future similar incidents	Incidents that could have or actually did result in a catastrophic release of a hazardous chemical need to be investigated by the employers' investigation team who shall produce a report that states the incident date, investigation date, incident description, contributing factors to the incident, incident findings and recommendations. The employer shall also establish a system for review reports with affected personnel, addressing recommendations and retaining the reports

(Continued)

TABLE 7.5 (Continued)

	CCPS risk-based 20 PSM elements (American Institute of Chemical Engineers Center for Chemical Process Safety, 2016; American Institute of Chemical Engineers Center for Chemical Process Safety, 2007)		OSHA PSM required 14 elements (OSHA., 2013)
LEARN FROM EXPERIENCE	18. Measurements and metrics	Establish a combination of leading and lagging indicators to monitor effectiveness and efficiency of performance	Do not apply
	19. Auditing	Conducted by a team of qualified internal and/or external personnel who evaluate whether management system is implemented and performing as expected. Audits should cover periodic evaluations of performance against all PSM elements to produce findings and recommendations for corrective actions	Employers shall ensure their performance is evaluated by a knowledgeable person(s) and a report of the audit findings along with appropriate responses are documented
	20. Management review and continuous improvement	A due diligence review performed by management to evaluate whether the management systems are performing as expected and producing the intended results as efficiently as possible	Do not apply
	Other	Do not apply	Trade secrets: Employers shall make all information available to responsible people that is necessary to perform process safety management activities regardless of the possible trade secret status of such information

7.8 Summary

In this chapter we have introduced the four types of health and safety risks, namely occupational health, human wellbeing, personal safety and process safety. We have also introduced several seminal case studies. These have highlighted the criticality of health and safety risks and highlight many key lessons to learn.

We have seen yet again how unwanted events occur through the failure of implementing known controls. This is evident for both health and safety risks. The reoccurrence of black lung in the Australian mining industry and the global growth in silicosis are both very clear illustrations of this. Ongoing regular major process safety incidents also reiterate our collective failure to learn from past events, and to implement known effective controls.

We have seen that safety covers both personal and process safety. Personal safety refers to high likelihood low consequence events such as slips, trips and falls. Process safety refers to catastrophic low likelihood but high consequence events. Most high hazard industries have invested time, resources and money into improving safety. This effort has resulted in significant improvements in personal safety outcomes with reductions in lost time injuries and total recordable injuries being reported by many companies over the last couple of decades. However, the same improvement has not been reported for process safety incidents, and it is now widely acknowledged that we need to employ different approaches to manage process safety.

The latter part of the chapter focussed on process safety in detail, and introduced key approaches to assess and treat process safety risks. Inherently safer design, DiD, resilience engineering, safety management systems and safety culture were all seen to play key roles in managing process safety risk. The effectiveness of these can be monitored by the use of lead process safety metrics.

Finally, this chapter has introduced safety management systems. These are the framework, processes, policies, resourcing, supporting information and communication processes that an organisation will implement and follow to ensure good health and safety outcomes.

Review questions

1. What are the four areas of health and safety that we are considering?
2. What is the difference between personal safety and process safety?
3. What is defence-in-depth (DiD)?
4. What is safety culture? Name some types of safety culture.
5. What is a process safety management system (PSMS) and why do we need them?
6. What are the principles of inherently safe design (ISD)?
7. Use the defence-in-depth (DiD) model to critique the Bhopal incident.
8. What is the difference between lead and lag process safety metrics? Provide two examples of each.
9. Select a contemporary process safety incident, review published incident investigations, and draft a concise summary of the incident, highlighting key lessons learnt. You may like to choose one of the incidents outlined in Table 7.2.
10. Describe the meanings of:
    a. Inherently safe design
    b. Defence in depth
    c. Resilience engineering

    How are they different and how are they complementary?

References

ABC News. (2016). *Mexico fireworks market explosion leaves at least 31 dead and scores injured, emergency services say*. ABC News. https://www.abc.net.au/news/2016-12-21/fireworks-market-explosion-leaves-31-dead,-dozens-injured-mexico/8137812.

American Institute of Chemical Engineers Center for Chemical Process Safety (2007). Guidelines for risk based process safety. Hoboken, New Jersey: Wiley.

American Institute of Chemical Engineers Center for Chemical Process Safety (2016). Guidelines for implementing process safety management (2nd ed). Hoboken, New Jersey: Wiley.

Anonymous (2014, June 19, 2014). *Neptune officially opens its new facility in Sherbrooke*. Information Hub News. http://sherbrooke-innopole.com/en/news/neptune-officially-opens-new-facility-sherbrooke/.

Ayala, D. (2014). *CSST issues report on fatal blast at Sherbrooke plant*. Montreal Gazette. http://montrealgazette.com/news/local-news/csst-issues-report-on-fatal-blast-at-sherbrooke-plant.

Baker, J. A. I., Leveson, N. G., Bowman, F. L. S., Priest, S., Erwin, G., Rosenthal, I. I., . . . Wilson, L. D. (2007). *The report of the BP U.S. refineries independent safety review panel*. sunnyday.mit.edu/Baker-panel-report.pdf.

Bagirova, N. (2015). *Azerbaijan says 29 missing after oil platform fire feared dead*. Reuters. https://www.reuters.com/article/us-azerbaijan-oil-fire-idUSKBN0TP0CL20151206.

BASF. (2016). *Update: Fire at the North Harbor in Ludwigshafen*. BASF. https://www.basf.com/global/en/media/news-releases/2016/10/p-16-359.html.

BEA. (2016). *Final Report. Accident on 24 March 2015 at Prads-Haute-Bléone (Alpes-de-Haute-Provence, France)*. Bureau d'Enquêtes et d'Analysespour la sécurité de l'aviation civile.

Bibby, N. (2016). *Preliminary findings of the cause of a blast that killed 21 in China is that a steam pipe bursting*. https://apfmag.mdmpublishing.com/preliminary-findings-of-the-cause-of-a-blast-that-killed-21-in-china-is-that-a-steam-pipe-bursting/.

BP. (2010). *Deepwater horizon accident investigation report*. http://www.bp.com/liveassets/bp_internet/globalbp/globalbp_uk_english/incident_response/STAGING/local_assets/downloads_pdfs/Deepwater_Horizon_Accident_Investigation_Report.pdf.

DARRP (Producer). (2022). *Refugio beach oil spill*. Santa Barbara County, California. May 2015. https://darrp.noaa.gov/oil-spills/refugio-beach-oil-spill.

Doyle, A. (2019). *Valero fined £5m following fatal Pembroke refinery explosion in 2011*. The Chemical Engineer. https://www.thechemicalengineer.com/news/valero-fined-5m-following-fatal-pembroke-refinery-explosion-in-2011/.

Elliott, D. (2012). *Fukushima: Impacts and implications*. Basingstoke: Palgrave Macmillan.

Eni. (2017). *Integrated annual report 2016*. Eni. https://www.eni.com/assets/documents/documents-en/Integrated-Annual-Report-2016.pdf.

Fu, G., Wang, J., & Yan, M. (2016). Anatomy of Tianjin Port fire and explosion: Process and causes. *Process Safety Progress*, *35*(3), 216–220. Available from https://doi.org/10.1002/prs.11837.

Gorman, S., & McWilliams, G. (2018). *Plains all American pipeline convicted in 2015 California oil spill*. Reuters. https://www.reuters.com/article/us-plains-all-amer-oilspill-court-idUSKCN1LN2SN.

Graham, B., Reilly, W. K., Beinecke, F., Boesch, D. F., Garcia, T. D., Murray, C. A., & Ulmer, F. (2011). *Deep water: The Gulf oil disaster and the future of offshore drilling*. USA.

Grupp, T. (2018). *Massive environmental and health damage from oil spill in Borneo*. https://www.climatescorecard.org/2018/05/massive-environmental-and-health-damage-from-oil-spill-in-borneo/.

Hackitt, D. J. (2018). *Building a safer future: Independent review of building regulations and fire safety: Final report*. London, England.

Hazardex. (2013). *South Korean chemical plant blast kills seven*. Hazardex. https://www.hazardexonthenet.net/article/56911/South-Korean-chemical-plant-blast-kills-seven.aspx.

Hazardex. (2020). *Plant manager detained after explosion and fire at chemical plant in China*. HazardEx. https://www.hazardexonthenet.net/article/178239/Plant-manager-detained-after-explosion-and-fire-at-chemical-plant-in-China.aspx.

Health and Safety Executive. (2020). *Chevron Pembroke Amine regeneration unit explosion 2 June 2011 - An overview of the incident and underlying causes*. https://www.hse.gov.uk/comah/chevron-pembroke-report-2020.pdf.

Hassall, M. E. (2013). *Methods and tools to help industry personnel identify and manage hazardous situations. Doctor of Philosophy thesis*. The University of Queensland.

Herink, T., Doskočil, J., Růžička, M., Henry, P. A., Sadowski, R. A., & Zámostný, P. (2022). Propylene column pressure relief valves chattering resulting in explosion and fire of the Steam Cracker unit. *Journal of Loss Prevention in the Process Industries*, *74*, 104658. Available from https://doi.org/10.1016/j.jlp.2021.104658.

Hoorelbeke, P., & Roosendans, D. (2019). *Industrial system for chemical inhibition of vapor cloud explosions*. Paper presented at the Process Safety Conference, Dordrecht, The Netherlands. https://pscongres.nl/wp-content/uploads/2019/05/02-Dirk-Roosendans-Total-Mitigation-of-Vapor-Cloud-Explosions.pdf.

Hudson, P. (2007). Implementing a safety culture in a major multi-national. *Safety Science*, *45*(6), 697–722. Available from https://doi.org/10.1016/j.ssci.2007.04.005.

IChemE (2015). *IChemE safety centre guidance: Lead process safety metrics – Selecting, tracking and learning*. IChemE. https://www.icheme.org/media/1092/safety-centre-metrics.pdf.

IMPEL - French Ministry for Sustaiable Development (2013). Overflow of a gasoline tank inside a refinery. 22 October 2011. Reichstett (Bas-Rhin) France. Report No. 41148. www.aria.developpement-durable.gouv.fr/wp-content/files_mf/A41148_ips41148_003.pdf.

Kátai-Urbán, L., & Csépló, Z. (2010). *Disaster in the AAjka red sludge reservoir on 04 October 2010.* Paper presented at the Sixth Meeting of the Conference of the Parties to the Convention on the Transboundary Effects of Industrial Accidents, The Hague.

Lakshmi, M. R., & Kumar, V. D. (2015). Anthropogenic hazard and disaster relief operations: A case study of GAIL pipeline blaze in East Godavari of A.P. *Procedia - Social and Behavioral Sciences, 189*, 198–207. Available from https://doi.org/10.1016/j.sbspro.2015.03.215.

Levy, G. (2016). *Break the silence on Azerbaijan oil workers' deaths.* Open Democracy. https://www.opendemocracy.net/en/odr/break-silence-on-azerbaijan-oil-workers-deaths/.

Marsh. (2020). *100 largest losses in the hydrocarbon industry 1974–2019.* https://www.marsh.com/au/industries/energy-and-power/insights/100-largest-losses-in-the-hydrocarbon-industry.html.

Melley, B. (2015). *Pipeline that leaked wasn't equipped with auto shut-off.* Pittsburgh Post-Gazette. https://www.post-gazette.com/business/powersource/2015/05/24/Pipeline-that-leaked-wasn-t-equipped-with-auto-shut-off-2/stories/201505240175.

Mexico News Daily. (2018). *2 years after explosion that killed 42, fireworks market reopens.* Mexico News Daily. https://mexiconewsdaily.com/news/tultepec-fireworks-market-reopens/.

Neptune Technologies & Bioressources Inc. (2012). *Press release: Neptune provides update on incident at its production plant.* Quebec, Canada: http://www.neptunebiotech.com/sites/default/files/20121112%20Neptune%20Press%20Release%20-%20Explosion%20-%20Press%20Release.pdf.

OSHA. (2013). *1910.119 - Process safety management of highly hazardous chemicals.* United States Department of Labor. https://www.osha.gov/laws-regs/regulations/standardnumber/1910/1910.119.

Parks, K. (2013). *Argentina storm sparks fire at YPF La Plata refinery.* Hydrocarbon Processing. https://www.hydrocarbonprocessing.com/news/2013/04/argentina-storm-sparks-fire-at-ypf-la-plata-refinery.

Parraga, M. (2013, 10 September, 2013). *Chronology: Pump collapse, leak caused Venezuela refinery blast.* Reuters. https://www.reuters.com/article/us-oil-venezuela-refinery-idUSBRE9880Z820130909.

Parker, D., Lawrie, M., & Hudson, P. (2006). A framework for understanding the development of organisational safety culture. *Safety Science, 44*, 551–562.

Reason, J. T. (1997). *Managing the risks of organizational accidents.* Aldershot, UK: Ashgate.

Safety Investigation Authority Finland. (2017). *Y2017-01 In-dus-trial fires in Sa-takunta in Jan-u-ary 2017.* https://turvallisuustutkinta.fi/en/index/tutkintaselostukset/other/tutkintaselostuksetvuosittain/2017/y2017-01teollisuuspalotsatakunnassatammikuussa2017.html.

Slater D. (2015). *Bhopal: A tragedy of unintended consequences.* The Chemical Engineer. January 2015. 32–36.

Stemn, E., Bofinger, C., Cliff, D., & Hassall, M. E. (2019). Examining the relationship between safety culture maturity and safety performance of the mining industry. *Safety Science, 113*, 345–355. Available from https://doi.org/10.1016/j.ssci.2018.12.008.

Stensland, J. (2014). *Settlement reached for 2010 Anacortes refinery explosion litigation.* Whidbey News-Times. Retried from https://www.whidbeynewstimes.com/news/settlement-reached-for-2010-anacortes-refinery-explosion-litigation/.

Swaby, N. (2019). *Chemical maker downplayed massive damage at facility before IPO, investors claim.* Courthouse News Service. https://www.courthousenews.com/chemical-maker-downplayed-massive-damage-at-facility-before-ipo-investors-claim/.

The Health Foundation. (2011). *Evidence scan: High reliability organisations.* https://www.health.org.uk/publications/high-reliability-organisations.

The National Diet of Japan. (2012). *The official report of the Fukushima nuclear accident independent investigation commission: Executive summary.* http://warp.da.ndl.go.jp/info:ndljp/pid/3856371/naiic.go.jp/en/report/.

Turi, D., Pusztai, J., & Nyari, I. (2013). *Causes and circumstances of red mud reservoir dam failure in 2010 at MAL Zrt factory site in Ajka, Hungary.* Paper presented at the International Conference on Case Histories in Geotechnical Engineering, Chicago. https://scholarsmine.mst.edu/cgi/viewcontent.cgi?article=3096&context=icchge.

U.S. Chemical Safety and Hazard Investigation Board. (2007). *Investigation report - Refinery explosion and fire - BP.* Texas City, March 23, 2005 (2005-04-I-TX).

U.S. Chemical Safety and Hazard Investigation Board. (2014). *Investigation report - Catastrophic rupture of heat exchanger (seven fatalities).* Tesoro Anacortes Refinery Anacortes, Washington, April 2, 2010. (2010-08-I-WA). https://www.csb.gov/tesoro-refinery-fatal-explosion-and-fire/.

U.S. Chemical Safety and Hazard Investigation Board. (2016a). *Case study - Williams Geismar Olefins Plant reboiler rutpure and gire Geismar, Louisiana - Incident date.* June 13, 3013 (2013-03-I-LA). https://www.csb.gov/williams-olefins-plant-explosion-and-fire-/.

U.S. Chemical Safety and Hazard Investigation Board. (2016b). *Investigation report - West Fertilizer Company fire and explosion (15 fatalities, more than 260 injured).* West Fertilizer Company, West, TX, April 17, 2013 (2013-02-I-TX). https://www.csb.gov/west-fertilizer-explosion-and-fire-/.

U.S. Chemical Safety and Hazard Investigation Board. (2017a). *Investigation report - Chemical spill contaminates public water supply in Charleston, West Virginia.* Frredom Industries Inc. Charleston, WV, January 9, 2014. (2014-01-I-WV). https://www.csb.gov/freedom-industries-chemical-release-/.

U.S. Chemical Safety and Hazard Investigation Board. (2017b). *Investigation report.* ExxonMobil Torrance Refinery.

U.S. Chemical Safety and Hazard Investigation Board. (2019). *Toxic Chemical Release at the DuPont La Porte Chemical Facility La Porte, Texas.*Incident Date: November 15, 2014 (2015-01-I-TX). https://www.csb.gov/dupont-la-porte-facility-toxic-chemical-release-/.

U.S. Chemical Safety and Hazard Investigation Board. (2021a). *Hydrogen Sulfide Release at Aghorn Operating Waterflood Station Odessa, Texas.* Incident Date: October 26, 2019 (2020-01-I-TX). https://www.csb.gov/arkema-inc-chemical-plant-fire-/.

Vaughan, A. (2018). *BP's Deepwater Horizon bill tops $65bn.* The Guardian. https://www.theguardian.com/business/2018/jan/16/bps-deepwater-horizon-bill-tops-65bn.

U.S. Chemical Safety & Hazard Investigation Board. (2018). *Organic Peroxide Decomposition, Release, and Fire at Arkema Crosby Following Hurricane Harvey Flooding.* U.S. Chemical Safety & Hazard Investigation Board. 2017-8-I-TX. May 2018. Available from https://www.csb.gov/arkema-inc-chemical-plant-fire-/.

U.S. Chemical Safety & Hazard Investigation Board. (2021b). *Chemical Reaction, Hydrogen Release, Explosion, and Fire at AB Specialty Silicones.* U.S. Chemical Safety & Hazard Investigation Board. 2019-03-I-IL. Sept 24, 2021. Available from http://www.csb.gov/assets/1/20/ab_specialty_investigation_report_final_(1).pdf.

U.S. Chemical Safety & Investigation Board. (2022). *TPC Port Neches Explosions and Fire. Popcorn Polymer Accumulation, Pipe Rupture, Explosions, and Fires at TPC Group Chemical Plant Butadiene Unit.* U.S. Chemical Safety & Investigation Board. 2020-02-I-TX. December 2022. Available from https://www.csb.gov/tpc-port-neches-explosions-and-fire/.

Waite, P. (2013). Recurring accidents: overfilling vessels. In *The Chemical Engineer*, (pp. 40–44). Institution of Chemical Engineers.

Weick, K. E., & Sutcliffe, K. M. (2007). *Managing the unexpected: Resilient performance in an age of uncertainty.* Hoboken, UNITED STATES: John Wiley & Sons.

Westrum, R. (2014). The study of information flow: A personal journey. *Safety Science, 67*, 58–63. Available from https://doi.org/10.1016/j.ssci.2014.01.009.

Wong, E. (2016). *Explosion at coal-fired plant in Central China kills at least 21.* The New York Times. https://www.nytimes.com/2016/08/12/world/asia/china-danyang-power-plant-blast.html.

Yang, H.-N., Chen, J.-H., Chiu, H.-J., Kao, T.-J., Tsai, H.-Y., & Chen, J.-R. (2016). Confined vapor explosion in Kaohsiung City – A detailed analysis of the tragedy in the harbor city. *Journal of Loss Prevention in the Process Industries, 41*, 107–120. Available from https://doi.org/10.1016/j.jlp.2016.03.017.

Zhang, Y., Jing, L., & Sun, C. (2018). Systems-based analysis of China-Tianjin port fire and explosion: A comparison of HFACS, AcciMap, and STAMP. *Journal of Failure Analysis and Prevention.* Available from https://doi.org/10.1007/s11668-018-0534-1.

조지현 (2013). *Daelim headquarters, factory raided over fatal explosion.* The Korea Herald. http://www.koreaherald.com/view.php?ud = 20130318001025.

CHAPTER

8

Environmental and social risk

OUTLINE

8.1 Introduction 194

8.2 What is environmental and social risk? 196

8.3 The project lifecycle 198

8.4 Impact assessment 200

8.5 Social licence to operate 202

8.6 Tools to identify, assess, treat and monitor environmental and social risks 204

- 8.6.1 *Preliminary stakeholder analysis and management* 204
- 8.6.2 *Stakeholder risk assessment* 207
- 8.6.3 *Social impact management plan* 208

8.7 Summary 210

Review questions 211

References 212

Fundamentals of Risk Management for Process Industry Engineers
DOI: https://doi.org/10.1016/B978-0-12-820320-0.00002-7

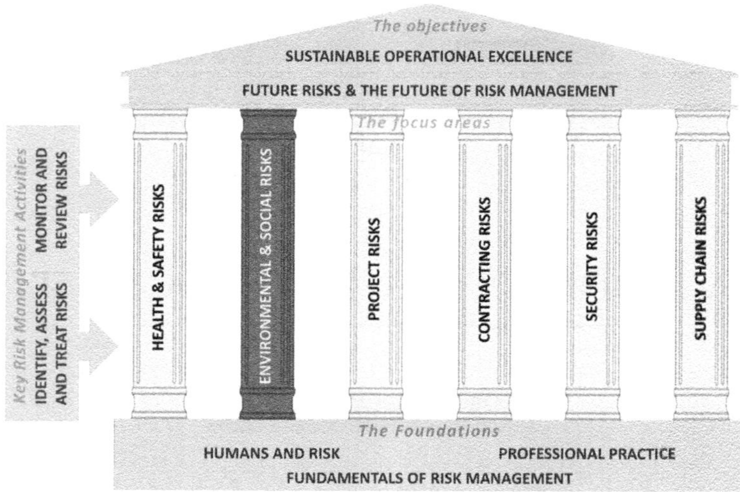

Failure is not an option but it is constantly a possibility. You manage your success by managing both the technical and the social risk ((Bratasanu, 2015), p. 62).

8.1 Introduction

Why are you studying or practicing engineering? Before proceeding, please list five answers to this question.

Were any of your answers related to social impact? If so, how often do you reflect on the social impact of your work?

We have decided to start with this question in order to emphasise a really critical point which will underpin the rest of this chapter. Namely, that positive social impact (societal benefit) is a core driver for most engineering projects, and likely a key reason for why you are pursuing an engineering career. So, rather than being a 'tangential topic' and peripheral to 'core' project engineering, we contend that positive social impact is actually our *raison d'etre*.

Recall the risk mindsets that we introduced in Chapter 2 – that is, the *loss reduction* and *system optimisation* mindsets (Table 2.4). We are arguing that our focus on positive social impact is consistent with the system optimisation mindset for managing risks. That is, we are implicitly approaching the challenges of managing environmental and social risk with the intent of managing downside threats whilst optimising the upside opportunities. Contemporary engineering practice is no longer solely about reducing the threat of environmental and social risks from impacting on the project. The goal of modern professional practice should be to integrate the environmental and social context into a project's design for outcomes that are beneficial to all stakeholders. Some projects will inevitably result in some stakeholders (which includes environmental stakeholders) being negatively impacted. In this case, the goal is to minimise these impacts and maximise the systemic benefits so that the environment and society benefits overall.

These principles are very nicely reflected in the World Bank Environmental and Social Framework (World Bank, 2017), which governs the conditions for borrowers from The Bank:

'The Bank's vision goes beyond "do no harm" to maximising development gains' (p. 2). The framework lists the following objectives for all projects:

- Avoid or mitigate adverse impacts to people and the environment.
- Conserve or rehabilitate biodiversity and natural habitats, and promote the efficient and equitable use of natural resources and ecosystem services.
- Promote worker and community health and safety.
- Ensure that there is no prejudice or discrimination toward project-affected individuals or communities and give particular consideration to Indigenous Peoples, minority groups, and those disadvantaged or vulnerable, especially where adverse impacts may arise or development benefits are to be shared.
- Address project level impacts on climate change and consider the impacts of climate change on the selection, siting, planning, design and implementation and decommissioning of projects.
- Maximise stakeholder engagement through enhanced consultation, participation and accountability.

(World Bank, 2017, p. 2).

Whilst the World Bank focus is primarily on financing projects in developing countries, this list of objectives is still highly pertinent for all industrial development projects, and you will see that these objectives underpin much of the content in this chapter. We will also see that failure to satisfy these objectives will quite likely result in failure to achieve social acceptance of the project.

Failure to achieve social acceptance of your project can mean that the project may never commence, meet problematic resistance, or may be forced to terminate early. There are many contemporary examples including: (1) the early closure of nuclear power plants in Germany in response to public concerns following the Fukushima disaster in Japan in 2011; (2) violent protests from local community members about the PNG LNG project in Papua New Guinea (Main, 2017); (3) the indefinite suspension of hydraulic fracturing (fracking) in the United Kingdom (2019); and (4) widespread public outrage and a government enquiry following the destruction of aboriginal heritage sites at Juukan Gorge in Western Australia as part of a Rio Tinto mining operation in 2020 (Parliament of Australia, 2020). In all of these cases, the project and/or industry either failed to obtain, or lost, their licence to operate. For each of these cases, it is worth considering which of the objectives above were not satisfied, and how they may have been better addressed.

As we saw in Chapter 1, environmental and social risks are high priority concerns for industry (Table 1.1). 'Licence to operate', 'environmental regulations' and 'environmental requirements' were all cited as 'top 10 risks' for the mining, manufacturing and oil and gas sectors. We also saw in Chapter 1 that businesses with more developed approaches to managing risk tend to perform better. This is certainly also the case for the management of social risks, where business benefits resulting from improved processes for assessing and managing social impacts include (Esteves et al., 2012):

- Greater certainty for project investments and increased chance of project success.
- Avoidance and reduction of social and environmental risks.
- Improved ability to identify problems at an earlier stage.
- A positive legacy beyond the life of the project.

- Improved quality of life for employees.
- Increased competitive advantage through enhanced social performance and corporate reputation.

On completion of this chapter, you should be able to describe what environmental and social risks 'look like', explain how ISO31000 informs how we manage environmental and social risk, and explain the processes by which we manage and monitor environmental and social risks in industry. Moreover, you should be able to explain why operational environmental and social risks matter, understand the importance of the social licence to operate, and understand the importance of asking the question: 'what is the environmental and social context for my project?'

In principle, environmental and social risk is managed like all other risks, and in accordance with ISO31000. As such, the risks are identified, analysed, evaluated and recorded in a risk register, and the appropriate treatment (controls) recommended. However, the management of environmental and social risk is also different in some ways. For example, in most jurisdictions, there is specific legislation in regard to the management of these risks. Also, the nature of environmental and social risks often means that the assessment and treatment processes that we employ are somewhat different to those that we might adopt to manage other risks.

The chapter starts by defining environmental and social risk, and then discusses the project and regulatory contexts within which we manage these risks. We will then introduce the Impact Assessment process, before examining the Social Licence to Operate. The final part of the chapter introduces tools to assist with stakeholder management, risk assessment and then treatment, via the use of social impact management plans (SIMPs).

8.2 What is environmental and social risk?

The aim of this section is to introduce environmental and social risk, and to discuss the role of engineers in managing it.

Why do we usually consider environmental and social risk together? It is simply because humans and society are intimately dependent on the environment in which we live. We are part of the environment and it is not separate from human society. Environmental harm often has strong interrelationships with social harm. As such, they are usually assessed as part of joint risk management processes. Indeed, in many jurisdictions around the world, it is legislated to perform environmental and social assessments together. This link will become more evident as we progress through the chapter, and in the many case studies presented.

What is environmental and social risk?

Environmental risk is commonly referred to as the threat of harm to an environmental value. In this book, we are proposing a slightly different definition in line with our broader definition of risk, namely:

Environmental risk is the uncertainty that matters for an environmental value.

What are environmental values? Environmental values are aspects of the environment that have value to the ecosystem and the broader environment – they are utilised in some way as part of a functioning and sustainable system. For example, in regard to water, environmental values are the qualities that make water suitable for supporting aquatic ecosystems and human use.

Similarly, we can define social risk as:

> Social risk is the uncertainty that matters for a human or social value.

What are social values? Social values are aspects of human life and society that are fundamental to health and safety, the ability to lead a fulfilling life and the structures and institutions that support a functioning society to progress in a direction of their choice. In regard to water, a social value may be the ability to safely use the water body for recreational activities such as fishing or swimming.

The links between environmental and social values are illustrated by coal bed methane extraction, whereby methane gas is extracted from coal beds after removing significance volumes of water. Ultimately, we want to extract the methane gas to support social values by providing affordable and effective energy services. However, the extraction of gas from coalbeds can directly impact on ground water (quantity and potentially quality). This can then impact surface water streams and dependent ecosystems. Water availability can also impact dependent human activities such as farming and access to drinking water. Such impacts lead directly to issues of human physical and mental health, ability to derive an income, and ultimately quality of life. This tension has been playing out in the public domain over the last decade, with many activist groups strongly protesting against the industry, resulting in prohibition in some regions and significant impact in many others (Ransan-Cooper et al., 2018).

Considering these broad definitions of environmental and social risk, it is easy to see that a lot of our work as engineers will have the risk of impacting on environmental and social values. We also live in an era where society is requesting higher levels of performance and transparency from process companies.

> Society rightly demands the highest standards from these high hazard industries. The risks created by these businesses must be managed effectively, because when things go wrong in this sector then the consequences are severe and can destroy lives, shatter local communities, and cause damage to the environment that can last for generations.... ***Gordon MacDonald, Director of Hazardous Installations, UK HSE. Commenting after the court proceedings into the Buncefield incident (2010).***

So we have established that environmental and social risks are critical, and that they have the potential to prevent projects from being approved or can be the reason for early termination. Whilst these are undoubtedly significant outcomes, these are strategic risks, and would typically be in the realm of the company board and senior executives (as discussed in Chapter 2).

So why does social and environmental risk management matter to engineers working in design and operations? Many environmental and social risks are *operational*, and are thus best managed by the engineering team. The engineers are also responsible for creating the environmental and social risk through their designs. It is only when the projects that engineers have designed are built, that the risk can become actual harm in the real world.

Thus the engineers are uniquely placed to remove or reduce the negative impacts, and to increase the positive benefits. Furthermore, we would argue that the engineers are trained to identify, assess and treat risks. As such, they have the toolkit and skills to identify, assess and treat the tactical and operational environmental and social risks.

However, despite this, engineers are often uncomfortable approaching environmental and social risks as they require a higher degree of qualitative analysis. Our challenge is to take what can be qualitative data and information, and enable objective decision-making and objective engineering design. This is not easy. The language is different, the data are often qualitative, and engineers are not trained to do this. Moreover, engineers are too often completely unaware of the environmental and social context of the projects in which they are working. They can become siloed, which can lead to major problems. But this can be overcome by asking simple questions such as 'what is the environmental context?', 'what is the social context?', 'what are the issues?'

8.3 The project lifecycle

It is useful to view environmental and social risk from the context of the project lifecycle. Projects have a lifecycle, starting from development studies through to closure and rehabilitation. As you will see in this chapter, environmental and social risk exists at all stages of the project lifecycle. For example, many projects fail to progress to construction because of a failure to achieve community acceptance. It is also interesting to note that several projects are experiencing major social acceptability challenges several decades after

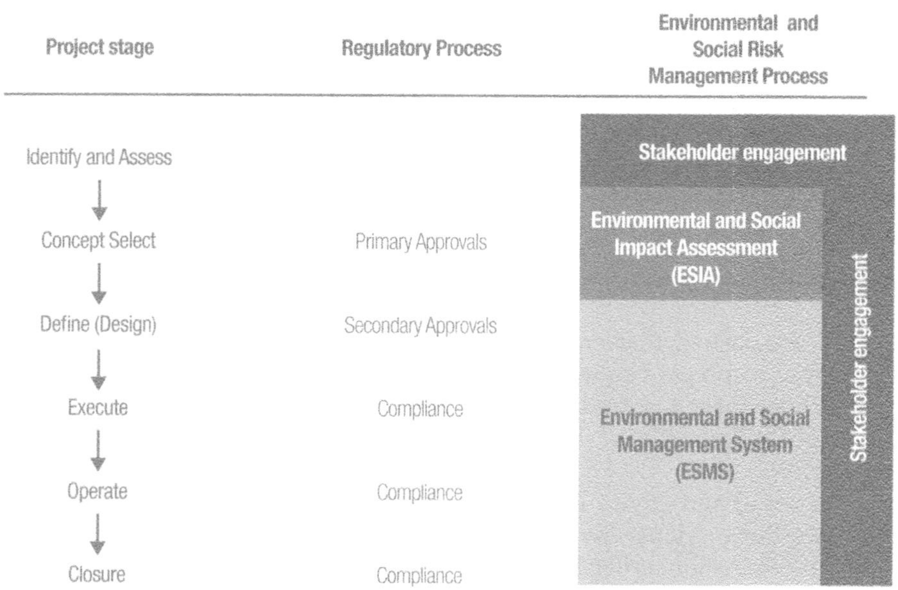

FIGURE 8.1 Project stages and lifecycle processes (Murphy, 2015). *Source: Adapted from Murphy, A. (2015).* Environmental risk. CHEE4002 impact and risk in the process industries course notes. *The University of Queensland.*

closure. These so-called legacy issues are becoming more prevalent, and need to be more actively considered much earlier in the project lifecycle (e.g. contaminated manufacturing sites (Sim, 2018); toxic and stockpiled by-products such as hexachlorobenzene (Rae & Brown, 2009)).

As Fig. 8.1 shows, the environmental and social management processes start at the project conception and continue through to project closure. This is very important. Many projects have struggled to achieve acceptance because the environmental and social management processes commenced too late.

As the project proceeds, the risk decision-making transitions from strategic to operational. Engineers are involved in managing the environmental and social risks at all project stages. The construction, operation and closure and rehabilitation (usually covering many decades) are focussed on operational risk and compliance.

A project goes through various phases – identify and assess, concept select, define, execute, operate and closure. In parallel with that, there is the regulatory process, which has three key phases – primary approvals, secondary approvals and then compliance. The primary approval refers to an overarching approval from the prime regulatory body in the jurisdiction of the project. For major projects, it is considered and issued on the basis of a large scale Environmental and Social Impact Assessment (ESIA) which examines the environmental, social and economic costs and benefits of the project, its alternatives, and how social and environmental risks will be managed for the life of the project and after its closure. The outcome from the assessment is the Environmental Impact Statement (EIS). A completed and accepted EIS leads to the issuing of a licence. This contains all the project-specific legal requirements and obligations that form the basis of the government's approval of the project. The primary approval is usually sought in parallel with the internal approvals process, but the primary approvals and the EIS can take years, even decades to prepare. Sometimes the engineering studies can take many years. Engineers are often very surprised to learn that environmental and social risk work can take many years and may involve many millions of dollars of expenditure, often exceeding the value of engineering work at the same time of project development.

Secondary approvals include a diverse range of permits, licences and other approvals that are only issued on the basis of an accepted EIS and, usually, existence of a primary licence. In some jurisdictions the primary licence is called an Environmental Authority. There can be dozens, even hundreds of secondary approvals which are contingent on the primary approval. Examples include water licences and vegetation clearance permits.

Construction, operation and finally closure and rehabilitation is all about compliance with legal obligations; both the general and also specific legal obligations arising from the primary and secondary approvals. Major projects can have hundreds, sometimes even thousands of licences, permits, obligations which are usually compiled into a compliance register. It is important to be aware that major projects, and major companies, should have a compliance register.

The Environmental and Social Management System (ESMS) is used to manage compliance with all the relevant legal obligations, regulations and commitments made by the company. It is based on an ISO standard – plan, do, check, act (ISO14001, 2015). Most major companies will have this sort of system and they're populated with documents and management tools based on the original EIS. The heart of this system is usually a risk register which provides a key summary of known risks and links to how they are managed. It is a live register in that it must always be revised to ensure that it captures new risks, and the management of known risks is adjusted to reflect actual experience. When working in an existing operation, this is where you can contribute to the management of environmental and social risk by getting familiar with what is in this system, and what aspects fall in your area of responsibility.

8.4 Impact assessment

Impact assessment more specifically refers to the process of performing an ESIA for a project. The names vary in different parts of the world. In many jurisdictions, the term 'Environmental Impact Assessment' is used to describe the overall impact assessment, including the social impact assessment. Throughout this section we will employ the more descriptive term, namely ESIA.

The ESIA process is now a legal requirement in many jurisdictions around the world. As such, we will focus this section on providing a slightly deeper understanding of the ESIA process. ESIA refers to the process whereby environmental, social and economic impacts of projects are assessed prior to development approval being granted. The practice of environmental impact assessment first started in the 1960s when it was introduced as part of US Environmental Policy. Subsequently, ESIA has been written into policy in many countries (UNEP, 2018), and it is also a requirement for project evaluation for organisations such as the UN and World Bank (World Bank, 2017). Not all projects need to undergo a full ESIA in order to obtain development approval. It is common to have a threshold test prior to requiring an ESIA. However, in most jurisdictions, it would be expected that any project which may have 'substantive environmental impact' will be required to perform an ESIA. This would certainly be the case for resource projects (such as mines, gas extraction and mineral processing facilities), chemical plants and energy facilities (e.g. refineries, power generation plants). In all of these cases, a thorough ESIA process would be required in order to obtain project approvals (Fig. 8.1).

ESIA is an internationally accepted and widely adopted process. It provides a level of consistency and credibility about how we manage environmental and social risk. It is a process of identifying, predicting, evaluating and mitigating risk. It is actually a specific field of specialisation, with many professional practitioners, professional associations (e.g. International Association for Impact Assessment) and specific academic journals. Most ESIAs conducted for industry would be performed by specialist contractors who could provide the broad range of expertise required for each specific impact area (e.g. noise, air pollution, water, aquatic ecology, cultural heritage).

There are four key stages to performing an ESIA. Not surprisingly, this is consistent with ISO31000 standard. The four stages are:

1. ***Baseline studies***: This involves a comprehensive investigation of the broad environmental and social context of a project, including a range of specialist areas such as flora and fauna, noise, dust, visual amenity, aquatic ecology and cultural heritage. This work can take years to complete for some projects.

2. ***Impact assessment***: This process assesses the project definition to identify the impacts that will be caused by the project on its environmental and social context. This is when engineers start to become involved in the process, as the impact assessment will contrast the design of the project with the baseline information.
3. ***Impact mitigation***: The development of avoidance, reduction and mitigation strategies to reduce impacts and to increase positive outcomes. This necessarily incorporates the technical and engineering teams who are developing the project's technical scope in an iterative and constructive manner. At this stage, it is possible that the engineering design work may be revisited multiple times to reduce, eliminate, and mitigate the impacts.
4. ***Decision gate***: Following completion of these three stages, the ESIA document is submitted to the relevant authority to seek approval for the development. At this stage, the project may not be approved, or approval may be given under specific conditions. The key outcome is the Environmental Licence which will authorise the project, stipulate conditions for the design, construction and operation, and specify the monitoring process and terms for review.
5. ***Monitoring and compliance***: This stage continues for the duration of the project. The aim is to continuously monitor the impacts and mitigation measures, and to monitor progress against the specified management plan, and ensure compliance with the Environmental Licence.

The Environmental Licence (note that the terminology changes for different jurisdictions) is the key document which results from the ESIA process. It defines the major legal obligations specific to a project. These are additional to the general legal obligations and don't necessarily include all of the obligations from secondary approvals. It also may not include commitments made to landowners and other stakeholders. This is why a compliance register is important, as it will bring all these obligations together in a single place to avoid people inadvertently relying solely on the Environmental Licence. But in the first instance, the project process engineers need to be very familiar with what is in the Environmental Licence, to ensure that all requirements are incorporated into the design and into operations.

As the outcome of the ESIA process includes both project approval and the establishment of legally binding obligations, it is usually an intensive activity where project proponents seek to ensure not only approval, but that the resulting obligations are manageable, and are not so onerous that they impact on the project's technical and financial viability. It is also a process where other stakeholders may hold an opposing view, where they hope to persuade the regulatory authority to not approve a project, or to impose conditions that meet particular goals. The process requires close collaboration in a multidisciplinary environment where a project design can be modified over many iterations. As such, early integration of the environmental and social context in the project design and development can significantly reduce both threats to the project, and the threats perceived by stakeholders.

Best practice in process and engineering design involves the development of environmental design criteria that are additional to technical design criteria. This is because environmental obligations are often presented in licence documents in ways that are not useful to other disciplines. Translating criteria that were specially developed for the management of environmental values into engineering criteria is a key step that can be overlooked. This

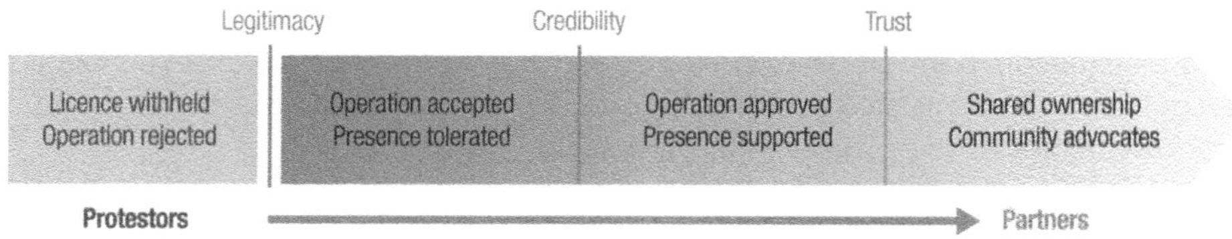

FIGURE 8.2 The social licence to operate (Boutilier and Thomson, 2011). *Source: Adapted from Boutilier, R. G., & Thomson, I. (2011).* Modelling and measuring the social licence to operate: Fruits of a dialogue between theory and practice. *Retrieved November 28, 2019, from https://socialicense.com/publications/Modelling%20and%20Measuring%20the%20SLO.pdf.*

can then lead to engineers designing according to their own process design criteria, leading to designs that do not comply with the Environmental Licence (primary approval).

8.5 Social licence to operate

As we have already discussed, community acceptance or support for the project can be critically important to ensure successful execution. The social licence to operate is a conceptual way to understand and foster support in the community for the presence of a project or ongoing facility. It is separate to any regulatory licence and, of course, there is no actual licence document. Whilst there is no agreed definition of what a social licence to operate is, it is broadly understood to mean *an informal agreement that infers an ongoing acceptance of a project by the affected stakeholders* (Gallois et al., 2017; Moffat & Zhang, 2014). It is a relatively new concept, originating in the mining industry in the 1990s, and much of the literature is focussed on the resources sector. However, the social licence to operate is equally relevant to other industries (Moffat et al., 2016).

The social licence to operate is a very powerful way to incorporate environmental and social risk into engineering design. It is a way to generate a shared understanding of issues and vocabulary that allows engineers to successfully communicate with environmental and social scientists. Fig. 8.2 shows a simple model that represents the social licence to operate along a continuous spectrum. It can be very useful to think of the SLO as a spectrum, as opposed to a binary concept, where it is either something that you have, or do not have. This figure was adapted from the original model proposed by Boutilier and Thomson (2011).

At the left end of the spectrum, the project is not legitimate in the eyes of society. The licence is withheld, and the operation is rejected without regulatory or community approval. Although it is possible for a regulatory licence to be granted without a social licence from the local community, this will likely result in sustained conflict and activism against the project. As Fig. 8.2 illustrates, the social licence to operate spectrum goes from rejection, to legitimate and the operation being accepted or tolerated, to where the project can be approved and actively supported, to the end of the spectrum where the community are widely and actively advocating for it, accepting it as their own. Strong social licences are valuable as they are more resilient to the inevitable issues that arise in any

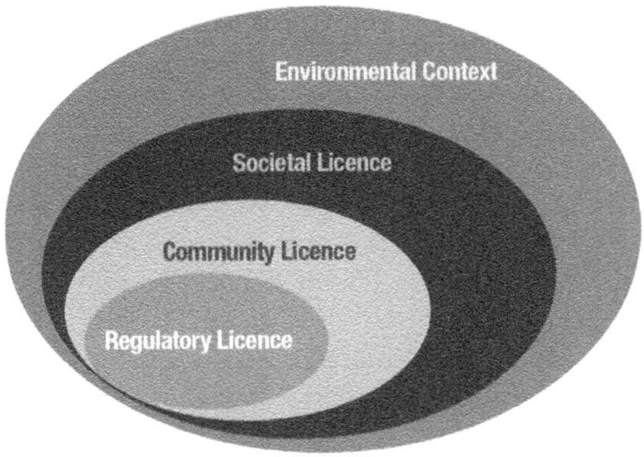

FIGURE 8.3 Layers of social licence (Murphy, 2015). *Source: Adapted from Murphy, A. (2015).* Environmental risk. CHEE4002: Impact and risk in the process industries course notes. *The University of Queensland.*

community and their businesses. It is this focus on maintaining a strong social licence that facilitates successful management of the environmental and social risks that arise as a company undertakes its activities.

The social licence to operate is made up of a number of key components. These can be considered as a set of dependent 'layers of license' as shown in Fig. 8.3. At the core of the social licence is the regulatory licence. There cannot be any sort of social license without regulatory approval from government. The community licence refers to those people who are immediately impacted by the project, and the societal licence refers to those people who might be remote, not immediately impacted, but have an interest in the project.

It is very important that both the community and broader society stakeholders are considered, and that ongoing acceptance, and ideally shared ownership, is achieved for both. The fossil fuel industry is currently providing some pertinent examples of tensions between local community and society stakeholders. There are several examples of contemporary projects that have acceptance and support from the local community, who can see local benefits in terms of jobs and income, but are strongly rejected by the broader society (typically people in the cities who are a long way from the project site) who are driven by global climate concerns. In such cases, the project may have the acquired regulatory approvals, and the support of the local community stakeholders, but still fails to have a social licence to operate.

Looking to the future, some companies such as BHP are shifting their business approach from a position of focussing on compliance to keep a social licence, to one of creating and adding social value to communities and others affected by an organisation's operations (Healy, 2019). It will be interesting to see if processing companies are able to evolve and deliver against this aspirational goal.

The following section will present some tools that can be used by the engineering team to ensure that environmental and social risks are given appropriate consideration and at the appropriate project stage.

8.6 Tools to identify, assess, treat and monitor environmental and social risks

The aim of this section is to present a number of tools that can help you to better understand and manage environmental and social risks. We have focussed on tools that you have previously encountered (Chapter 5), but which have been extended to consider environmental and social risks more thoroughly. The first step is to consider the stakeholders. We will then introduce a risk register approach to document the risk assessment, and a modified risk assessment matrix to assist with risk analysis. The risk register will also include a set of recommended actions for treating each of the identified risks. For high priority risks, it may then be required to develop a more detailed management plan. For this, we will introduce the social impact management plan (SIMP).

Importantly, the stakeholder engagement, communication and management plans as well as the social and environmental risk assessments and management plans should be a 'live' documents. They should be updated as time progresses and as more information emerges and as changes occur over the lifetime of a project.

8.6.1 Preliminary stakeholder analysis and management

Stakeholder engagement is the process of establishing a dialogue with stakeholders so that different perspectives can be identified, understood and incorporated into the project design and management. Projects that are welcomed by their stakeholders are likely to have a strong social licence to operate. It is now well established that public participation is a key component for achieving best practice for both environmental impact assessment (UNEP, 2018) and social impact assessment (Esteves et al., 2012).

Stakeholder identification and analysis is essentially a process of 'walking in other people's shoes'. In our engineering disciplinary environment, we can easily get siloed in our thinking and lose our ability to connect with other world views. The process can be uncomfortable, but

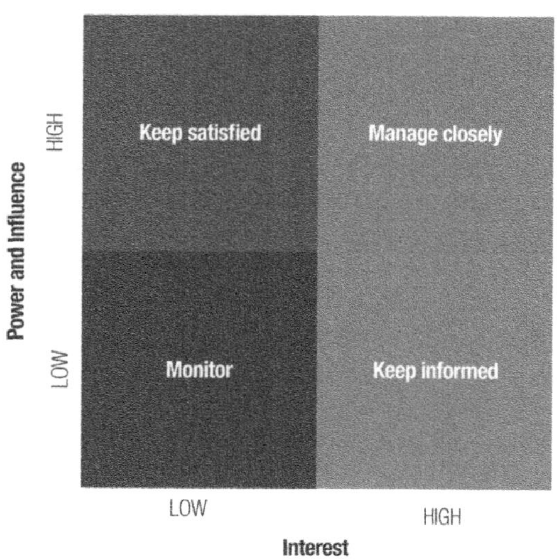

FIGURE 8.4 Stakeholder analysis and management (Mendelow, 1981). *Source: Adapted from Mendelow, A. L. (1981). Environmental scanning – The impact of the stakeholder concept. ICIS 1981 Proceedings. Paper 20.*

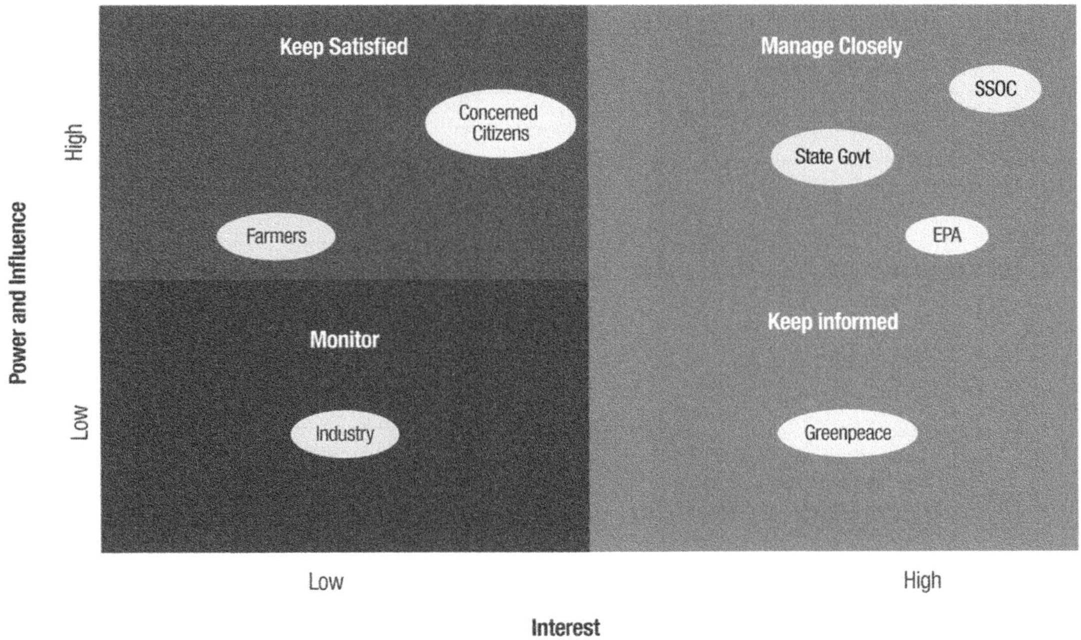

FIGURE 8.5 Stakeholder analysis and management grid for a hypothetical unconventional oil project.

it is a very important, and very valuable, skill to develop. The process begins with a broad analysis of the context of a project. Social engagement specialists, and engineers with experience, can begin to identify logical groupings of stakeholders, and potentially key individuals.

The power-interest grid is a very commonly used tool to assess stakeholders in many fields (Fig. 8.4). It seems to have originated from Mendelow (1981). Each stakeholder is assessed on two major attributes:

- The degree of influence and power they can exert over the project.
- Their level of interest in the project (or the extent to which they are impacted by the project)

Based on these attributes, a plan of engagement, communication, dialogue and management can be developed:

- Low power and low interest stakeholders should be ***monitored*** in case their situation changes. There is no need to engage these stakeholders.
- Low power and high interest stakeholders should be ***kept informed*** about the project. Engagement with these people can be useful for identifying problems and/or opportunities.
- High power and low interest stakeholders should be kept ***satisfied***. They should be informed and engaged so that they are satisfied. Avoid over-communication and boring them.
- High power and high interest stakeholders must be ***managed closely***. These stakeholders must be fully engaged and effort must be made to fully satisfy them.

TABLE 8.1 Stakeholder risk and opportunity assessment template.

Stakeholder Information			Potential Impacts (health/safety, environment, financial, societal)		
Name	**Description**	**Classification (Fig 8.05)**	**Threats/Issues**	**Impact (Table 8.03)**	**Recommended Actions (Figs 8.06 & 8.07)**
Stakeholder	Stakeholder description. Who are they and what are their interests in the project.	Manage Closely; Keep satisfied; Keep informed; Monitor	Threat 1	eg B4	Mitigation plans
			Opportunities/Benefits	**Impact (Table 8.03)**	**Recommended Actions (Figs 8.06 & 8.07)**
			Opportunity 1	eg B2	Enhancement plans

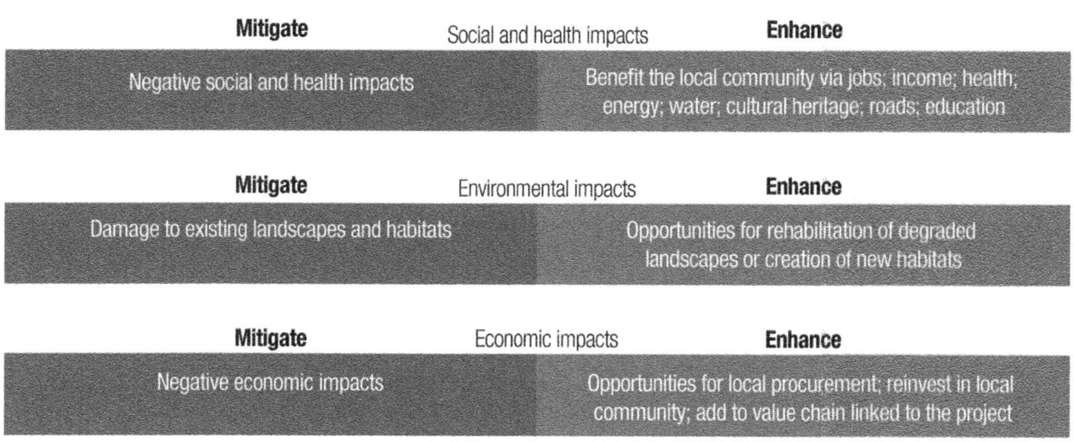

FIGURE 8.6 Enhancing positive impacts in addition to mitigating negative impacts (João et al., 2011). *Source: Adapted from João, E., Vanclay, F., & den Broeder, L. (2011). Emphasising enhancement in all forms of impact assessment: introduction to a special issue. Impact Assessment and Project Appraisal, 29(3), 170–180. https://doi.org/10.3152/146155111X12959673796326.*

TABLE 8.2 Social opportunity and threat risk assessment matrix.

		E – Rare Rarely occurs at either a specific stage of the project or more broadly	D - Unlikely Unlikely to occur at either a specific stage of the project or more broadly	C - Possible Possible to occur at either a specific stage of the project or more broadly	B - Likely Likely to occur either at a specific stage of the project or more broadly	A - Almost certain Very likely to occur either at a specific stage of the project or more broadly
5 – Serious/ Significant Impact	Irreversible serious impact and the stakeholder no capacity to adapt and cope with change. Significant opportunities from the project. Impact is expected to continue indefinitely.	E5	D5	C5	B5	A5
4 - Major Impact	Major long-term threat with recoverable impact. Stakeholder has limited capacity to adapt and cope with impact. Major opportunities emanating from the project. The impact is long-term.	E4	D4	C4	B4	A4
3 – Moderate Impact	Significant medium-term threat with recoverable impact. Stakeholder has some capacity to adapt and cope with impact. Significant opportunities emanating from the project. Probably a medium-term impact.	E3	D3	C3	B3	A3
2 – Minor Impact	Minor short-term threat with recoverable impact. Minor opportunities emanating from the project. Probably a short-term impact.	E2	D2	C2	B2	A2
1 – Negligible Impact	Local, small-scale, easily addressed threat with reversible impact. Very small-scale opportunities that can readily pursue and capitalise on.	E1	D1	C1	B1	A1

Low social impact or opportunity | Moderate social impact or opportunity | High social impact or opportunity | Extreme social impact or opportunity

It is important that the stakeholder management plan remains a 'live' document, and that it is updated over the lifetime of a project. This recognises that stakeholders may change or change their perspective of a project over time.

As an example of a stakeholder mapping and management exercise, consider a new proposed project to investigate the development of an unconventional oil facility, whereby kerogen in oil shale is converted into shale oil, which may then be further refined into conventional hydrocarbon fuels. The project proponent is State Shale Oil Company (SSOC). The project would be the second attempt at producing unconventional oil from oil shale in the designated region. The first project was terminated early, and had experienced many issues with the local community due to air quality concerns and water management practices. A stakeholder analysis and management grid is presented for the following stakeholders: the local industry (including large chemical and aluminium refining sites), local farmers, concerned citizens, Greenpeace, EPA, Government, SSOC as shown in Fig. 8.5.

8.6.2 Stakeholder risk assessment

Following the risk assessment approach presented in Chapter 5, and consistent with ISO31000, after risk identification we perform risk assessment and analysis. We introduced the risk register in Chapter 5 to represent this analysis. Similarly, we are proposing a

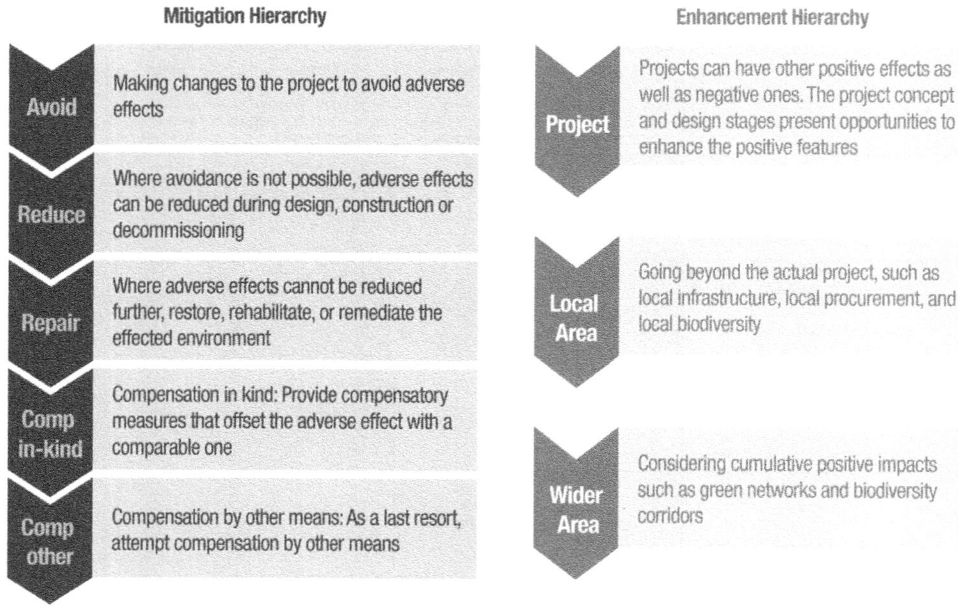

FIGURE 8.7 Mitigation and enhancement hierarchies (João et al., 2011). *Source: Adapted from João, E., Vanclay, F., & den Broeder, L. (2011). Emphasising enhancement in all forms of impact assessment: introduction to a special issue. Impact Assessment and Project Appraisal, 29(3), 170–180. https://doi.org/10.3152/146155111X12959673796326.*

Stakeholder risk register to collate the assessment and analysis as shown in Table 8.1. This risk register is extended to explicitly consider threats and opportunities. For each stakeholder, consider the threats that they may be concerned about, and thus require mitigating, and identify opportunities to provide benefit, which you would like to enhance. Fig. 8.6 may be helpful to assist you to consider threats to mitigate and opportunities to enhance.

Following identification of the risks, we advocate the use of a modified risk matrix to assist with the risk analysis as shown in Table 8.2. Having analysed the risk, it needs to be evaluated to determine if it will be accepted as it is, or whether it needs proactive treatment in order to mitigate or enhance it. Analogous to the risk treatment hierarchy presented in Chapter 5 (Figure 5.14), we can employ a hierarchy of *mitigation* measures and *enhancements* to inform the recommended risk treatment actions (Fig. 8.7). This analysis can then be compiled in the stakeholder risk register. Note that, consistent with the hierarchy of controls, avoidance is the preferred option for impact mitigation.

8.6.3 Social impact management plan

A SIMP is very simply just a management tool for addressing the social impacts of projects. SIMPs are developed either as part of, or in addition to, the EIS. They are still a relatively new component of the overall Impact Assessment process. Franks and Vanclay (2013) present a review of the history and development of SIMPs, and discuss their development in both the private and public sector. The International Finance Corporation (World Bank) has been a major driver behind the development of SIMPs, requiring borrowers to prepare Environmental and Social Management Plans as a condition of

TABLE 8.3 Social impact management plan template.

<table>
<tr><td>Impact Category (e.g. air quality)</td><td colspan="2"></td></tr>
<tr><td>Impact strategy (e.g. Minimise adverse impacts on air quality)</td><td colspan="2"></td></tr>
<tr><td colspan="2">Summary of Potential Threats/Opportunities that could affect Impact Category (e.g. construction activity, transport)</td><td>Impact Assessment</td></tr>
<tr><td colspan="2" rowspan="2">Threats

Opportunities</td><td>Anticipated social impact/opportunity without strategy:</td></tr>
<tr><td>Anticipated social impact/opportunity after strategy:</td></tr>
<tr><td colspan="2" rowspan="2"></td><td>Key Stakeholders</td></tr>
<tr><td></td></tr>
<tr><td>Action Plans</td><td>Timeframes</td><td>Monitoring and Stakeholder Engagement Activities</td></tr>
<tr><td></td><td></td><td></td></tr>
<tr><td>Accountabilities:</td><td colspan="2">e.g. General Manager, Community Engagement</td></tr>
</table>

financing since 2006 (Franks & Vanclay, 2013). Whilst they were initially developed with a focus on large resource projects, we contend that SIMPs can be an integral part of social risk management for all processing businesses which have social impacts.

Many SIMPs for large projects are available in the public domain. Some examples include the Trans-Adriatic Pipeline (TAP, n.d.) the Surat Gas Project (URS, 2012) and Arafura Resources Nolans Project (Michels Warren Munday, 2016). As you will see, SIMPs currently vary in scope, format and content. Holm et al. (2013) proposed the following principles, from a government perspective, to guide the development and implementation of SIMPs:

- *Adopt a risk management approach – The SIMP should clearly prioritise impacts, and the SIMP should only be developed after the significance of impacts is well understood.*
- *Reaffirm a hierarchy of controls – Only when the hierarchy of controls is applied can it said that social impact considerations have been appropriately incorporated into the project design an execution.*
- *Separate out opportunity realisation from social impact management – SIMPs should focus on impact management and benefit sharing should be addressed under other frameworks.*
- *Re-evaluate the requirement to reach agreement – The level of stakeholder engagement required to develop a SIMP is sufficient to ensure that it is aligned with community expectations, and a formal requirement for agreement is not necessary.*
- *Reaffirm the role of government in providing social infrastructure and services - Many SIMPs identify mitigation measures to address impacts on the demand for social infrastructure and services usually provided by government. SIMPs are an excellent tool to reaffirm the role of government in providing these services to people impacted.*

Holm et al. (2013) – page 219. (Slightly adapted).

A template for a summary SIMP is shown in Table 8.3, and an example SIMP is presented for the case study in the next section.

In Appendix B we work through a case study to demonstrate the use of the tools introduced in the preceding sections. The case study is the CarbonCap project, a proposed carbon capture and storage project. This case study is hypothetical, although based on real projects.

8.7 Summary

This chapter has introduced environmental and social risk. We have seen that environmental and social risks are high priority concerns for industry, and that there are several significant benefits for businesses that effectively manage environmental and social risks. These include reduced uncertainty, improved identification of problems, positive project legacies, and increased competitive advantage.

In Chapter 1 we referred to the importance of focussing on both the upside and downside of risk. This has become very clear in this chapter, where we have seen that the management of environmental and social risks often offers opportunities to both enhance project benefits, as well as mitigating the potential negative impacts of the project.

We looked at how environmental and social risks vary throughout the project lifecycle, discussed the environmental and social risk processes employed throughout the project, and the accompanying regulatory processes and approvals. For major projects, ESIAs must be conducted prior to obtaining the primary approval for the project. The outcome from the ESIA is the EIS. A completed and accepted EIS leads to the issuance of a licence for the project.

In addition to the formal assessment and approval processes, we have seen that the informal community acceptance of a project – the Social Licence to Operate – is also critically important. The SLO is best considered as a continuum, progressing from having no community approval for the project, through to shared ownership and community advocacy of the project.

Finally, we introduced a number of tools to assist with identification, analysis, assessment and treatment of environmental and social risks. Whilst the language and context are somewhat different, these tools are analogous to the tools that were introduced in Chapter 5. A major extension is that we also need to identify, understand and classify project stakeholders prior to identifying and assessing the risks.

Managing environmental and social risks is challenging for engineers. The language is different, and much of the analysis is qualitative. However, as we have seen, it is critically important for process companies to effectively manage these risks, and the operational risks are often best managed by the engineering team. As a junior engineer, we encourage you to start by asking 'what is the social context, what is the environmental context, and what are the risks?'

Review questions

1. What is environmental risk?
2. What is social risk?
3. List some examples of social values.
4. What is the difference between strategic and operational risk?
5. At what stage of the project should stakeholder engagement occur?
6. What are the four stages of ESIA?
7. What do we mean by the term *social licence to operate*?
8. What is the stakeholder power-interest grid?
9. What is the strategy for managing stakeholders with a high interest and low power?
10. Is there a hierarchy of controls for social risk?
11. List some examples of how a project may be enhanced to address identified social risks (opportunities).
12. Is the approach for managing environmental and social risks consistent with ISO31000?
13. Considering coal bed methane projects (for example), explain why stakeholder interests really matter.
14. If a project has a regulatory licence to permit operation, then the opinions of the wider community are not important. Discuss this statement, citing contemporary case studies in your answer.
15. Write a short reflection on two cotemporary projects that have been impacted by poor management of environmental and/or social risk. What are the key lessons learnt?

References

Boutilier, R.G., & Thomson, I. (2011). *Modelling and measuring the social licence to operate: Fruits of a dialogue between theory and practice*. Retrieved November 28, 2019, from https://socialicense.com/publications/Modelling%20and%20Measuring%20the%20SLO.pdf.

Bratasanu, D. (2015). *Engineering success: The true meaning of leadership and team building*. CreateSpace Independent Publishing Platform.

Esteves, A. M., Franks, D., & Vanclay, F. (2012). Social impact assessment: The state of the art. *Impact Assessment and Project Appraisal, 30*(1), 34–42, March 2012.

Franks, D. M., & Vanclay, F. (2013). Social impact management plans: Innovations in corporate and public policy. *Environmental Impact Assessment Review, 43*, 40–48.

Gallois, C., Ashworth, P., Leach, J., & Moffat, K. (2017). The language of science and social licence to operate. *Journal of Language and Social Psychology, 36*(1), 45–60.

Healy, G. (2019). BHP social value briefing. Geoff Healy, Chief External Affairs Officer. *October, 8*, 2019.

Holm, D., Richie, L., Snyman, K., & Sutherland, C. (2013). Social impact management: A review of current practice in Queensland, Australia. *Impact Assessment and Project Appraisal., 31*(3), 214–219.

ISO 14001. (2015). *ISO 14001:2015 Environmental management systems—Requirements with guidance for use*. International Organization for Standardization. Geneva.

João, E., Vanclay, F., & den Broeder, L. (2011). Emphasising enhancement in all forms of impact assessment: Introduction to a special issue. *Impact Assessment and Project Appraisal, 29*(3), 170–180. Available from https://doi.org/10.3152/146155111X12959673796326.

Main, M. (2017). Papua New Guinea gets a dose of the resource curse as ExxonMobil's natural gas project foments unrest. *The Conversation*, March 9 2017.

Mendelow, A. L. (1981). Environmental scanning – The impact of the stakeholder concept. *ICIS 1981 Proceedings. Paper*, 20.

Michels Warren Munday. (2016). Arafura resources: Social impact management plan - Nolans Project. Retrieved March 12, 2021, from https://www.arultd.com/images/EIS/DOCUMENTS/Volume6/Nolans_EIS_Appendix_X_K_web.pdf.

Moffat., & Zhang. (2014). The paths to social licence to operate: an integrative model explaining community acceptance of mining. *Resources Policy, 39*, 61–70.

Moffat, K., Lacey, J., Zhang, A., & Leipold, S. (2016). The social licence to operate: A critical review. *Forestry, 89*, 477–488.

Murphy, A. (2015). *Environmental risk. CHEE4002: Impact and risk in the process industries course notes*. The University of Queensland.

Parliament of Australia. (2020). *Juukan Gorge – Interim report*. Parliament of Australia, December 3, 2020.

Rae, I., & Brown, P. (2009). Managing the intractable: Communicative structures for management of hexachlorobenzene and other scheduled wastes. *Journal of Environmental Management, 90*(4), 1583–1592.

Ransan-Cooper, H., Ercan, S. A., & Duus, S. (2018). Getting to the heart of coal seam gas protests – it's not just the technical risks. *The Conversation*, December 4 2018.

Sim, M. (2018). Industrial aftermath: Project ends toxic pesticide legacy. *The Chemical Engineer, Environment Features*, August 24 2018.

TAP. (n.d.). Trans adriatic pipeline: Environmental and social management plan. TAP AG Doc. no.: CAL00-PMT-601-Y-TTM-0006. Rev. no. 2. Retrieved March 12, 2021, from https://www.tap-ag.com/sustainability/lenders-requirements/environmental-and-social-management-plan/$2683/$2684/$2685.

UNEP. (2018). *Assessing environmental impacts – A global review of legislation*. ISBN No: 978-92-807-3679-3.

URS. (2012). Surat gas project: Social impact management plan. Arrow energy. Document no.: 42626746/SIMP/1.0. Retrieved March 12, 2021, from https://www.arrowenergy.com.au/__data/assets/pdf_file/0003/28713/Attachment2000620-20Social20Impact20Management20Plan.pdf.

World Bank. (2017). *World bank environmental and social framework*. Washington, DC: World Bank.

CHAPTER

9

Project risks

OUTLINE

9.1 Introduction 214

9.2 What are projects? 214

9.3 Project life cycle 217

9.4 Project risks 220

9.5 Managing project risks 221

9.6 Summary 224

Review questions 224

References 225

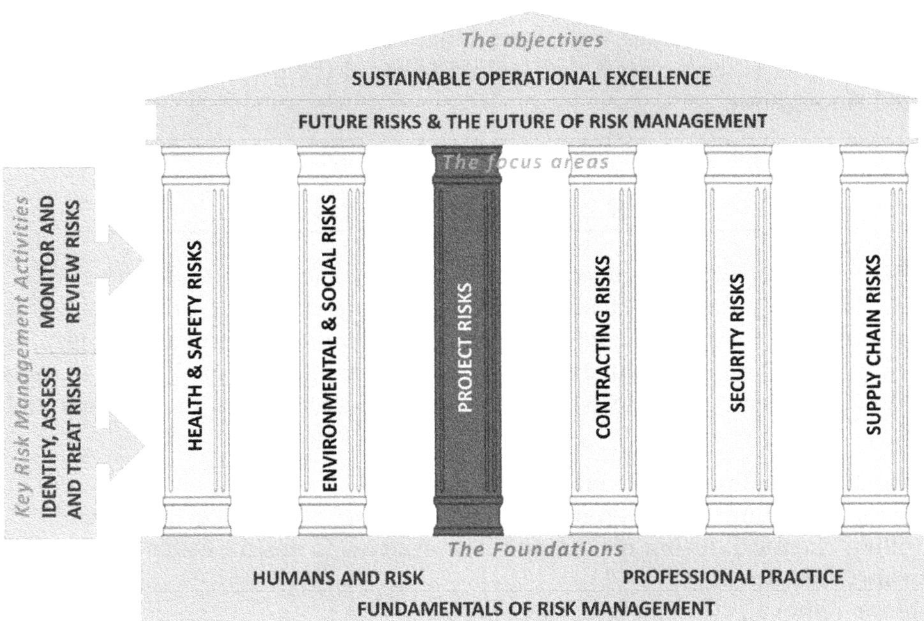

Fundamentals of Risk Management for Process Industry Engineers
DOI: https://doi.org/10.1016/B978-0-12-820320-0.00006-4

Getting something done is an accomplishment; getting something done right is an achievement.
Anonymous

9.1 Introduction

A lot of engineering work is conducted within the context of a 'project'. Projects are often undertaken to help organisations achieve their objectives. Projects are initiated to deliver benefits associated with a set of deliverables that help organisations sustain or improve operational performance (Hillson, 2009). Project success is heavily reliant on effective management of risks throughout the project life cycle. The framework and tools used to manage operational risks within projects are the same as previously introduced. However, it is important to be familiar with project structure and life cycle stages, as the risks, costs of risk control, risk management tools employed, and engineering tasks undertaken, vary throughout the project.

In this chapter we will provide an overview of what projects are and provide examples of projects undertaken in the process industries. We will then describe project risks and demonstrate the potential significance of these risks and the many lessons learned with case study examples. These will highlight the criticality of identifying, correctly assessing and effectively controlling project risks. The chapter concludes with options for approaching risk management activities across the different phases of a project's life cycle.

The scope of some large projects (cost and lifetime) means that many of the risks are strategic, and the key decision makers will be senior executives. However, engineers play critical roles at all project stages, and operational risk management is a key feature for all stages of projects. It is thus critical that you understand the project life cycle and how risk is managed throughout.

9.2 What are projects?

A project is a *'temporary endeavour to achieve one or more defined objectives'* (ISO 21500, 2020, p. 3). Similarly, the Project Management Institute defines a project as a *'temporary endeavor undertaken to create a unique product, service or result'* (sourced from http://www.pmi.org). Within the process industries, projects can range from small scale to mega scale (nominally >$1 billion). Examples of small-scale projects from the process industries include:

- Selecting and installing a more efficient pump.
- Testing the capabilities of different air-mist nozzles to cool a hot slab of metal.
- Controlling chemical dosing of cooling water systems to ensure water quality remains within specified limits.
- Membrane removal, cleaning and reinstallation.
- Various maintenance projects which will require management of change (e.g. managing hot works on site, removal and replacement of pipes and/or equipment).

Examples of some medium-scale projects could include:

- Managing the shutdown, maintenance and recommissioning of an operating unit within a processing facility.
- Investigating and eliminating causes of process disturbances.
- Redesign and upgrade of a section of plant within an industrial site (e.g. wastewater treatment plant).
- Optimising energy use per operating hour or per tonne of product produced.

Examples of large-scale or megaprojects include:

- Designing, building and operating a new processing plant. Contemporary examples include floating LNG [e.g. Prelude (Shell)], large integrated chemical complexes (e.g. Sadara, Saudi Arabia), nuclear power plants (approx. 50 under construction in 2020), large integrated biorefineries (e.g. UPM Leuna Biorefinery, Germany). These are known as greenfield projects.
- Undertaking a major expansion of an existing operation. For example, expansion of the Pluto LNG project in Western Australia.
- Decommissioning and repurposing industrial sites. For example, decommissioning nuclear power facilities.

Project complexity, innovation, size and location are some of the major factors that determine project risk as shown in Fig. 9.1. Consider building something that has been built before and in the same country. For example, if you were to build a compression facility for shale gas in the United States, where there are a lot of them already, then the project complexity is low and the location is mature. Hence the project is a 'standard project', where the risk is low. In contrast, if there was a project to build a whole new type of

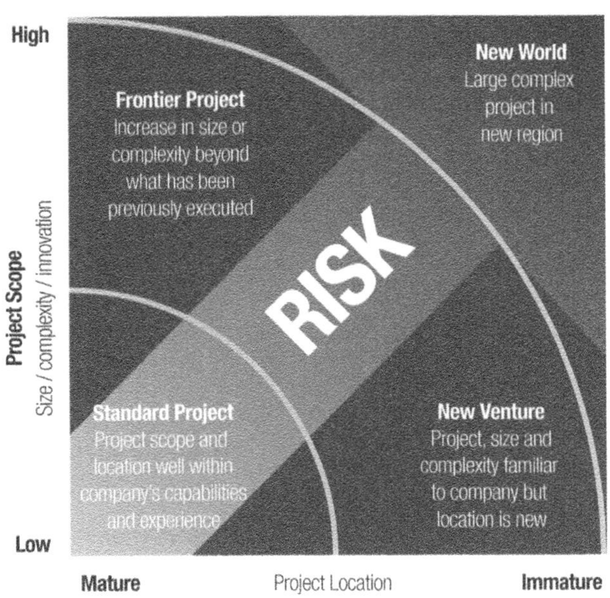

FIGURE 9.1 Project risk (Killick, 2016). *Source: Adapted from Killick, C. (2016).* Project & process safety. *CHEE4002 course lectures. The University of Queensland.*

power station in a developing country that has no experience with that sort of technology, then the project is designated as high risk, and that defines the level of effort required to manage the risks.

The different projects and level of risk should also impact the consultation processes and resources used to inform risk-based decision-making processes as shown in Fig. 9.2.

This figure illustrates how the engineering decision-making process varies with the project risk. For 'standard projects' (Type A), where something is being built that is not new, is not unusual and has been done often before, where there's lots of available data and no major stakeholder issues, then much of the decision-making can be made based on complying with codes and standards, good practice and engineering judgement.

Codes and standards become less important for Type B decisions, where the project is beyond established practice. Whilst it is always important to comply with codes and standards, they reflect historical practice, and may not be appropriate for projects which are not 'standard'. Good practice and engineering judgement are important, but there will also be a need for better risk-based analyses, such as Quantitative Risk Assessment and Cost Benefit Analysis. This will require analyses to justify stepping outside of what is normally done.

The transition from technology to values-based decision-making is completed with Type C projects, whereby the major factors guiding decision-making are company and societal values. This transition is accompanied with a concomitant shift from operational to strategic decision makers. These are typically 'new world' high-risk projects in new

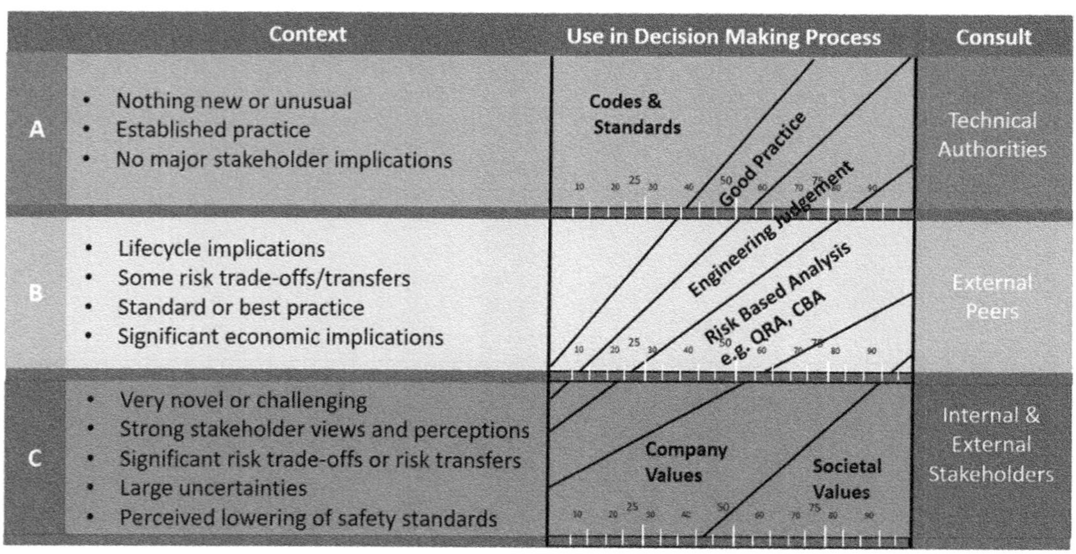

FIGURE 9.2 Relationship between project context and decision and consultation processes (United Kingdom Offshore Operators Association, 1999). *Source: Adapted from United Kingdom Offshore Operators Association (UKOOA). (1999).* A framework for risk related decision support. *UKOOA Publication.*

locations with a range of different stakeholders to work with. There are significant trade-offs and risk-based analysis tools are employed, but decision-making really comes down to ethics and values. The decisions made must reflect what is right, both in engineering terms and also what is right for society. Genetically modified food and the nuclear fuel cycle are two good examples of projects with Type C decision-making.

9.3 Project life cycle

Due to the temporary nature of projects, they have a defined life cycle comprising a number of phases. The scale and specific industry practices can influence the number and naming of the phases.

Most industry projects may be described using the generic life cycle stages shown in Fig. 9.3. This is the 'standard' stage-gate project development and delivery process (Barshop, 2016). There are key decision gates after the first three stages, with the final investment decision (FID) made after the 'Define' phase. Depending on the scale of the project, each of these stages may be less or more extensive, and the critical decision gate process may differ. For example, for a small project (e.g. selecting and installing a more efficient pump), the 'Identify and Assess' and 'Concept Select' stages may be straightforward and only comprise a summary report, and the investment decision will be made by an operations manager. Alternatively, for a greenfield mega project (e.g. floating LNG), the overall project timeline may be many years, and the FID process will only come after an extensive design process (Define) and will be approved by company senior executives. Barshop (2016) states median duration for projects to reach a strong level of definition to be 3 months for $100,000 projects up to 23 months for $1b projects.

Projects exist to generate value. It is thus important to consider how value varies throughout the project life cycle as shown in Fig. 9.4. Good project definition is critical for achieving optimum value. As such, it is critical that sufficient rigorous and critical analysis is conducted in early project stages (i.e. 'Identify and Assess', 'Concept Select' and 'Define'). It is in these early project stages where the costs of change are low and the ability to implement effective risk controls is high. It is critical to understand that whilst project definition often represents small fractions of the overall project timeline and costs, it is crucial for ultimately achieving good value from the project. Early in the project when the design, economics and even location are not yet selected, there is often large amounts of uncertainty with whether the project will proceed and result in a successful outcome. During this phase, the cost of making changes to the selected technology, location and input streams and proposed outputs is relatively low as shown in Fig. 9.4. As the project progresses into more detailed designs, decisions about plant location and layout, process flows and technologies, project timing and costs are made that reduce the uncertainty. As the activities ramp up and continue to progress, any proposed changes to location, process design and technologies are exponentially more costly to implement due to the increasing span and detail of activities, documentation and procured goods and services involved in the project. Thus early and ongoing risk assessments should be conducted within a project to enable the early prediction

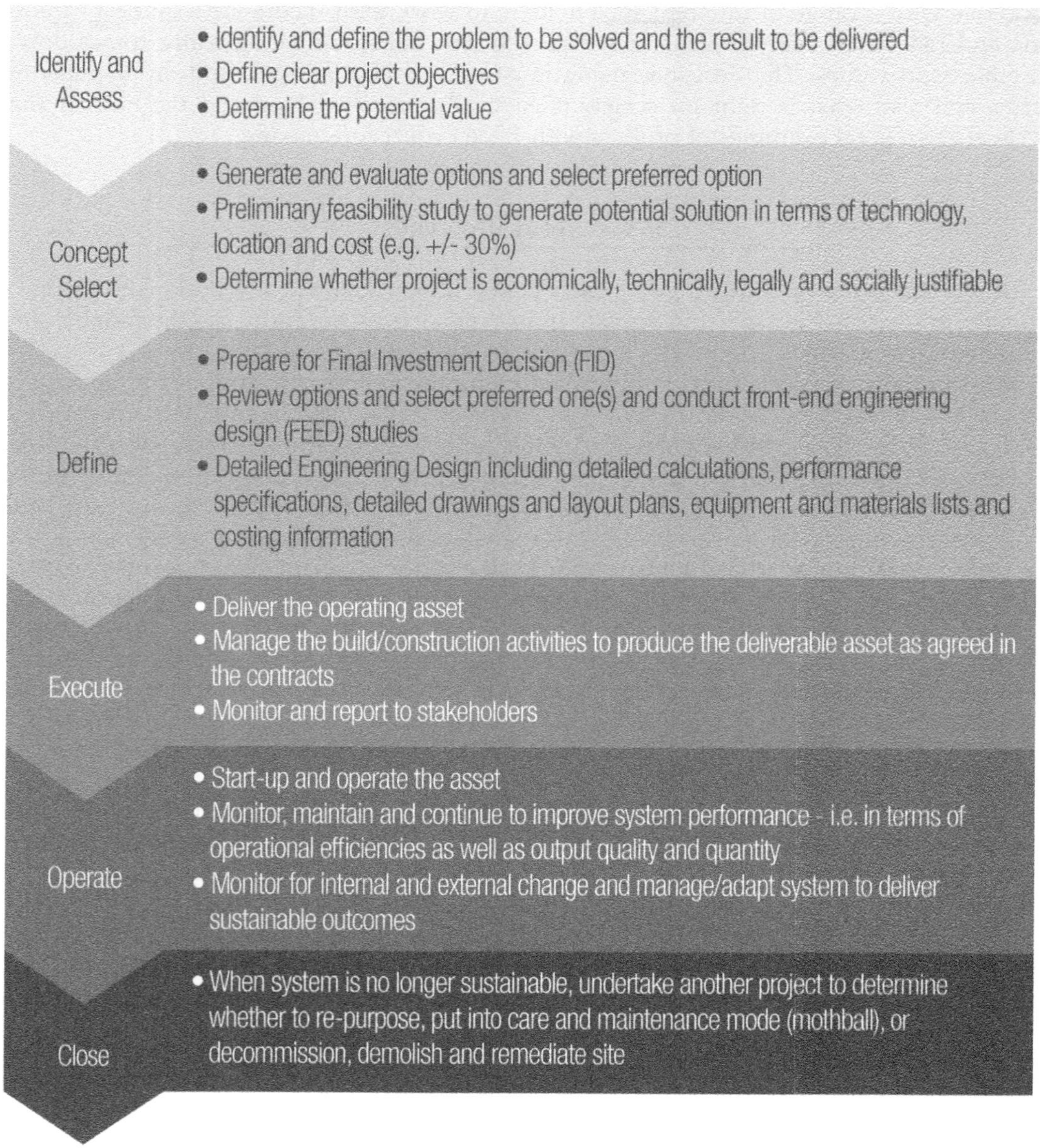

FIGURE 9.3 Life cycle phases associated with larger process industry projects.

and identification of risks that could impact the project so these can be managed effectively and in a timely manner.

Projects often fail to deliver the predicted value. In a study of 431 industrial projects, it was found that the average project delivers 22% less net present value than was forecast

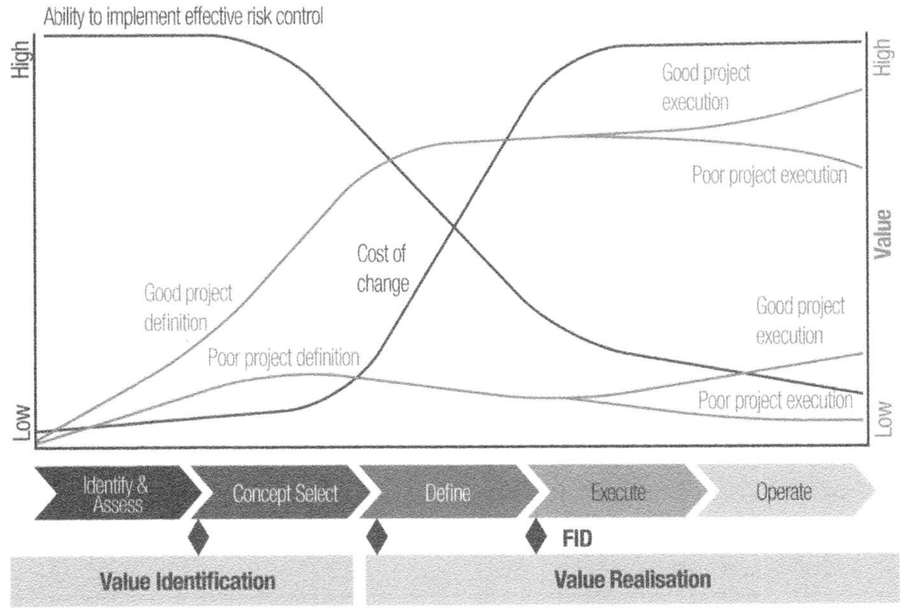

FIGURE 9.4 Life cycle phases, value addition and cost of change (Killick, 2016). *Source: Adapted from C Killick, (2016).* Project & process safety. *In CHEE4002 course lectures. The University of Queensland.*

when the project was funded (Barshop, 2016). Barshop (2016) states the three main reasons for value erosion were (1) lower demand for the product than expected; (2) cost and/or schedule overrun; and (3) the facility did not operate as expected. The risk of failure is greater for large projects. In a survey of 381 industrial megaprojects, 65% of projects were classified as failures (Merrow, 2011) because of either cost overruns (>25%), poor cost competitiveness (>25%), slips in the execution schedule (>25%), poor schedule competitiveness (>50%) or significantly reduced production in year 2 of operations. It is pertinent to note that the extensive study reported by Merrow (2011) did not include project safety outcomes as measures of success or failure because 'Too many of the safety numbers for megaprojects are not credible' (Merrow, 2011, p. 45). Oil and gas sector megaprojects had a 78% failure rate and resource sector projects had a 70% failure rate.

According to Flyvbjerg et al. (2003), one of the reasons for projects failing to deliver the forecast value is that only a handful of projects 'contain "thoughtful" risk analysis showing "good practice"' (p. 76). Flyvbjerg et al. (2003) go on to state *'Risks cannot be eliminated from major projects, but they can be acknowledged and their impacts reduced through careful identification and by allocation of risks to those best suited to manage them'* (p. 84). These authors advocate that a full risk analysis be conducted during the feasibility study to identify the most risky parts of the project and the worst-case scenarios that might occur, and to use this information to develop a risk management plan that eliminates the risk by dropping that part of the project or reduces the risks to a level appropriate to all involved. The type of risks that might be associated with process industry projects are discussed next.

9.4 Project risks

As illustrated in Fig. 9.4, failure to determine sufficient project details early in the project, and failure to identify, assess and treat risks early in the project life cycle, may result in significant changes later in the project, increasing the likelihood of overruns and cost blow outs.

Projects involve people, processes, technology and information that develop, build, operate and close processing facilities within a context. The context comprises the external environment, external stakeholders, and both the input and output supply chains.

Essentially, project risks can stem from the following:

1. The risks that the project can impose on itself.
2. The risks that external entities can impose on the project.
3. The risks that the project can impose on external entities.

Examples of these risks are listed in Table 9.1.

TABLE 9.1 Examples of project risks.

Risk category	Types of risks	Example events that impact projects either during execution and/or operational phases
Risks that the project can impose on itself		
To the project itself	Poorly defined or changing scope and/or schedule Deployment of obsolete or poorly designed technology or noncompetitive/ nonsustainable designs Unsafe and/or poor-quality construction work practices	Deepwater Horizon drilling project turned into a disaster, at least in part, because of delays to the schedule and pressure to complete the Macondo well (Hopkins, 2012). Imperial Sugar Company dust explosion occurred after a project was conducted to cover the conveyors to confine the dust but this created explosive levels of dust in the conveying system (CSB, 2009).
Risk to project from external entities		
From external environment	Natural disasters such as hurricanes/ cyclones, tsunamis, bush/wild fires Biological disruptors such as pandemics Economic conditions such as financial booms and crises Lack of supply, oversupply or poor-quality supplies Loss of market or profit for outputs	COVID-19 resulted in 36% of water/sewer construction projects in the United States being cancelled and 40% experiencing delays or disruption due to shortages in people, materials, equipment or parts (AGC The Construction Association, 2020). Hurricane Harvey caused the loss of refrigeration at the Arkema organic peroxide plant which led to decomposition of the organic peroxides which subsequently caught on fire and leaked (CSB, 2018). The styrene gas leak at LG Polymers India was thought to be caused by negligence in maintenance activities during COVID-19 lockdown (Srivastava, 2020).

(Continued)

TABLE 9.1 (Continued)

Risk category	Types of risks	Example events that impact projects either during execution and/or operational phases
From external stakeholders	Public commentary or protests for/against project or project owners/constructors Changes to funding provisions or withdrawal of funds Legal and regulatory risks Political change	Protests against fracking have hampered gas extraction projects in the United Kingdom (U.K. Government indefinitely suspended fracking in November 2019), United States and Australia (Ransan-Cooper et al., 2018). Residential recycling schemes in some Australian States have not been approved or been decommissioned prior to the end of their life due to public health concerns and changes in government funding (West, Kenway, Hassall, & Yuan, 2016). The extension of the Keystone XL pipeline from Canada to United States was abandoned in 2021 after President Biden revoked its permit due to protests and environmental concerns. Resolution copper mine in the United States comprises a significant and large deposit of copper sited below sacred Apache ground – a land swap was proposed based on an environmental impact statement which has been done but the project has been subject to lawsuits from the local Apache tribe and environmental groups. The US Agriculture Department withdrew approval for the proposal in 2021.
Risk of project to external entities		
To environment	Potential contamination or disruption to air, water, ground Clearing or disturbances of flora and fauna Carbon emissions	The Deepwater Horizon drilling project disaster caused one of the worst oil leaks and environmental contamination experienced in modern times (Beyer et al., 2016). The operation of many industries, especially fossil fuel producers, fossil fuel-based power producers, transportation, iron and steel making, cement manufacturing.
To society	Increased traffic, noise, people in local area Impact on jobs Impact on local services and supplies Impact on sites of cultural significance	The construction of many new processing plants often creates jobs, but also brings increased traffic, demand for housing and on local goods and service providers. The destruction of the Juukan Gorge (Western Australia) caves by Rio Tinto in 2020 was an 'act of cultural vandalism' (Hopkins & Kemp, 2020). The caves were destroyed as part of an approved iron ore mining project.

9.5 Managing project risks

The object of a project is to deliver the required plant, process or product that meets the agreed performance or functional specifications within the specified timeframe and budget. Therefore the identification of project risks focuses on the 'things' that could cause deviations in the quality, timeliness or cost associated with delivering the project. The project risks should be managed in accordance with the appropriate standards and regulations. As a minimum, this should include setting the scope, assessing the risks and determining how best to control the risks as discussed previously (Chapter 2). The project risks should be identified, assessed and the unacceptable risks managed. The outputs of these processes are often documented through the development of a risk register (see Chapter 5). The processes used to identify, assess and treat the risks should be drawing on past experience, expert opinion and available literature. These processes can use checklists, flowcharts, case studies, SWOT analyses and interviews to produce qualitative and/or quantitative risk assessments (Shevtshenko et al., 2008). Decisions are then made as to the acceptability of the risk and how unacceptable risks are going to be addressed. For project risk these can be addressed by eliminating the risk (e.g. doing the project in another location), transferring the risk (e.g. passing it to contractors and/or insurers), or treating the risk by implementing actions that limit exposures, reduce likelihood and mitigate potential impacts associated with the risk (Chapter 7). Developing a risk register captures the potential risks to a project, the assessment of the severity of those risks, as well as the preventative actions to be implemented to reduce the likelihood of occurrence and potential severity of the risk. The risk register should also capture current status of the preventative and mitigating actions and the person responsible for monitoring, maintaining and communicating about the risk and actions. An example of a project execution risk register is shown in Fig. 9.5.

Project feasibility and execution or project management risks should be managed first with impact assessments. Early in the project there should also be feasibility-based risk assessment that identifies future states of costs, revenues, social, regulatory and other stakeholder requirements in order to estimate and do a sensitivity analysis on the potential viability and sustainability of the project. If the project is deemed feasible then the application of risk management approaches to project activities can help deliver overall project success, keep the project on track with respect to project schedule, budget, and plan and it can help achieve customer satisfaction (Raz & Michael, 2001).

The preferred risk management tools and methods employed can vary through the project life cycle as shown in Fig. 9.6.

After the project has been executed, there will be operational risks associated with the resultant process, plant or product. These risks should be regularly assessed iteratively throughout the project life cycle using a range of techniques. An illustration of the preferred techniques, and when they might apply throughout the project life cycle, is shown in Fig. 9.6. For process engineers, the aim of these types of risk management activities is to identify the possible means by which there could be a loss of control of, or a loss of containment from, the process that could result in adverse health, safety, environmental, asset, financial and/or reputational outcomes. This type of risk assessment should focus on identifying early in the project the most inherently safe design that meets specifications. Maximising the inherent safety of the design (at Define stage) will minimise risks, the reliance of layers of protections and operational procedures to sustain effective, efficient and safe operations. Identifying inherently safer design options early in the project should also minimise the changes and associated costs of design modifications in later stages of the project (see Fig. 9.4). Once the inherently safe design is established, then risk-based

Risk description	Likelihood of the risk occurring	Impact if the risk occurs	Risk Rating *Based on impact & likelihood.*	Potential causes	Preventative action *Actions to address cause and reduce the likelihood of risk occurring.*	Potential impacts	Mitigating action *Action to be taken if the risk happens to mitigate impacts.*	Owner *Person who will manage the risk.*
Stakeholder action delays project start or halts project.	Low	High	High	Misunderstanding about project	Identify stakeholders, analyse power and influence and create and action stakeholder engagement plan.	Protests or curtailment/loss of license to operate	Comply with appropriate authorities' internal procedures. Seek independent expert help.	Project Sponsor
Project overruns schedule	High	Medium	Medium	Unrealistic schedule or scope changes Schedule not understood. External disruption (e.g. natural disaster	Add contingency in schedule Hold scheduling workshops to detect and correct potential deviations.	Financial penalties	Clauses in contract that require schedule review for scope changes Late penalties in subcontractor and supplier contracts. Force majeure clause in contracts	Project Manager
Project exceeds budget	High	Medium	Medium	Unrealistic budget or scope changes Supply/market issues	Add contingency in budget Hold budget meetings to detect and correct potential deviations.	Loss of profit	Clauses in contract that require budget review for scope changes. Goods and service subcontracts based on fixed price agreements. Force majeure clauses in contracts	Project Manager
Accident on the job	Medium	High	High	Issues with condition of work site, equipment, execution of tasks.	Safety review of project design to eliminate/minimise hazards Daily inspection of work site and worker conditions Only plant and equipment authorised as safe to be used Only competent, fit for duty and authorised personnel allowed	Worker fatalities or severe injuries. Asset damage	Provision and use of personal protection and rescue equipment Provision of emergency response personnel and equipment	Project Manager
Theft of materials, intellectual property or equipment.	Low	Medium	Medium	Sabotage Illegal activity and unsecured assets	Install physical security measures (e.g. fencing with guards and security cameras). Employ cyber security measures	Loss of assets. Loss of commercially sensitive information.	Involve police Insurance.	Project Manager
Customer refuses or delays approval of deliverables.	Medium	Medium	Medium	Deliverable definition/ process unclear or incomplete, customer not available Poor quality job	Preapprove deliverable process, milestones and resourcing and sign off as part of contract prior to starting process. Quality assurance performed prior to asking for approval	Long rework/ defects list Potential overruns	Agreed process for challenging and dealing with rework/ defects list Subcontractor penalties for poor quality work.	Project Manager

FIGURE 9.5 Example of a project risk register (Morphy, 2008). *Source: Adapted from Morphy, T. (2008).* Free risk register of common risks. *Retrieved from https://www.stakeholdermap.com/risk/register-common-project-risks.html#download.*

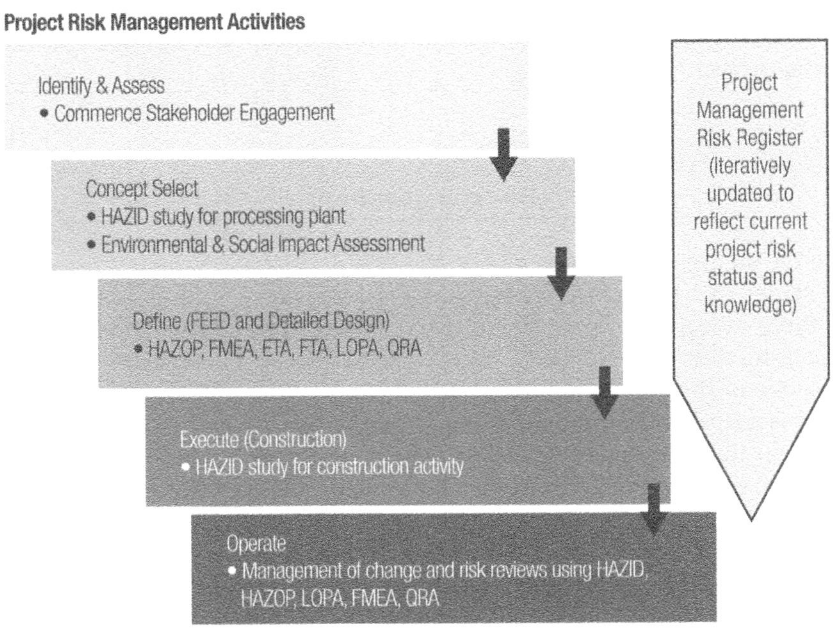

FIGURE 9.6 Different risk assessment processes used throughout a project life cycle.

approaches should be deployed to understand the additional layers of protection, resources and other tasks required to monitor, detect, assess, treat and review operational risks. This should be conducted on an ongoing basis.

9.6 Summary

A lot of engineering work is conducted within the context of a 'project'. It is thus very important for process engineers to understand project structures and systems, the project life cycle, and the associated risk management approaches. A project is a *'temporary endeavour to achieve one or more defined objectives'* (ISO 21500, 2020, p. 3), and projects are used extensively in the process industries to achieve company objectives. Projects vary enormously in their scope, ranging from small projects (nominal budget of a few thousand dollars) through to megaprojects (nominally >$1b).

We have seen that project complexity, innovation, size and location are some of the major factors that determine project risk, and that the decision-making process varies with the project risk, transitioning from technical (e.g. codes and standards) to values-based decision-making as the project risk increases.

Beware! The evidence shows us that on average, industry projects fail to deliver the forecast value, and a significant majority of large projects fail. It is thus critical to identify, assess and treat the project risks appropriately. As we have seen in this chapter, it is crucial that this commences early in the project life cycle, thus reducing the risk of costly changes and overruns.

It is very important to adopt a structured stage-gate framework for managing projects. Such a framework has been introduced in this chapter. As part of this, it is important to incorporate an appropriate risk management system, one which incorporates a live risk register, and which stipulates the deployment of the appropriate risk tools at the appropriate stage of the project.

Review questions

1. What are the stages of a project?
2. At what stages of a project do we require a risk register?
3. Provide three examples of process industry projects and specify the likely scale for each (small, medium, mega).
4. At what stage of the project should stakeholder engagement commence?
5. Why is project definition important?
6. We have seen that projects often fail to deliver the predicted value. Why?
7. At what stage of a project is the ability to implement effective risk control significantly reduced?
8. What is the approximate failure rate for megaprojects?
9. At what stage in the project life cycle is the FID?

10. Cite two examples of projects that have had significant negative impacts on society and the environment. In each case, what happened, why did it happen and what are the key lessons to learn?

References

AGC The Construction Association. (2020). *AGC Coronavirus survey results (March 23–26).* https://www.agc.org/sites/default/files/AGC%20Coronavirus%20Survey%202.pdf.

Barshop. (2016). *Capital projects. What every executive needs to know to avoid costly mistakes and make major investments pay off.* John Wiley & Sons, Inc.

Beyer, J., Trannum, H. C., Bakke, T., Hodson, P. V., & Collier, T. K. (2016). Environmental effects of the Deepwater horizon oil spill: A review. *Marine Pollution Bulletin, 110*, 28–51.

CSB. (2009). *Investigation report: Sugar dust explosion and fire.* Imperial Sugar Company. US Chemical Safety and Hazard Investigation Board. Report Number 2008-05-I-GA.

CSB. (2018). *Organic peroxide decomposition, release, and fire at Arkema Crosby following Hurricane Harvey flooding.* US Chemical Safety and Hazard Investigation Board. Report Number 2017-08-I-TX.

Flyvbjerg, B., Bruzelius, N., & Rothengatter, W. (2003). *Megaprojects and risk. An anatomy of ambition.* Cambridge: Cambridge University Press.

Hillson, D. (2009). *Managing risk in projects.* Farnham: Ashgate Publishing Ltd.

Hopkins, A. (2012). *Disastrous decisions. The human and organisational causes of the Gulf of Mexico Blowout.* Australia: CCH, Ltd.

Hopkins, A., & Kemp, D. (2020). Corporate dysfunction on Indigenous affairs: Why heads rolled at Rio Tinto. *The Conversation, 11*, September.

ISO 21500. (2020). *Project, programme and portfolio management—Context and concepts.* Geneva: International Organization for Standardization.

Killick, C. (2016). *Project & process safety. CHEE4002 course lectures.* The University of Queensland.

Merrow, E. W. (2011). *Industral megaprojects. Concepts, straegies, and practices for success.* John Wiley & Sons, Inc.

Morphy, T. (2008). *Free risk register of common risks.* Retrieved from https://www.stakeholdermap.com/risk/register-common-project-risks.html#download.

Ransan-Cooper, H., Ercan, S.A., & Duus, S. (2018). *Getting to the heart of coal seam gas protests – it's not just the technical risks.* The Conversation Paperpress, December 4, 2018.

Raz, T., & Michael, E. (2001). Use and benefits of tools for project risk management. *International Journal of Project Management, 19*(1), 9–17. Available from https://doi.org/10.1016/S0263-7863(99)00036-8.

Shevtshenko, E., Karaulova, T., & Kramarenko, S. (2008). *Risk factors in project management life cycle.* In: 6th International DAAAM Baltic conference industrial engineering at Estonia, Tallinn.

Srivastava, S. (2020, May 12, 2020). *Prioritise safety in industry during the COVID-19 era.* https://www.downtoearth.org.in/blog/governance/prioritise-safety-in-industry-during-the-covid-19-era-71067.

United Kingdom Offshore Operators Association (UKOOA). (1999). *A framework for risk related decision support.* UKOOA Publication.

West, C., Kenway, S., Hassall, M., & Yuan, Z. (2016). Why do residential recycled water schemes fail? A comprehensive review of risk factors and impact on objectives. *Water Research, 102*, 271–281. Available from http://www.sciencedirect.com/science/article/pii/S0043135416304821.

CHAPTER

10

Contracting risks

OUTLINE

10.1 Introduction 228

10.2 Contracting in the process industries 229

10.3 Classification of contractors 230

10.4 Contract agreements 233

10.5 Contractor selection and management 236

10.6 Managing contractor risks 237

10.7 Case studies 238

10.7.1 Grosvenor coal mine gas explosion 238

10.7.2 Management of contracted work: hot work, confined spaces and working at height 239

10.7.3 Donaldson Enterprises Fireworks Explosion Hawaii 239

10.8 Summary 239

Review questions 240

References 241

Fundamentals of Risk Management for Process Industry Engineers
DOI: https://doi.org/10.1016/B978-0-12-820320-0.00003-9

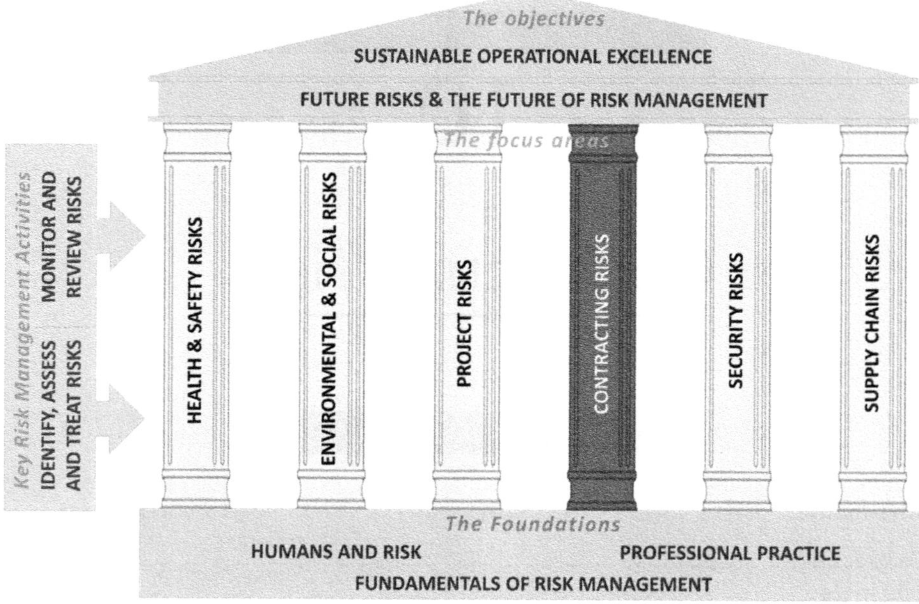

Nothing we do is more important than hiring people. At the end of the day, you bet on people, not strategies. Lawrence Bossidy

10.1 Introduction

As we have just seen, a lot of process engineering work is conducted in projects. It is quite common that many of the people working on these projects are not employees of the company, but actually staff of other businesses that are contracted to deliver certain tasks (i.e. contractors). In this chapter we will outline the types of work and agreements associated with contractors in the process industries. We will then describe the three major types of risks associated with contractors, those associated with (1) classifying contract work and contractors, (2) the terms and structure of the contract agreement, and (3) the contractors performing the work as highlighted in Fig. 10.1.

Often, operational situations make it necessary to bring in external resources to sustain or maintain operations. The businesses providing those resources are called contractors. The contractor provides labour and/or materials as per the written contract. In the process industries, contractors are typically engaged to perform specialist tasks (e.g. tasks requiring unique technical skills and/or equipment not possessed by the client company), as top-up labour (e.g. to augment employee numbers during peak work periods such as shutdowns) or supplement employee numbers when there are shortfalls due to recruitment delays. Depending on the stage of a project, contractors may also be engaged to provide independent expert advice, peer reviews and auditing.

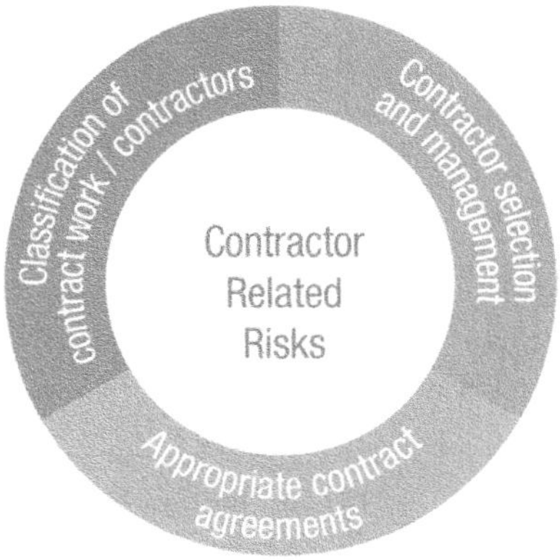

FIGURE 10.1 Categories of contractor risks.

Due to the added complexity associated with contracting work, additional and significant risks can be created when using contractors. These risks comprise risks associated with both the use of contract personnel as well as the nature of the contracting agreements. They can impact the operational performance and safety, occupational health and safety of personnel, environmental and social performance, and also supply chains. Thus contractor risks are important risks that should be identified, assessed and managed where unacceptable – from the perspective of the client as well as from the perspective of the contractor.

10.2 Contracting in the process industries

The contract lifecycle typically comprises the following phases:

- Determining the scope of work to be executed and the skills, timing and cost constraints. This information, combined with an understanding of the work-related hazards, informs a risk-based determination of who should perform the work – that is employees or contractors.
- Deciding the most appropriate contract for managing the work. This involves selecting both the type and mode of contract. Types of contracts include fixed cost, cost plus and unit cost-based contracts. Modes of contracts describe the management system requirements that the contractors need to comply with.
- Selecting contractors and managing the contract. This phase of work comprises tendering, awarding, mobilisation, execution, demobilisation and finalisation of the contract.

Contractors are typically hired to help companies reduce risks by allowing them to select the right skillsets for jobs for the duration that the requisite skillsets are required, without the overheads associated with directly employing workers. In the process industries such scenarios include:

- Hiring contractors to manage the entire project (i.e. turnkey projects).
- Hiring contractors to temporarily top-up the workforce or to fill-in for workers while they are unavailable, to manage peak workloads (workloads above what the permanent

employees can typically handle) or while the permanent recruitment process is being undertaken.
- Hiring contractors to do specialist, usually high-risk, work such as conveyor belt replacements, spent catalyst removal, oil/gas well or water bore drilling, smelter vessel refractory lining, hot works and working in confined spaces.
- Hiring expert consultants to provide technical advice, facilitate risk management processes, provide peer review and audits of business-critical processes and reports.
- Companies contracted to provide goods and materials – these entities are often called suppliers. Risks specific to provision of supplies are discussed in Chapter 12.

Depending on the size of the scope of work in the contract, there may be primary contractors, subcontractors, subsubcontractors, and suppliers. In these instances, the operational organisational structure often comprises contractors and subcontractors as shown in Fig. 10.2. Examples from the process industries include contractors employed to run oil rigs, mine sites, or major projects.

When done well, hiring contractors can reduce liabilities, improve financial results, produce higher-quality deliverables, facilitate the exchange of best practices, help develop internal and subcontractor capabilities, and improve risk awareness and management activities (Choma, 2008). However, introducing contractors into an operation, or even outsourcing whole operations to contractors, can also increase risks. Understanding what these risks are, and how they can and should be controlled, is crucial in achieving successful outcomes for both clients and contractors. We will look at these issues in the following sections.

10.3 Classification of contractors

People working in the process industries typically do so under some form of legal agreement (i.e. a Contract). They can be deemed employees or contractors who supply services and/or materials. Different jurisdictions have different laws that distinguish contractors from employees. Misclassifying workers as contractors creates a significant risk if it contravenes the law because of the legal and financial penalties, social impacts and reputational issues. Relevant legislation includes employment laws, workplace health and safety laws and even immigration laws. Some factors that distinguish employees from contractors are presented in Table 10.1.

Understanding the key difference between the classification of workers will help inform decisions about whether to do work in-house (i.e. with employees) or to outsource it (i.e. with contractors). Using employees can be advantageous especially when tasks require a certain level of competence-based training and in-house experience to allow them to safely, efficiently and effectively execute the task, while managing site-based hazards and complying with site requirements. However, it can be more time-consuming, costly and even risky to appoint employees to do work that is ad-hoc or temporary, or that requires specialist skills, or needs an independent perspective. Employing contractors also involves risks associated with whether the requisite knowledge and skills will be provided, whether their services will comply with the requisite health, safety, quality and

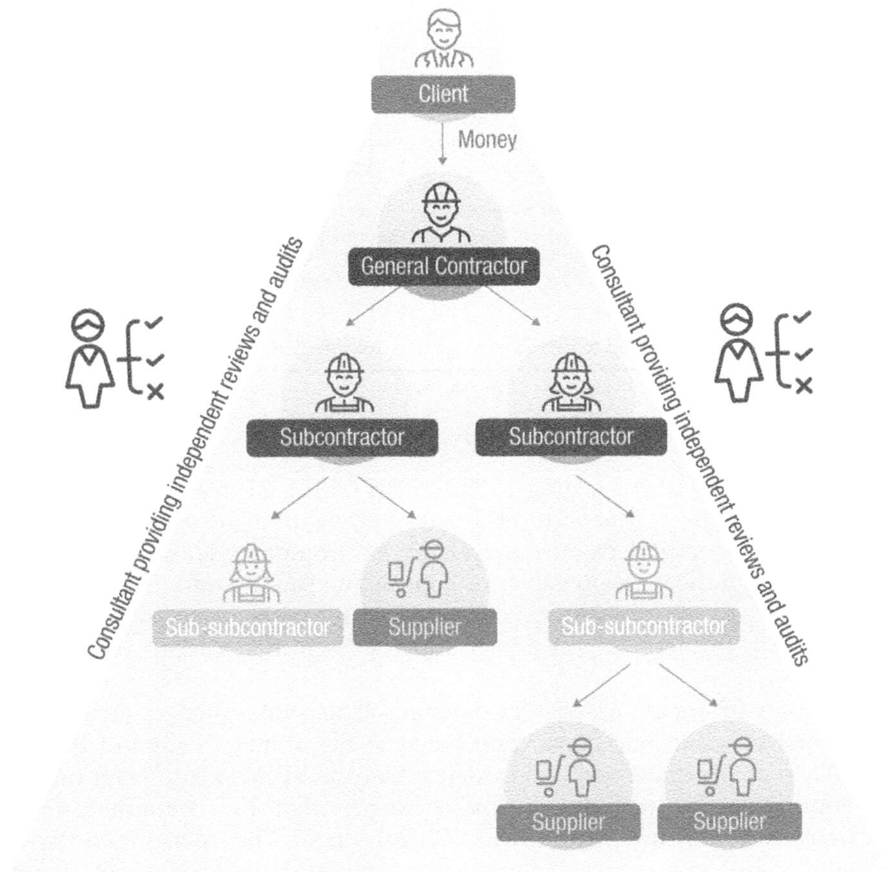

FIGURE 10.2 Example of different types of contractors associated with major works. *Source: Adapted from https://www.levelset.com/blog/difference-between-contractors-and-subcontractors/.*

other site-based standards, and whether the engagement may incur conflicts and disputes. Therefore in order for client companies, contractors and workers to make informed choices about the correct type of formal agreement they should work under, these entities should identify and assess the risks associated with the different classes of work arrangements to determine the most appropriate one for their circumstances.

The importance of selecting the right contracts and contractors has unfortunately been highlighted by many seminal industry case studies (see Section 10.7). One contemporary example is The Deepwater Horizon drilling rig disaster which killed 11 people and led to the release of millions of barrels of oil into the Gulf of Mexico, causing significant environmental harm. The Deepwater Horizon was owned and operated by Transocean who were contracted by BP to drill the Macondo well (Transocean, 2011). Another contractor, Halliburton, was responsible for supervising cement operations that were done to encase

TABLE 10.1 Factors that can distinguish employees from contractors (Hensaw, n.d.).

Employee	Independent contractor
Required to attend work at certain times on certain days. Little or no control over when and how work is performed	Has control over days and hours worked. Has control when and how work is performed
Must wear the employer's uniform or otherwise appear like they are providing services on behalf of the employer	Adopts their own branding, such as their own business logo or uniform
The employer provides the equipment required to perform their work	Must provide their own equipment. The equipment they provide is expensive or specialised
Prohibited from subcontracting and expected to provide all services personally	Can subcontract their work out to other people to perform
Paid an hourly wage and is not required to invoice	Issues invoices and is paid per task completed

Adapted from Hensaw, A. (n.d.). Contractor or employee: What's the risk? *Velocity Legal. https://www.velocitylegal.com.au/contractor-or-employee-whats-the-risk/.*

the well (Transocean, 2011). According to BOEMRE (2011), BP was ultimately responsible for conducting operations at Macondo in a way that ensured the safety and protection of personnel, equipment, natural resources and the environment. Transocean, the owner of the Deepwater Horizon, was responsible for conducting safe operations and for protecting personnel onboard. Halliburton, as a contractor to BP, was responsible for conducting the cement job, and, through its subsidiary (Sperry Sun), had certain responsibilities for monitoring the well.

The report found BP 'made a series of decisions that complicated cementing operations, added incremental risk, and may have contributed to the ultimate failure of the cement job' (BOEMRE, 2011, p. 2) but there was 'no evidence that the BP Macondo team fully evaluated ongoing operational risks, nor did it find evidence that BP communicated with the Transocean rig crew about such risks' (BOEMRE, 2011, p. 6). The Panel found that 'BP failed to protect health, safety, property, and the environment by (1) performing all operations in a safe and workmanlike manner; and (2) maintaining all equipment and work areas in a safe condition' (BOEMRE, 2011, p. 6) and that BP and its contractors 'failed to take necessary precautions to keep the well under control at all times' thereby 'creating conditions that posed unreasonable risk to public health, life, property, aquatic life, wildlife, recreation, navigation, commercial fishing, or other uses of the ocean' (BOEMRE, 2011, p. 7).

This position was supported by the Chief Counsel's report which found

> BP and its various subcontractors appear to have lost sight of that danger, compartmentalizing information that would have been useful to other companies carrying out their respective tasks. The onus fell on BP to ensure that its contractors were providing all of relevant information to the respective decision makers (Bartlit et al., 2011, p. 238).

This report also found that BP did not adequately supervise or review the work of its Macondo contractors.

> many of BP's contractors were unduly deferential toward BP's design decisions. . . . In several instances, BP's contractors expressed private reservations about the plans and procedures at Macondo but did not more forcefully communicate to BP that there were better ways to do things (Bartlit et al., 2011, p. 238).

10.4 Contract agreements

This section describes the range of risks that should be considered when deciding what class and mode of contract is best offered or accepted for a specified scope of work. Clarifying and selecting the most appropriate contract for the work can lessen the risks associated with using contractors. Thus the process of choosing which contract to offer, and which to accept, should consider the risk factors and their significance as well as legal and ethical requirements. This process should follow the steps outlined in ISO31000 whereby the risks associated with the different options are identified and assessed to select the preferred option. Then the risks associated with the preferred option should be reviewed to determine how best these risks should be treated and monitored throughout the contractor lifecycle.

Contract agreements are the formal articulation of the contracted terms of work and are intended to be legally binding. An agreed contract is where two or more parties – the one that makes the offer (i.e. the client) and the others that receive and respond to the offer (i.e. the contractor(s)) – accept to abide by the contents of the contract. If one or more parties fail to deliver on what was agreed, then the other is entitled to, and can, seek legal redress.

In general, in the process industries, the most commonly used types of contracts are:

- Fixed price or lump sum contracts
- Cost-reimbursable contracts
- Unit price contracts

Each contract has different risks associated with it. For example, with fixed price contracts, there is little risk to the client of cost overruns, however with such contacts, the contractors often add contingency costs to the price to allow for and mitigate the risk of any deviations. This increases the risk of the client paying higher than required prices. It also leaves the contractor carrying the risk of incurring, at their own expense, any deviations that fall outside those covered by the contingencies. Thus these types of contracts are best suited for jobs where the scope of work, work plan and schedule are clearly defined and well understood.

Cost-reimbursable, or cost plus, contracts are where the client pays the contractor for the actual cost of the materials and service, plus agreed amounts for contractor overheads and for profit. The profit arrangement could be a fixed fee, per cent markup on costs and/or some form of incentive or bonus/penalty payment. These types of contracts are typically used when the scope of work, type of labour or amount and types of materials are not well known or are open ended. The risk to the client is cost and time overruns. The risk to the contractor is loss of bonuses or even being required to pay penalties.

Time and materials or unit price contracts are where the client pays the contractor for materials and an agreed labour rate for work done. The unit prices often contain an allocation for overheads and profits. These types of contracts can expose clients to cost and time overruns as there is no contract-based incentive for the contractor to complete the job in the least amount of time nor with the least amount of materials.

In addition to these three commonly used contract types, there are other adaptations that may also be used. Examples include fixed price plus incentive, and fixed price with allowable variations type contracts.

All contracts, regardless of the type, should contain details about the scope of work, timelines, the deliverables, responsibilities, liabilities, communication and reporting requirements, as well as information about dispute and penalty processes.

Selecting the most appropriate contract type for the job can lessen risks for both the client and contractor. Table 10.2 lists some of the factors to consider when deciding on a contract type. However, it is best to seek specialist advice when it comes to selecting, issuing and accepting contracts. It is also important to ensure that contracts are approved by appropriate advisers and signed by all parties prior to any work commencing, as this protects all parties and workers.

It is, of course, critical to consider how contractor occupational, health and safety will be managed within a contract. Contractors often have higher incident and accident rates than employees (Valluru et al., 2017). Notably, contractors appear to have significantly increased risk of major injury and fatality in comparison with employees of the client company. This could be due to them being hired to do high-risk work (e.g. drill oil and gas wells, working in confined spaces, working with explosives). It could also be due to contractors being less experienced and less qualified (due to the temporary nature of jobs) and the fact that they often do not receive the same level of site-based training as employees. In an extensive study of 157,000 mine workers between 1998 and 2007, Muzaffar et al. (2013) determined that the odds of fatal versus nonfatal injury were 2.8 times higher for contractors than for operators (employees). Similarly, Concawe annually reports extensive safety data from the European downstream oil industry. The reporting Concawe members represent 98% of the European downstream oil industry capacity. In the period 2010–19,

TABLE 10.2 Summary of contract types and risks (Amado et al., 2011).

Contract type	Scope type	Cost certainty and risk	Risks associated with meeting milestones (time and/or quality)
Fixed price	Very certain and well defined	Cost paid is fixed – contractor takes on all risks associated with deviations	No/low contract-based incentive
Cost plus	Some uncertainty	Cost paid has some uncertainty but this is usually bounded and risk/reward is shared	Financial incentives to meet and potentially exceed milestone requirements
Time and materials	Open-ended	Cost paid is uncertain – client takes on all risks associated with overruns	No/low contract-based incentive

Adapted from Amado, M., Ashton, K., Ashton, S., Bostwick, J., Clements, G., Drysdale, J., Francis, J., Harrison, B., Nan, V., Nisse, A., Randall, D., Rino, J., Robinson, J., Snyder, A., Wiley, D., & Anonymous. (2011). Project management for instructional designers. *Press Books. http://pm4id.org/. Licensed under a Creative Commons Attribution-NonCommercial-ShareAlike (BY-NC-SA) license.*

the industry reported 72 fatalities, 62 of which were contractors (86%).[1] Contractor safety thus should be a very important consideration in all contracts.

The nature of the contract agreement typically contains information about who is responsible for management of contractors' safety. According to the International Association of Oil and Gas Producers (2017), contracts can fall under one of three categories which are described as per the following:

- Mode 1 contract: The contractor provides people and materials and executes the contract under the oversight and instructions, and in compliance with, the client's safety management system. The contractor needs to provide assurance that their personnel are qualified and fit for work and that the procedures, tools, equipment and materials are fit for purpose, in serviceable condition and suitable for the contract. Under this mode of contract, the contractor reports health and safety performance, including incidents and accidents, to the client. Typically, Mode 1 contracts are applied for minor scopes of work such as minor modification projector or on maintenance/turnaround jobs that are undertaken on a client owned/operated site.
- Mode 2 contract: The contractor provides people and materials and executes the contract under the oversight and instructions and in compliance with its own safety management system. The contractor still needs to interface with the client's safety management system and report on health and safety performance and accidents and incidents to the client. For Mode 2 contracts, clients are responsible for ensuring that both health and safety management systems – the clients and the contractors are compatible. Clients are also responsible for assuring that the contractor's controls for managing health and safety risks are implemented and effective. This includes how they are interfacing with subcontractors. Mode 2 contracts can apply to client, contractor or third-party operating sites.
- Mode 3 contract: The contractor provides people and materials and executes the contract in compliance with its own safety management system. There is no interfacing with, or reporting into, the client's safety management system. This type of contract typically applies to work done on contractor owned/operated or third-party sites and/or when the contract is for the supply of materials. Examples include the fabrication and delivery of replacement parts of plant, the supply of parts or other goods, and the contracted use of transport (e.g. public airlines, taxis). If Mode 3 contracts can still pose a risk that impacts the client (e.g. the reliability, safety and costs of using airlines) then these contractors should still be assessed and monitored by the client.

It is important to note that when considering which mode is most appropriate, consideration should be given to the workplace health and safety laws of the jurisdiction within which the work is conducted. Different jurisdictions can impose different obligations on clients and on contractors depending on who bears the risks and who actually has control over the work and working environment.

[1] Determined from the 10 2011–2020 annual Concawe reports entitled 'European downstream oil industry safety performance.' Concawe. Brussels.

10.5 Contractor selection and management

Contractor selection and management involves activities associated with the tendering, awarding, mobilisation, execution, demobilisation and finalisation of the contract. The tender process involves specifying the scope, deliverable and timeline of the work to be contracted. The tender also should detail the minimum requirements that the bidders need to meet to be considered. These minimum requirements might include qualifications and previous experience with the work, safety systems and history as well as liability and insurance information. Ensuring the tender accurately reflects the work required and the minimal requirements and expectations of those executing the work is key to minimising the risk of having things go wrong.

The award process is often split into vetting, prequalification and award. Vetting is performed by checking potential contractors against a list of general and specific requirements, including (Govindasamy, 2018; McLeod, 2012):

- assessing financial condition
- checking certificates licences, and training
- verifying safety record, insurances and previous job-specific experience
- ensuring they are trained in safety
- checking references by contacting previous employers

Contractors that satisfy the vetting criteria then prequalify to quote for jobs. This process of vetting and prequalification minimises the risk of hiring new contractors that do not have the capacity or capability to successfully perform the work (McLeod, 2012). However, there can be still issues during award. Awarding contracts involves selecting the preferred contractor from the list of prequalified and vetted contractors based on their response to the tender. Typically, responses to the tender contain information about the personnel involved, how the job will be executed, relevant historical performance on similar jobs, and the proposed cost of the job. The most significant risk about the awarding process has historically come from awarding the lowest bidder irrespective of their ability to meet the selection criteria and historical performance. The significance of this risk was explicitly highlighted in the official CSB investigation report into the Xcel Energy hydroelectric plant penstock fire in 2007, in which five contract workers were killed (U.S. Chemical Safety and Hazard Investigation Board, 2010). The CSB (U.S. Chemical Safety and Hazard Investigation Board, 2010, p.16) stated, as per the following quotes, that:

- Xcel's prequalification process for determining which potential contractors were allowed to participate in the Cabin Creek bid process considered only the contractors' financial capacity and did not disqualify bidders based on unacceptable past safety performance.

- RPI [the chosen contractor] received the lowest score, "zero," in the safety category, which, according to Xcel's evaluation form, meant that the proposal should have been automatically rejected.

- RPI's proposal received the highest ranking in the evaluation process, based primarily on the low price.

Mobilisation and demobilisation activities will vary depending on the size and scope of the contracted work. Mobilisation starts after the contract has been awarded. It may include procurement of goods and subcontracts. It could also involve relocation to site (for remote work) and site setup (i.e. installation of temporary offices, workshops, laydown and storage areas, etc.). Demobilisation includes the removal from site of the contractors' personnel, equipment and temporary facilities. Mobilisation and demobilisation activities can impact project risks (e.g. timing and cost), operational risks (e.g. these activities can disrupt operations as well as introducing potential health and safety risks) and other risks (e.g. the social impact of mobilisation and demobilising large amounts of people and equipment). The BP Texas City incident in 2005 dramatically highlighted risks associated with mobilisation and demobilisation of contractors. All 15 fatalities were contractors located in temporary work trailers located near to the explosion site. This led to the CSB issuing urgent safety recommendations in regard to the safe siting of trailers and temporary facilities away from hazardous areas of process plants (U.S. Chemical Safety and Hazard Investigation Board, 2007). This was the precursor to significant global changes in regard to protocols for siting and constructing worker offices on processing plants.

A range of risks can emerge during contract execution. A key example is not-fit-for-purpose materials, equipment and contractors arriving on the job. Sites often use quality assurance checks before allowing materials and equipment onsite. These checks are done to ensure all incoming goods comply with site standards and requirements. Personnel are often given inductions or other relevant training so they know how to work safely and in compliance with the requisite standards while on site. They can also be subjected to random and/or routine drug and alcohol testing and on-job work performance checks. Other common risks include labour shortage and/or productivity issues, increased vulnerabilities to health and safety hazards, failure to manage subcontractors, cost overruns, dealing with changes to contracted work or on-job conditions and the risk of receiving no warranty for work (Choma, 2008).

10.6 Managing contractor risks

In the preceding sections, we outlined contractor risks, the threats and consequences, and discussed the types of controls that might be considered to prevent and mitigate contract-related risks. Consistent with the generic approach for managing risks introduced in earlier chapters, a risk register may be used to document contractor risks and controls. It can also be helpful to include each entity's responsibilities (i.e. client, contractor, subcontractor). In addition, it is worth performing a bowtie analysis for high-risk scenarios to ensure that there are adequate and effective controls to prevent and mitigate the risk. It can also be beneficial to develop an overarching or contractor governance bowtie to ensure the management of all contractors is conducted in a manner that provides sufficient oversight and delivers safe and successful work. An example of an overarching bowtie is shown in Fig. 10.3. This bowtie shows threats, consequences and controls associated with the loss of control of contracted work. Such a bowtie should be accompanied by an AcciMap analysis which highlights controls, control supports and entities responsible for verifications and auditing of controls and supports.

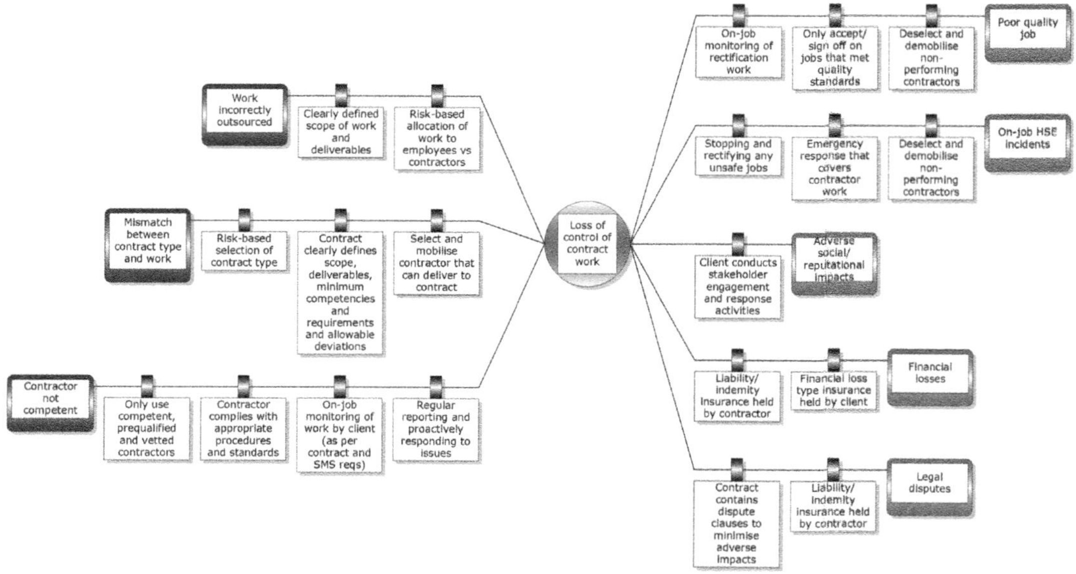

FIGURE 10.3 Example of high-level bowtie for managing contracted work.

10.7 Case studies

We have already seen that poor management of contract and contractor risks was a significant contributing factor in the Deepwater Horizon, Xcel Energy penstock and BP Texas City accidents. There are many other case studies that highlight the importance and impact of contractor risks within the process industries. Below is a sample of some other case studies.

10.7.1 Grosvenor coal mine gas explosion

In 2019 there was a gas explosion at the Grosvenor underground coal mine in Queensland, Australia. An enquiry (Martin & Clough, 2021) into the explosion and gas exceedances in other Queensland Coal mines found:

> 'Since the 1990s there has been a substantial increase in the proportion of labour hire workers and contractors at Queensland coal mines, although the proportions vary between sites. At Grosvenor, 76% of its total site workforce were contractors and labour hire workers engaged in mining tasks' (p. 34).
> 'The Board's conclusion is that there is a perception among coal mine workers that a labour hire worker or contractor who raises safety concerns at a mine might jeopardise their ongoing employment at the mine. It has not been possible to assess how widespread that perception might be. However, the existence of a perception, no matter how widespread, creates a risk that safety concerns will not always be raised' (p. 34).

'Production and safety bonuses largely based on lag safety performance indicators are not a reliable means of improving safety outcomes and may in fact lead to under-reporting of safety incidents and injuries' (p. 34).

10.7.2 Management of contracted work: hot work, confined spaces and working at height

There are numerous incidents where contractors have been killed while performing hot work in or near process vessels containing flammable liquids and gases. Examples include:

A contractor performing welding and grinding work on a Briar Chemicals vessel died when toluene residue left inside was ignited and exploded (Doyle, 2021). The vessel was supposed to be empty.
The tank explosion at the Avonmouth Wastewater treatment plant which killed four people including a contractor. The workers were carrying out hot work on top of a tank containing methane (Aviram, 2021).
A foul condensate tank exploded during a maintenance shutdown at a Packaging Corporation of America containerboard mill in February, 2017. A flammable vapour mixture in the tank was ignited by sparks from contractors welding a water pipe directly above the tank. The explosion killed 3 and injured 7 workers. All 10 workers were contractors (U.S. Chemical Safety & Hazard Investigation, 2018).

There are also many incidents of contractor fatalities in confined spaces and from falls from heights. These cases highlight the importance that both the operator or contractors need to take appropriate risk identification and control precautions.

10.7.3 Donaldson Enterprises Fireworks Explosion Hawaii

A fire and explosion in a seized fireworks magazine killed five workers who were disposing of the fireworks. According to the (U.S. Chemical Safety and Hazard Investigation Board, 2013), there was no safety-related review of the company involved prior to the awarding of the contract for fireworks disposal. The company changed procedures several times without proper hazard analysis, training or oversight. As a result, on 17 January 2013, a large accumulation of explosive material was placed just inside the magazine entrance and it exploded. The CSB found that this was due, at least in part, to insufficient contractor selection and oversight requirements especially for those involved in hazardous activities. Based on this incident, the CSB recommended the use of experts to assist with contractor selection and ensuring that the contractor work is overseen to ensure that it is being conducted safely.

10.8 Summary

A lot of process engineering work is conducted in projects, and contractors are often key actors throughout the project lifecycle. The contractor is an external business that provides labour and/or materials, as per the written contract, to support the project to achieve its objectives.

Contractors are typically hired to help companies reduce risks by allowing them to select the right skillsets for jobs, for the duration that the requisite skillsets are required. However, due to the added complexity associated with contracting work, additional and significant risks can be created when using contractors. We have seen that contractors may have significantly increased risk of major injury and fatality when working on projects. In two extensive studies of industry data, we saw that the odds of fatal versus nonfatal injury were 2.8 times higher for contractors than for operators on mine sites, and that 86% of the fatalities reported in the European downstream oil industry between 2010 and 2019 were contractors. Thus contractor risks are important risks that should be identified, assessed and where unacceptable, managed properly.

Using a bowtie analysis, we identified the major threats leading to the 'loss of control of contracted work' as: (1) work incorrectly outsourced; (2) mismatch between contract type; and (3) contractor not competent. Key controls involved in ensuring good contract and contractor management include: (1) selecting the appropriate type and mode of the agreement; (2) good contractor selection and management; and (3) appropriate mobilisation and demobilisation activities.

As a means of emphasising the importance of contractor risk, we have shown how poor contractor management has been identified as a significant contributing factor in many serious incidents, including the devastating consequences observed in the Deepwater Horizon, BP Texas City refinery and Xcel Energy Hydroelectric Tunnel incidents.

Review questions

1. In what circumstances do companies prefer to use contractors?
2. Contractors are typically engaged to help companies reduce risks. Contracts and contractors thus result in low-risk work. Discuss.
3. State some differences between employees and contractors.
4. What are the three major types of risks associated with contracts and contractors?
5. Why do we need to consider contractor risks separately?
6. What are the phases of a contract lifecycle?
7. What causal factors contribute to contractors having a higher risk of injury when working on projects?
8. What are the three different types of contracts that have been presented and how do they differ?
9. What are the different modes of contract that are used for the managements of contractor safety?
10. What processes should be used to identify and address contractor risks?
11. What major threats may lead to 'loss of control' of contracted work?
12. What controls can we implement to ensure good contract and contractor management?
13. During shutdown, contractors are often engaged to perform maintenance work comprising hot work (e.g. welding) and working in confined spaces. Because the plant

is 'shutdown', then this work is low risk. Comment. Cite some case studies in your answer.

14. Review the Xcel Energy Hydroelectric Tunnel fire and explosion case study. The CSB investigation report (U.S. Chemical Safety & Hazard Investigation, 2010) and accompanying video are excellent resources. Summarise the key lessons learnt and highlight the recommendations in regard to managing contractor risks.

15. Review the BP America Texas City Refinery explosion case study. The CSB investigation report (U.S. Chemical Safety and Hazard Investigation Board, 2007) and accompanying video are excellent resources. Summarise the key lessons learnt and highlight the recommendations in regard to managing contractor risks.

References

Amado, M., Ashton, K., Ashton, S., Bostwick, J., Clements, G., Drysdale, J., Francis, J., Harrison, B., Nan, V., Nisse, A., Randall, D., Rino, J., Robinson, J., Snyder, A., Wiley, D., & Anonymous. (2011). *Project management for instructional designers*. Press Books. Licensed under a Creative Commons Attribution-NonCommercial-ShareAlike (BY-NC-SA). http://pm4id.org/.

Aviram, A. (2021). The Avonmouth explosion was horrific. But it wasn't unprecedented. *The Bristol Cable*, 17 February, 2021. Available from https://thebristolcable.org/2021/02/the-avonmouth-explosion-was-horrific-but-it-wasnt-unprecedented-wessex-water-silo-bristol/.

Bartlit, F. H., Sankar, S. N., & Grimsley, S. C. (2011). *Macondo the gulf oil disaster*. Chief Counsel's Report. National Commission on the BP Deepwater Horizon Oil Spill and Offshore Drilling.

BOEMRE. (2011). *Report regarding the causes of the April 20, 2010 Macondo well blowout*. September 14, 2011. The Bureau of Ocean Energy Management Regulation and Enforcement.

Choma, A. A. (2008). *How to reduce risks in contractors' management*. Paper presented at the PMI® Global Congress 2008, North America, Denver.

Doyle, A. (2021). *Briar Chemicals fined £1m following fatal accident*. The Chemical Engineer. 27 September 2021.

Govindasamy, A. (2018). *Minimising risks when hiring contractors*. https://conserve.com.au/minimising-risks-when-hiring-contractors/.

Hensaw, A. (n.d.). *Contractor or employee: What's the risk?* Velocity Legal. https://www.velocitylegal.com.au/contractor-or-employee-whats-the-risk/.

International Association of Oil and Gas Producers. (2017). *HSE management – guidelines for working together in a contract environment*. Report No. 423. April 2017.

Martin, T. S. C., & Clough, A. (2021). *Queensland Coal Mining Board of Inquiry Report Part II*. May 2021. State of Queensland (Queensland Coal mining Board of Inquiry) 2021.

McLeod, D. (2012). Prequalifying subcontractors has become crucial, but the vetting process can prove to be difficult. *Business Insurance*, *11*(2012), October.

Muzaffar, S., Cummings, K., Hobbs, G., Allison, P., & Kreiss, K. (2013). Factors associated with fatal mining injuries among contractors and operators. *Journal of Occupational and Environmental Medicine*, *55*(11), 1337–1344.

Transocean. (2011). Macondo well incident - Transocean investigation report. *Transocean*, June 2011.

U.S. Chemical Safety and Hazard Investigation Board. (2007). Investigation report: Refinery Explosion and Fire. *Report No. 2005-04-I-TX*. U.S. Chemical Safety and Hazard Investigation Board.

U.S. Chemical Safety and Hazard Investigation Board (2010). *Investigation report: Xcel Energy Hydroelectric Plant Penstock Fire*. Report No. 2008-01-I-CO. U.S. Chemical Safety and Hazard Investigation Board.

U.S. Chemical Safety and Hazard Investigation Board. (2013). Investigation report: Donaldson Enterprises, Inc. Fireworks Disposal Explosion and Fire. *Report No.: 2011-06-I-HI*. U.S. Chemical Safety and Hazard Investigation Board.

U.S. Chemical Safety and Hazard Investigation Board (2018). *Non-Condensable Gas System Explosion at PCA DeRidder Peper Mill*.Report No.: 2017-03-I-LA. U.S. Chemical Safety and Hazard Investigation Board.

Valluru, C. T., Dekker, S., & Rae, A. (2017). How and why do subcontractors experience different safety on high-risk work sites. *Cognition, Technology & Work*, *19*, 785–794.

CHAPTER

11

Security risks

OUTLINE

11.1 Introduction 244

11.2 Physical security in process industries 245

11.3 Cyber security in process industries 248

11.4 Summary 253

Review questions 253

References 254

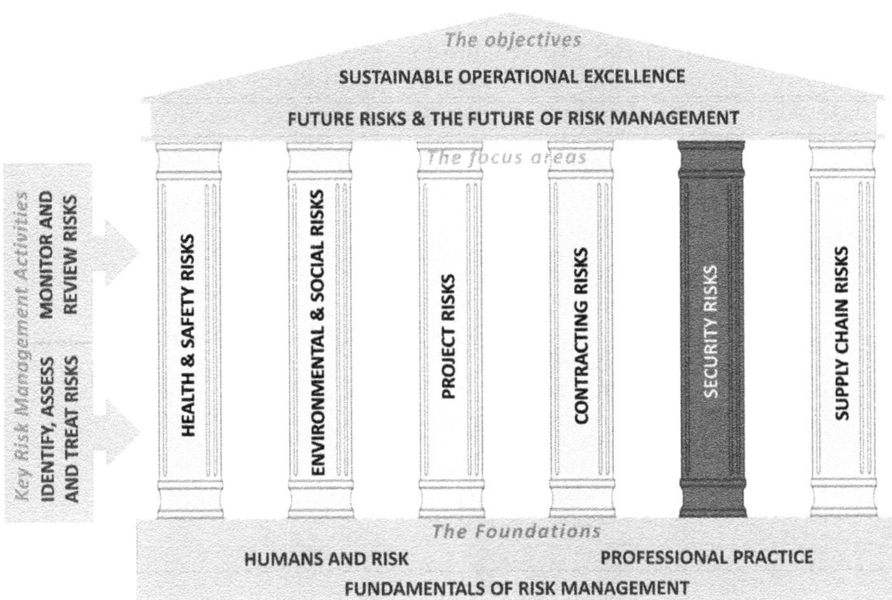

Fundamentals of Risk Management for Process Industry Engineers
DOI: https://doi.org/10.1016/B978-0-12-820320-0.00014-3

There are risks and costs to a program of action, but they are far less than the long-range risks and costs of comfortable inaction John F Kennedy

11.1 Introduction

Contemporary processing plants are cyber-physical systems, comprising computer systems and information technology (IT) integrated with the physical system or the operational technology (OT). Security risk refers to risks posed by external threats to both the OT and IT. We will refer to these risks as physical security and cyber security risks, respectively. The identification and management of security risks—both physical and cyber—is becoming increasingly important for process engineers in order to assure sustainable operations. Also, as shown in Fig. 11.1, it is important to also understand that these risks often overlap.

As introduced in Chapter 1, security risks are now high priority for business. Physical and/or cyber security risks rated in the top 10 risk concerns for manufacturing (BDO, 2017a), mining and metals (Ernst & Young, 2018), oil and gas (BDO, 2017b) and business overall (Allianz, 2021). In a recent industry survey of 400 managers and senior executives from companies with critical distributed infrastructure (e.g. manufacturing, telecommunications, energy), 58% of organisations had experienced a security breach of their control systems in the last 12 months (Fortinet, 2021).

In this chapter, we will provide examples of different types of physical and cyber security threats that the process industries have experienced. We will outline the significance of these risks and some lessons learned with case-study examples. We include discussions of how those in the process industries can identify, assess and effectively control security risks. The chapter concludes by highlighting the importance of remaining up-to-date with potential security threats and leading practices for preventing or mitigating them.

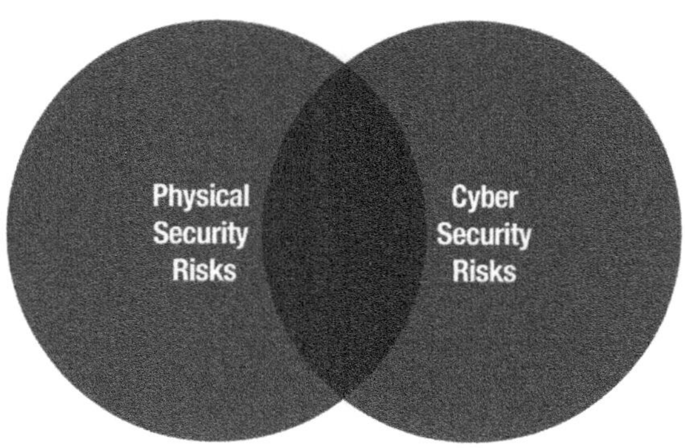

FIGURE 11.1 Processing plant security risks.

11.2 Physical security in process industries

Physical security can be defined as the protections from external threats to the OT. The threats could emerge from people, goods, the environment and neighbouring activities. These might include:

- Unauthorised persons accessing processing sites or process equipment.
- Animals accessing processing sites or process equipment.
- Noncompliant or contrabanded goods coming onto site and into restricted areas within a site. Such contraband could also include infiltration of an operation by drones, robots or other similar technologies.
- External man-made events, for example emissions from an external plant or damage from external accidents, vehicle crashes, utility failures.
- Extreme weather events that produce projectiles or compromise site facilities, for example via flooding, earthquakes, cyclones.
- Introduction of biological hazards such as viruses.
- Opportunistic crime (e.g. theft), organised crime (e.g. counterfeiting), civil unrest, acts of sabotage and terrorism.

The determination of which of these risks are relevant to a given process operation should be derived from review of historical events—locally, nationally and internationally—as well as from consultation with relevant stakeholders to capture current and future risk scenarios of significance that haven't yet resulted in an event.

Unauthorised access of persons into a processing plant or processing plant infrastructure has resulted in numerous unwanted events. For example, Berbotto and Chainey (2021) found that USD$133 billion dollars a year is lost globally due to oil being stolen. The stealing of oil has also resulted in significant fires and explosions. Examples include the deaths of 73 people in January 2019 when residents tried to steal gasoline from a pipeline outside Mexico City (Sieff, 2019). Other examples are associated with breaking and entering and stealing chemicals. Incidents include the theft of highly dangerous chemicals including cyanide, sulphuric acid, hydrochloric acid and nitric acid from a Northhamptonshire chemical plant in the United Kingdom (BBC, 2015) and phosphine from a Perth grain terminal (WAtoday, 2014). Further examples include the stealing of 137 L of chemicals from Hong Kong university laboratories (Xinhua, 2019), and the stealing of 400 pounds of palladium from a company in Georgetown County, SC, USA (WMBF News Staff, 2019).

Unwanted animal access to work sites has also caused issues. For instance, in 2014, an Alberta oil sands worker was attacked and killed by a black bear (Farshbaf, 2014) and in 2017 a Siberian oil worker was killed by a brown bear (Stewart, 2017). Another example of negative impacts from unwanted animal access includes animals causing power outages. Snakes have caused numerous and significant power losses including in Jacksonville in 2015 (Bereola, 2015), Greenville County in 2017 (Felicien, 2017), Denton, North Carolina in 2021 (Doliner, 2021). Other problem animals can include rodents, birds and animals whose bites can cause anaphylactic shock. As such, process industry risk analyses should also include the identification and assessment of potential fauna hazards like the ones mentioned.

Both authorised and unauthorised people can introduce risks associated with restricted goods and contraband. For example, many processing sites have volatile chemicals which could easily ignite causing fire and explosion. One of the controls used to manage these types of risks is associated with excluding ignition sources. As such, these sites only allow intrinsically safe equipment into these areas and prohibit any other sources of ignition (e.g. personal lighters, matches, mobile phones, car key fobs). Without such controls, events like the Pike River Mine disaster (Royal Commission on the Pike River Coal Mine Tragedy, 2012) and BP Texas City Refinery explosion (CSB, 2007) may happen. Other processing industries have similar contraband rules. For example, aluminium smelting operators forbid glass containers that could contain liquid. This is done to prevent molten metal explosions that can occur when the contained liquid is vapourised (expands rapidly) within the molten metal causing a primary explosion which in the right conditions can atomise the molten aluminium and cause it to exothermically oxidise in the air leading to a secondary more intense explosion.

Advanced technologies such as drones and robots are another type of contraband that an unauthorised person may introduce to an operational site. This might be an accidental infiltration from a recreational user. For example, a recreational drone was crashed into the White House in 2015 (Grossman & Nicas, 2015), and another was flown into the World Trade Centre in 2021 (Hambling, 2021). It could also be a deliberate attack such as the September 2019 drone attacks on Saudi Aramco's oil and gas facilities. The drone strikes targeted Abqaiq and Khurais and 'knocked out 5% of the world's oil production capacity' (Franck, 2019).

Another type of physical security risk is that posed by external man-made events. Such events include external fire and explosions harming internal operations. For instance, at the Port of Tianjin, a fire and explosion in one company's warehouse spread to other storage facilities at the port (Bloor et al., 2022). Loss of services due to disruptions to supplies and industrial accidents can also harm downstream processing operations. For example, 21% of unscheduled shutdowns in U.S. refineries were caused by electrical problems of which 53% were due to loss of power supply from third-party suppliers (Hydrocarbon Publishing, n.d.). Another example is the Esso Longford disaster where an explosion shut down the gas plant resulting in 2 weeks of disrupted gas supplies to other industry in Victoria (Dawson & Brooks, 1999). Similarly, accidents associated with loss of control of vehicles can harm operations. For example, a laser factory in Connecticut was hit by a plane in 2021 (Manno, 2021) and a fertilizer plant in Napier New Zealand caught on fire after a car crashed into it (Fuller, 2020).

Other external events that can disrupt processing operations include extreme weather events. In August 2017, flooding caused by Hurricane Harvey caused the Arkema Crosby chemical plant to lose power. As a consequence, organic peroxide products stored in a refrigerated warehouse had to be moved to refrigerated trailers, but these too subsequently lost power which caused the organic peroxide products to decompose and catch on fire. Approximately 160 tonnes of organic peroxide combusted and 200 nearby residents were evacuated (CSB, 2018). Similarly, Superstorm Sandy resulted in Phillips 66's Bayway New Jersey refinery being shut down for 3 weeks due to salt water flooding causing extensive equipment damage which resulted in $650 m in revenue losses and $56 m in other costs (Hydrocarbon Publishing, n.d., Reuters

Staff, 2012). Another example is the Fukushima tsunami. In 2011, a tsunami triggered by an offshore earthquake surged over the walls of the Fukushima Nuclear Power plant disabling the cooling system power supply and backup power generators which led to a reactor meltdown and significant radiation releases into the atmosphere causing more than 100,000 people to be evacuated (World Nuclear Association, 2022). Earthquakes and bushfires can also disrupt process operations. In 2014, a bushfire in Victoria (Australia) infiltrated and ignited the Hazelwood coal mine which continued to burn for 45 days (Hazelwood Mine Fire Inquiry, 2015).

Pandemics and epidemics can also disrupt processing operations. Some processing operations may be ramped up in response to a pandemic. For example, sanitiser and face mask manufacturers had to work overtime to meet demand during the COVID-19 pandemic. In contrast, other processing operations may be shut down totally as was also experienced in the COVID-19 pandemic. The shutting down and restarting of operations due to these types of disruptions can be hazardous, as evidenced with the LG Polymers chemical factory gas leak in Visakhapatnam in India in 2020 which killed 11, left hundreds hospitalised and impacted thousands (Burke, 2020).

Processing operations can be vulnerable to counterfeiting of supplies and products. For example, counterfeit cosmetics and perfumes can be laced with dangerous ingredients (ABC7.com staff, 2018; Summers, 2017). Counterfeit lithium batteries are becoming more prevalent and can result in fires and explosions (Kong et al., 2022). Counterfeit drugs are also increasing in prevalence, and these not only have economic impacts for manufacturers but also potentially serious health implications for the public (Blackstone et al., 2014). The European Crop Protection Association has also previously issued warnings about fake pesticides as they can cause crop damage and compromise food safety (Sanderson, 2006). They can also cause brand and reputation issues.

Lastly, processing operations can be subject to community unrest as well as acts of sabotage and terrorism. Examples include:

- The explosion of terrorist bombs at a petrochemical plant near Marseille which resulted in a full tank of petrol and another full tank of naphtha catching on fire (ABC, 2015).
- Catalent, a pharmaceutical company, had to halt production of its Beinheim plant for 4 months in 2015 to tighten security after suspected sabotage resulted in out-of-placed capsules being found (Palmer, 2015, 2016).
- The 2015 failed terrorist bombing of the Air Products & Chemicals plant in France (Trager, 2015).
- The 2016 sabotage of the Dakota Access Pipeline in the United States (Flock, 2021).
- The 2016 terrorist attack on a gas processing plant near Baghdad which resulted in 14 people being killed (Brindle, 2016).

The process used to assess the physical security risks associated with a processing operation should be compliant with ISO31000. Thus the first step is to determine the scope. The scope should include all external threats and unauthorised or illegal internal activities that could compromise the achievement of the organisation's objectives. It should consider not only the processing operations but also the provision of services and supplies as well as the distribution of products and disposal of wastes. Then according to ISO31000, risks need to be identified, assessed and then risk treatments implemented to prevent loss, theft, damages or other consequences that compromise assets or interrupt operations. With regard to risk treatment, inherently safer design options should be considered firstly, followed by defence in depth options.

For plant physical security, inherently safer design options should include consideration of plant location and the siting of equipment within the plant. Considerations that might impact plant location are the degree of unrest and crime in a given area, and the vulnerability of the area to adverse environmental events such as earthquakes, tsunami's, hurricanes, cyclones and volcanic activity. Inherently safe design approaches would consider options for locating away from areas where the environmental and security risks are high. Other security risks to consider for site location include supply chain risks for both supplies into the site, and for product and waste from the site. Considerations relating to environmental events should consider historical records (dating back to indigenous information if available) and future climate forecasts to determine the safest design based on consequence analyses.

Selecting and locating equipment within the plant is another option for creating an inherently safe design. For example, the backup generators at the Fukushima nuclear power plant could have been placed above, not below, the predicted tsunami flood levels. Other inherently safe designs might include eliminating, minimising or substituting the storage inventory quantities of valuables (e.g. explosives, gold bullion, valued chemicals) at processing facilities so as to not attract thieves.

Once the site and equipment design has been made as inherently safe as possible, then defence in depth options should be consider for unacceptable physical security risks. Examples include boundary fencing or barricades with security gate access. The fencing or barricade should prevent incursions of pedestrians, vehicles and other projectiles, fire, and smoke that might come from external man-made activities. Additional security surveillance, both electronic and manual systems, can also be deployed to monitor authorised access and activity around the perimeter of the site. Such surveillance would seek to detect physical objects or people. But if required, it could also be designed to detect other hazards such as volatile or toxic gases or smoke. Similarly, weather monitoring alert and response systems could be used to detect and respond to adverse external threats.

In summary, the physical security risks that can impact a processing operation can vary depending on the nature and location of the operation. However, identification and treatment of these risks can still be managed with the risk assessment and risk treatment processes described in Chapter 5. An example of a hazard analysis with a focus on security risks, both physical and cyber, is shown in the risk register illustrated in Fig. 11.2. An example of a bowtie for the knot 'unauthorised access of person' is shown in Fig. 11.3.

11.3 Cyber security in process industries

Cyber security risks are increasingly emerging due to the digitisation of the industry and the increased interconnectedness of equipment and sites. Digital control systems, digital enterprise wide procurement, maintenance and other management systems, use of smart sensors, computerised models and computer-based work are becoming increasingly

Description of Hazard	Risk scenario	Consequences	Maximum Potential Impact	Estimate of Likelihood	Overall Risk Rank	Assessment/recommendation including comments on risk treatment options	Responsible person and bowtie ref
Terrorist/ saboteur / Disgruntled employee	Unauthorised openning of drain valves	Drain valves opened on tanks and bunds that release tank contents which can ignite or form vapour cloud and explode killing people and destroying assets	5 Catastrophic	Rare	Very High	Full risk treatment analysis to identify ISD and prevention and mitigation DiD and LOPA that produces bowtie, control specification and control verification information	Site Executive: *Unauthorised access bowtie* info
Hacker	Unauthorised access of person into control system(s)	Hacker overrides high level SIS systems resulting in overfilling tanks. The overflowing petrol forms pools of liquid which ignite and catch on fire and/or vapourise to form cloud that explodes killing people and destroying assets	5 Catastrophic	Likely	Very High		
	Loss of personnel and commercially sensitive data	Identity theft from employee data. Loss of contracts from others having commercial data	3. Moderate	Likely	Significant		
Ransomware or other type of computer virus	Access into control system(s) blocked by randsomware	Full site shutdown resulting in loss of revenue and additional costs to deal with randsomware	3. Moderate	Likely	Significant		
Tourist, unsolicited visitor	Person struck by moving vehicles/ equipment	Visitor gets in way of and is struck by moving equipment causing injury or fatality	4. Major	Rare	Medium		
Protestors	Physical access to site blocked	Delivers to and from site halted resulted in lost revenue	2. Minor	Unlikely	Medium	Risk treatment analysis to determine adequate levels of safety critical supplies to be held on site in case of supply chain disruptions.	Engineering Manager
Disgruntled employee	Contamination of fuels (e.g. fuel in wrong tank)	Contaminated project that cannot be shipped and will need to be reprocessed	2. Minor	Moderate	Medium	Risk treatment analysis that monitors fuel quality and detects and alerts when contamination may or has occurred	Engineering Manager

FIGURE 11.2 Examples of security risks from tank farm risk register.

more common as is the interconnection of these systems both on-site and to off-site entities through the Industrial Internet of Things (IIoT).

When these systems are infiltrated by unauthorised actors then this can adversely impact on operations, personnel and ability of the business to achieve its objectives. Such security threats and actual infiltrations have been defined by Kissel (2013):

- Cyber security threats which entail 'any circumstance or event with the potential to adversely impact organisational operations (including mission, functions, image or reputation), organisational assets, individuals, other organisations or the nation through an information system via unauthorised access, destruction, disclosure, modification of information, and/or denial of service' (p. 198).
- Actual cyber security attacks which are 'an attack, via cyberspace, targeting an enterprise's use of cyberspace for the purpose of disrupting, disabling, destroying, or maliciously controlling a computing environment/infrastructure; or destroying the integrity of the data or stealing controlled information' (p. 57).

Cyber security risks are now high priority for industry, and a recent industry survey of 320 worldwide professionals found that 77% of companies surveyed ranked cyber security as a major priority and that the consequences of 'major concern' included damage to products and services (69%), employee death or injury (64%), damage to company reputation (62%) and violation of regulations (62%) (Schwab & Poujol, 2018).

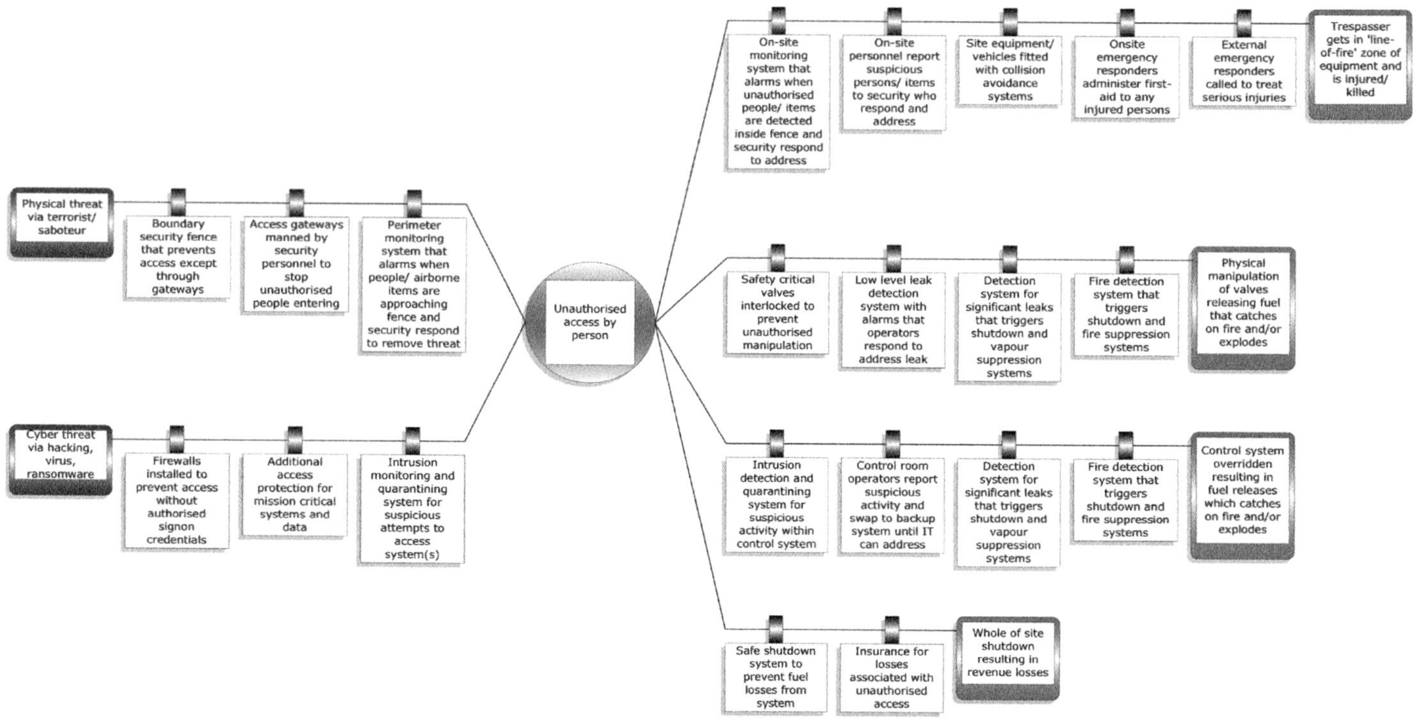

FIGURE 11.3 Example bowtie for unauthorised access by person on tank farm.

Iaiani et al. (2021) compiled a database of 82 cyber security reported incidents that affected industrial facilities (chemical, energy, petrochemical, water) between 1982 and 2019. Seventy-seven incidents occurred after 1999. That work concluded that the impacts of cyber security incidents can be very significant:

> Some attacks resulted in very severe consequences for the companies affected, as in the transportation of hydrocarbons by pipeline, where two major events (explosions) were induced by remote manipulations of the OT system. Although no fatalities were recorded, huge economic losses, shutdown of process units, loss of process control and monitoring were caused when the control and supervision network was affected by the attackers. Although based on a relatively limited number of CSIs, the analysis carried out and the lessons learnt point at defense-in-depth and multiple IT countermeasures as a key point to hinder such attacks (Iaiani et al., 2021, p. 29).

The cyber security risks faced by the process industries include: (1) external entities taking over control systems; (2) ransomware infiltration of computer systems to extort money; and (3) theft, destruction or modification of data. Examples of each of these are briefly introduced in the following.

External entities taking over control systems:

- In 2008 hackers infiltrated Turkey's Baku-Tbilisi-Ceyhan crude oil pipeline surveillance cameras. They then got into the internal network to manipulate the pipeline pressure to superpressurise the crude oil in the pipeline, causing an explosion. The ensuing fire continued to burn for more than 2 days (Hazardex, 2014; Iaiani et al., 2021).
- In 2011 hackers gained access to a U.S. Water utility control system and caused a water pump to turn on and off repeatedly until it burned out (RISI, 2011).
- An attack on a German steel company in 2014 involved hackers gaining sign-on information via booby-trapped emails, and subsequently accessing and reprograming parts of the control system, forcing the blast furnace into an unscheduled shutdown and causing significant damage (BBC, 2014).
- A power grid attack by the 'Dragonfly Group' in Ukraine took control of the workstations and Supervisory Control and Data Acquisition (SCADA) systems in 2015 and 2016 and crashed the grid causing widespread power outages affecting hundreds of thousands of people (Symantec, 2017).
- In 2021 hackers changed the dosing level of sodium hydroxide in the Tampa Bay area water supply from the safe 100 parts per million to the dangerous 11100 parts per million (Robles & Perlroth, 2021). Fortunately, this attack was noticed and addressed before harm was inflicted on residents (Robles & Perlroth, 2021).
- The 2021 cyber attacks on JBS – the largest meat processing company in the world– impacted operating plants in Canada, the United States and Australia (Geiger, 2021).

Ransomware infiltration of computer systems used to extort money:

- In May 2021, The Colonial Pipeline (USA) was shutdown for 5 days as a consequence of a ransomware attack. The operator, the Colonial Pipeline Company, reportedly paid a ransom to cyber attackers (Jasi, 2021).
- Tanner et al. (2021) present several ransomware attacks including: (1) Wannacry ransomware disrupted U.K. National Health Service, Honda, Nissan, Renault, Fedex

and Boeing in 2017 which cost an estimated US$4 billion; (2) Not-Petya ransomware impacted the operations of many companies including Fedex, Merck & Co, Modelez International, Reckitt Benckiser and Maersk Line in 2017 costing an estimated US$10 billion; and (3) Locker-Goga ransomware impacted organisations including Norsk Hydro costing more than $US60 million.

Theft, destruction or modification of data or information via direct stealing:

- In 2019 a cyber attack of an Indian nuclear power plant involving the DTrack malware was designed for stealing data (Hazardex, 2019).
- In March 2020, Royal Dutch Shell announced that hackers had targeted a file transfer application and stolen files containing company data and personal files (Mathews, 2021).

Whilst offering some unique and specific challenges, cyber security risks should be managed using the generic risk management framework that was introduced in Chapter 2, and consistent with ISO31000. However, the competencies required will need to extend to incorporating cyber security expertise to:

- Identify which risks are relevant
- Assess the significance of the relevant risks
- Determine the best way to manage the significant risks (e.g. what is ALARP standard?)
- Implement and sustain effective monitoring and review systems

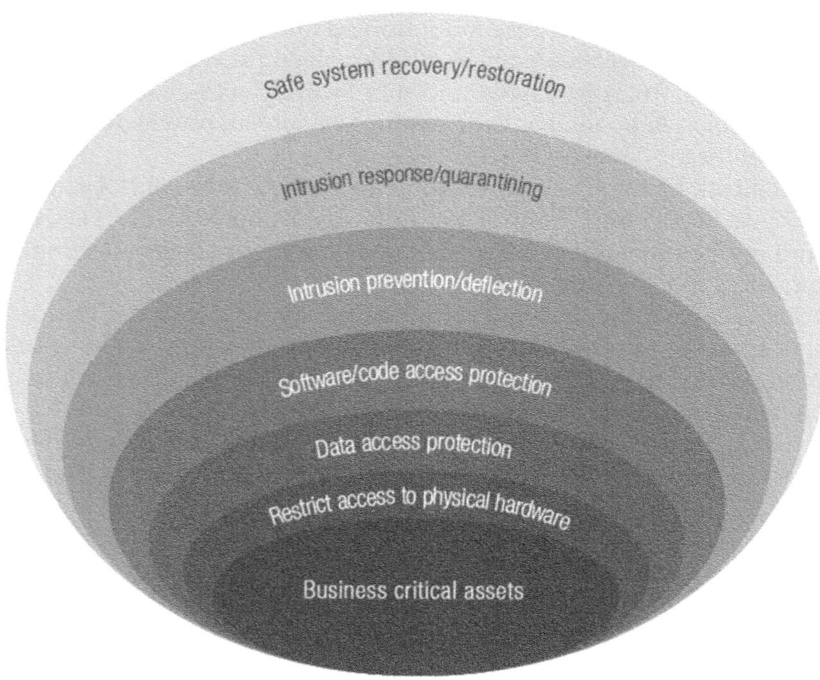

FIGURE 11.4 Example layers of defence that can be deployed against cyber security threats.

The identification and assessment of cyber security risks can also be done using the hazard analysis process that produces risks for a risk register as shown in Fig. 11.2. An example of some risk treatment options for the very high risks associated with unauthorised assess is shown in Fig. 11.3.

The management of cyber security threats is often done with a defence-in-depth approach. An example of what these layers might entail is shown in Fig. 11.4. The layers of defence model can then be used to guide bowtie (or similar) analysis to determine what controls are going to be added into each layer to address both internal and external threats. The bowtie analysis would also require a control support analysis to determine the support and verification activities required to ensure the controls are maintained to deliver effective prevention and mitigation against cyber threats. It is also critically important that the controls (e.g., access protections and virus detection software) are always being improved to address dynamic changing and emerging threats.

11.4 Summary

Security risk refers to risks posed by external threats to both the OT (physical security) and IT (cyber security). Security risks are now high priority for businesses. Recent industry surveys have highlighted that 77% of companies surveyed ranked cyber security as a major priority, and 58% of organisations surveyed had experienced a security breach in their control system in the previous 12 months.

Physical security risks include: (1) unauthorised persons accessing sites; (2) noncompliant goods coming onto site; (3) external man-made events; (4) extreme weather events; and (5) crime and civil unrest. Cyber security risks include: (1) external entities taking over control systems; (2) ransomware infiltration of computer systems for extortion; and (3) theft, destruction or modification of data.

We have presented many contemporary examples of security risks that have resulted in significant consequences for process companies. Consequences reported include fires and explosions, fatalities, product losses, major asset damage and huge economic losses.

Whilst materially very different, the approaches adopted to manage physical and cyber security risks should be consistent with the ISO31000 framework. That is, the scope must be specified, the risks identified and assessed, and then the risk treatments implemented. We have shown that for both physical and cyber security risks, risk treatment can be informed by adopting inherently safer design and defence-in-depth principles. We have also illustrated that tools such as the risk register and bowtie can also be very helpful for assisting in the assessment and treatment of security risks.

Review questions

1. What do IT and OT refer to?
2. State three examples of physical security risks.
3. Security risks are becoming more relevant, but still not high priority for companies in the process industries. Comment.

4. What types of approaches should be employed for designing control treatments for security risks? Hint: Consider approaches introduced in Chapter 5.
5. Write short summaries of two contemporary incidents that have happened as a result of security risks.

References

ABC. (2015). *France 'terrorist' attack on military foiled; chemical plant explosions a 'criminal act' with no link established.* ABC News. July 16, 2015. https://www.abc.net.au/news/2015-07-16/france-terrorist-attack-on-military-foiled/6623300.

ABC7.com staff. (2018). *Bacteria and feces found in fake cosmetics similar to popular makeup brands.* ABC13 Eyewitness News. April 15, 2018. https://abc13.com/lapd-los-angeles-police-department-makeup-counterfeit/3342874/.

Allianz. (2021). *Allianz risk barometer: Identifying the major business risks for 2021.* Allianz. https://agcs.wufoo.com/forms/download-the-allianz-risk-barometer-2021-report/.

BBC. (2014). *Hack attack causes 'massive damage' at steel works.* BBC News. December 22, 2014. https://www.bbc.com/news/technology-30575104.

BBC. (2015). *'Dangerous' chemicals stolen in Northamptonshire raid.* BBC News. April 8, 2015. https://www.bbc.com/news/uk-england-northamptonshire-32216115.

BDO. (2017a). *2017 BDO Manufacturing Riskfactor Report.* BDO USA. https://www.bdo.com/getattachment/07362692-3d80-4640-ac0a-d310bc60b946/attachment.aspx?2017-Manufacturing-Riskfactor-Report-Brochure_WEB.pdf.

BDO. (2017b). *2017 BDO Oil & Gas Riskfactor Report.* BDO USA. https://www.bdo.com/getattachment/a1bf67be-1beb-42b1-8f0c-f3db2446c6ed/attachment.aspx?2017-Oil-Gas-Riskfactor-Report-Brochure-WEB.pdf.

Berbotto, A. A., & Chainey, S. (2021). Theft of oil from pipelines: An examination of its crime commission in Mexico using crime script analysis. *Global Crime*, 1–23. Available from https://doi.org/10.1080/17440572.2021.1925552.

Bereola, L. (2015). *See what happens when a snake causes a power outage.* https://www.actionnewsjax.com/news/local/see-what-happens-when-snake-causes-power-outage/27093839/.

Blackstone, E. A., Fuhr, J. P., Jr, & Pociask, S. (2014). The health and economic effects of counterfeit drugs. *American Health and Drug Benefits, 7*(4), 216–224, PMID: 25126373; PMCID: PMC4105729.

Bloor, M., Boyle, T., & Chintapatla, T. (2022). Failures, repeated - the Tianjin explosion. *Loss Prevention Bulletin IChemE, 286*, 17–18. Available from https://www.icheme.org/media/18714/lpb286_pg17.pdf.

Brindle, F. (2016). *Taji gas processing facility back online. Hydrocarbon Engineering.* May 18, 2016. https://www.hydrocarbonengineering.com/gas-processing/18052016/gas-processing-facility-in-taji-iraq-restarts-production-after-attacks-3338/.

Burke, J. (2020). *India's chemical plant disaster: Another case of history repeating itself.* The Guardian. May 8, 2020. https://www.theguardian.com/world/2020/may/07/indias-chemical-plant-disaster-another-case-of-history-repeating-itself.

CSB. (2007). *Investigation report: Refinery explosion and fire.* Report No. 2005-04-I-TX. US Chemical Safety and Hazard Investigation Board.

CSB. (2018). *Organic peroxide decomposition, release, and fire at Arkema Crosby Following Hurricane Harvey flooding.* Report No. 2017-08-I-TX. Chemical Safety and Hazard Investigation Board.

Dawson, D. M., & Brooks, B. (1999). *The Esso Longford Gas Plant Accident: Report of the Longford Royal Commission.* Melbourne: Parliament of Victoria.

Doliner, A. (2021). *Entire town nearly left without power after snake started electrical fire.* https://www.newsweek.com/entire-town-nearly-left-without-power-after-snake-started-electrical-fire-1629986.

Ernst & Young. (2018). *Top 10 business risks facing mining and metals in 2019-20.* EYGM Ltd. https://assets.ey.com/content/dam/ey-sites/ey-com/en_gl/topics/mining-metals/mining-metals-pdfs/ey-top-10-business-risks-facing-mining-and-metals-in-2019-20-v2.pdf.

Farshbaf, N. (2014). *Bear attacks, kills oil sands worker in Canada.* CNN. May 8, 2014. https://edition.cnn.com/2014/05/08/world/canada-fatal-bear-attack/index.html.

Felicien, T. (2017). *Snake causes power outage for thousands in Greenville Wednesday morning*. Greenville News. July 5, 2017. https://www.greenvilleonline.com/story/news/crime/2017/07/05/thousands-without-power-greenville-wednesday-morning/450835001/.

Flock, E. (2021). *The pipeline saboteurs*. The Economist. November 11, 2021. https://www.economist.com/1843/2021/11/11/the-pipeline-saboteurs.

Fortinet. (2021). *Independent study finds that security risks are slowing IT-OT convergence. Fortinet*. May 23, 2021. Forrester and Fortinet report pinpointing SCADA & ICS Security Risks.

Franck, A. (2019). *Saudi Arabia – An object lesson in the importance of guarding against low-tech threats*. Hazardex. September 19, 2019. https://www.hazardexonthenet.net/article/173858/Saudi-Arabia-An-object-lesson-in-the-importance-of-guarding-against-low-tech-threats.aspx).

Fuller, C. (2020). *Car crashes into Awatoto fertilizer factory, starting large sulphur fire*. Hawke's Bay Today. December 13, 2020. https://www.nzherald.co.nz/hawkes-bay-today/news/car-crashes-into-awatoto-fertiliser-factory-starting-large-sulphur-fire/6P3EA6XBFJ7UP7IL4OOQDNS5PE/.

Geiger, G. (2021). *Cyber attack sabotages world's largest meat processor*. Motherboard Tech by Vice. https://www.vice.com/en/article/m7eb7a/cyber-attack-sabotages-worlds-largest-meat-processor.

Grossman, A., & Nicas, J. (2015). *White House drone crash said to be 'recreational'*. Wall Street Journal. January 26, 2015. https://www.wsj.com/articles/drone-crashes-on-white-house-grounds-1422286164.

Hambling, D. (2021). *The next 9/11 could be a mass drone attack. Here's how we can stop it*. Forbes. September 10, 2021. https://www.forbes.com/sites/davidhambling/2021/09/10/why-the-next-911-could-be-a-mass-drone-attack-and-how-we-can-stop-it/?sh = 79d524975878.

Hazardex. (2014). *Russian hackers now thought to have caused 2008 Turkish oil pipeline explosion*. Hazardex. December 21, 2014. https://www.hazardexonthenet.net/article/88497/Russian-hackers-now-thought-to-have-caused-2008-Turkish-oil-pipeline-explosion.aspx.

Hazardex. (2019). *Indian nuclear power plant detects cyber attack*. Hazardex. November 1, 2019. https://www.hazardexonthenet.net/article/175519/Indian-nuclear-power-plant-detects-cyber-attack.aspx.

Hazelwood Mine Fire Inquiry. (2015). *Hazelwood mine fire inquiry report 2015/2016 Volume I – Anglesea mine*. Hazelwood Mine Fire Inquiry. Government of Victoria.

Hydrocarbon Publishing. (n.d.). *Refinery power outage mitigations*. https://www.hydrocarbonpublishing.com/ReportP/power.pdf.

Iaiani, M., Tugnoli, A., Bonvicini, S., & Cozzani, V. (2021). Analysis of cybersecurity-related incidents in the process industry. *Reliability Engineering & System Safety*, *209*, 107485. Available from https://doi.org/10.1016/j.ress.2021.107485.

Jasi, A (2021). Major US pipeline restarts following cyberattack. *The Chemical Engineer*, 14th May 2021.

Kissel, R. (Ed.), (2013). *Glossary of key information security terms. National Institute of Standards and Technology (NIST) U.S.* Department of Commerce. https://nvlpubs.nist.gov/nistpubs/ir/2013/nist.ir.7298r2.pdf.

Kong, L., Das, D., & Pecht, M. G. (2022). The distribution and detection issues of counterfeit lithium-ion batteries. *Energies*, *15*(10), 3798. Available from https://doi.org/10.3390/en15103798.

Manno, A. (2021). *Four people die after $3M private jet crashes into industrial laser factory in Connecticut - sending flames into the sky and sparking evacuation of surrounding area*. Daily Mail Australia. September 3, 2021. https://www.dailymail.co.uk/news/article-9951729/Small-planes-crashes-industrial-factory-Connecticut-sending-flames-shooting-sky.html.

Mathews, L. (2021). *Oil Gian Shell Victimized IN December 2020 Hack*. Forbes. March 23, 2021. https://www.forbes.com/sites/leemathews/2021/03/23/oil-giant-shell-victimized-in-december-2020-hack/?sh = 4f4169b514f2.

Palmer, E. (2015). Regulator halts Catalent production at French plant where product sabotage suspected. *Fierce Pharma*. Available from https://www.fiercepharma.com/manufacturing/regulator-halts-catalent-production-at-french-plant-where-product-sabotage-suspected.

Palmer, E. (2016). Catalent France plant that was victim of sabotage getting back online. *Fierce Pharma*. Available from https://www.fiercepharma.com/partnering/catalent-france-plant-was-victim-of-sabotage-getting-back-online.

Reuters Staff (2012). *NJ Phillips Bayway refinery still idle, assessing damages from Sandy*. Reuters. November 4, 2012. https://www.reuters.com/article/us-storm-sandy-phillips-bayway-idUSBRE8A20A320121103.

RISI. (2011). *Water utility hack destroys pump. The respository of industrial security incidents*. https://www.risidata.com/Database/Detail/water-utility-hack-destroys-pump.

Robles, R., & Perlroth, N. (2021). *'Dangerous stuff': Hackers tried to poison water supply of Florida Town*. The New York Times. https://www.nytimes.com/2021/02/08/us/oldsmar-florida-water-supply-hack.html.

Royal Commission on the Pike River Coal Mine Tragedy. (2012). *Royal Commission on the Pike River Coal Mine Tragedy Volume 1 and Overview*. Wellington, New Zealand: Government of New Zealand.

Sanderson, K. (2006). *Fake pesticides pose threat*. Nature. November 5, 2006. https://www.nature.com/articles/news061030-14.pdf.

Schwab, W., & Poujol, M. (2018). *Trend study – The state of industrial cybersecurity 2018*. Kaspersky Lab & CXP Group. https://ics.kaspersky.com/media/2018-Kaspersky-ICS-Whitepaper.pdf.

Sieff, K. (2019). *Attempts to steal fuel from Mexian pipeline set off massive fireball, killing at least 73*. Washingtonpost.com. January 19, 2019. https://www.washingtonpost.com/world/the-americas/attempts-to-steal-fuel-from-mexican-pipeline-sets-off-massive-fireball-killing-at-least-66/2019/01/19/a952b606-1bf0-11e9-b8e6-567190c2fd08-story.html.

Stewart, W. (2017). *Eaten alive: Oil worker is disembowelled by bear that buried the remains and returned to feast on the body*. The Scottish Sun. September 19, 2017. https://www.thescottishsun.co.uk/news/1583274/oil-worker-is-disembowelled-by-bear-that-buried-the-remains-and-returned-to-feast-on-the-body/.

Summers, N. (2017). *Code rouge: Cracking the case of the counterfeit makeup*. Bloomberg Asian Edition. https://www.bloomberg.com/features/2017-counterfeit-makeup/.

Symantec. (2017). *Dragonfly: Western energy sector targeted by sophisticated attack group*. Symantec. October 20, 2017. https://symantec-enterprise-blogs.security.com/blogs/threat-intelligence/dragonfly-energy-sector-cyber-attacks.

Tanner, J., & Hunter, T. (2021). Digital risk management: It's everyone's responsibility. *The Chemical Engineer*, 19–20, June 2021. The Institution of Chemical Engineers.

Trager, R. (2015). *Failed terror attack raises alarms about chemical plant security*. Chemistry World. https://www.chemistryworld.com/news/failed-terror-attack-raises-alarms-about-chemical-plant-security/8708.article.

WAtoday. (2014). *Dangerous chemicals which could prove fatal, stolen from Perth terminal*. https://www.watoday.com.au/national/western-australia/dangerous-chemicals-which-could-prove-fatal-stolen-from-perth-terminal-20140409-36cp8.html.

WMBF News Staff. (2019). *$300K worth of chemicals stolen from Georgetown facility*. https://www.wmbfnews.com/2019/08/19/report-k-worth-chemicals-stolen-georgetown-facility/.

World Nuclear Association. (2022). *Fukushima Daiichi accident*. World Nuclear Association. Updated May 2022. https://world-nuclear.org/information-library/safety-and-security/safety-of-plants/fukushima-daiichi-accident.aspx.

Xinhua. (2019). *Hong Kong police seize 59 bottles of dangerous chemicals stolen by rioters from university laboratories*. Xinhuanet. http://www.xinhuanet.com/english/2019-12/06/c_138611184.htm.

CHAPTER

12

Supply chain risks

OUTLINE

12.1 Introduction 258

12.2 Process industry supply chain risks 258

12.3 Managing supply chain risks 259

12.4 Case study examples 261

12.5 The changing paradigm of supply chain risk management 266

12.6 Summary 267

Review questions 267

References 268

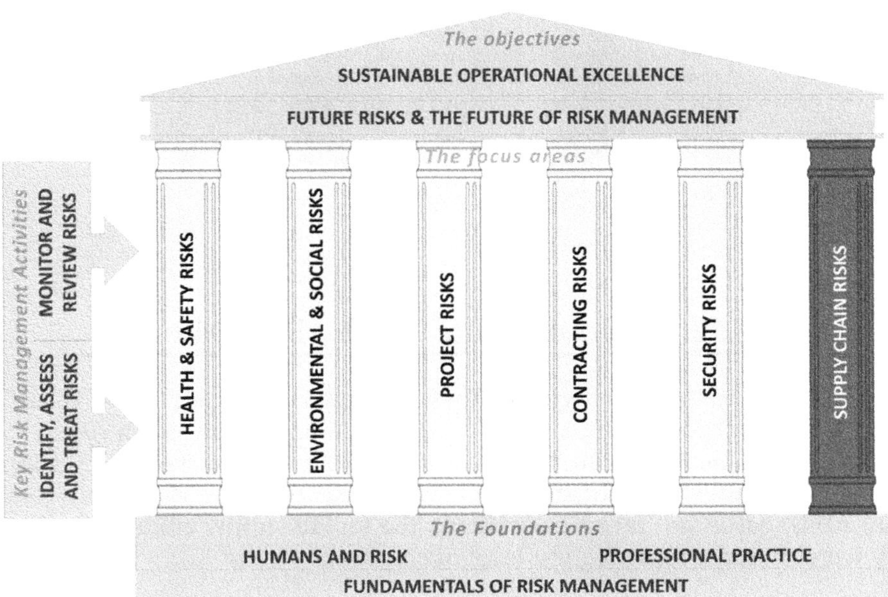

Fundamentals of Risk Management for Process Industry Engineers
DOI: https://doi.org/10.1016/B978-0-12-820320-0.00015-5

You will not find it difficult to prove that battles, campaigns, and even wars have been won or lost primarily because of logistics Dwight D. Eisenhower.

12.1 Introduction

A supply chain is a system comprising the people, plant, processes and subsystems required to transform raw materials into finished products, and to deliver them to the customer (Fig. 12.1). Process companies (manufacturers) are pivotal players in complex global supply chains. They are both customers, receiving materials, people, goods and equipment, and also suppliers providing products to customers and consumers. Some companies are also the distributor as they transport their goods to customer sites.

We saw in Chapter 1 that supply chain risk is a high priority for industry, recently ranking in the top 10 risks for both the manufacturing (BDO, 2017a) and oil and gas (BDO, 2017b) sectors.

In this chapter we first discuss the range of risks that could threaten the achievement of supply chain objectives. We then present some case study examples to illustrate the types of impacts these risks can have for processing companies. We conclude by discussing how these risks can be managed.

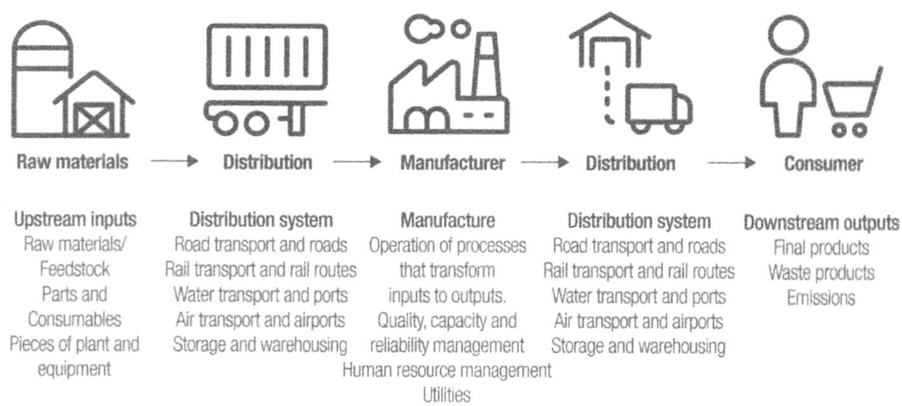

FIGURE 12.1 Overview of a supply chain.

12.2 Process industry supply chain risks

Supply chains are complex and global. They are exposed to many risks. Table 12.1 uses the PESTLE framework to identify a range of contemporary supply chain risks.

The Supply Chain Institute from the University of Tennessee surveyed 150 supply chain executives across multiple industries and identified the top 10 supply chain risks (Dittmann, 2014). Based on 'rating of concern', the top 10 supply chain risks identified in that study were (in order from highest level of concern):

- Quality

TABLE 12.1 Example of supply chain risks.

Category	Examples
P: Political factors	Changes to taxes/tariffs, sanctions, border closures. Global trade wars.
E: Economic factors	Impact on prices due to economic conditions (e.g. recession, inflation) and instabilities. Circular economy risks. Solvency of contracted suppliers.
S: Social factors	Changes in customer demand. Availability and cost of utilities, e.g. energy, water, waste disposal, etc. Activist/Terrorist threats. Not meeting UN Global Compact's Ten Principles or using ethical suppliers – ones that do not use child labour, slavery or contravene employees human rights. Cargo theft. Security risks.
T: Technology factors	Supplier and customer shutdowns and disruptions. Transport losses. Cyber risks. Quality of supplies and products.
L: Legal factors	Changes to regulations, e.g. labour laws, for transportation etc. Contract issues.
E: Environmental factors	CO_2 emission impact of supplies and suppliers. Management of waste packaging and generated wastes. Natural disaster risks to supply chains (e.g. floods, cyclones/hurricanes, super storms, earthquakes, bushfires, extreme weather events, volcanic eruptions).

- Inventory
- Natural disasters
- Economics
- Transit loss
- New product delay
- Cyber security
- Intellectual property
- Political instability
- Customs
- Terrorism

12.3 Managing supply chain risks

The management of supply chain risks should follow the risk management approaches presented in earlier chapters. While the risks may change, and in this case may be complex and have global influences, our approach for addressing them is unchanged. You should now be very familiar with this framework, namely: (1) set the scope; (2) assess the risk; (3) treat the risk; and (4) review and report.

In regard to setting the scope for supply chain risks, consideration should be given to suppliers, customers, and the manufacturing business. Consideration should also be given to short-term acute risks and long-term chronic legacy risks as illustrated in Table 12.2.

The identification of the risks should reference existing internal and external information, relevant incidents, as well as involve a foresighting exercise to identify relevant emerging risks. These risks can then be ranked using a risk assessment matrix to determine the risks of concern. The risks can then undergo risk treatment analysis as discussed in Chapter 5.

It is important to highlight that supply chain risks can result in both adverse and positive outcomes. This was very clearly illustrated during the global COVID-19 pandemic. They can also can be unknown and/or result from unexpected events. As Kerry Ingalls, VP of Amgen a pharmaceutical manufacturing company, stated after Hurricane Maria:

> Crises don't fol>low the script. Weird things happen. (**Jarvis, 2018**).

It thus follows that for supply chain risks, resilience initiatives should also be considered (Fig. 12.2). A resilient supply chain has the 'adaptive capability of the supply chain to prepare for unexpected events, respond to disruptions, and recover from them by maintaining continuity of operations at the desired level of connectedness and control over structure and function' (Ponoarov & Holcomb, 2009, p. 131).

The importance of considering supply chain resilience has been highlighted by McKinsey (2020) who found that 93% of global supply chain leaders are planning to increase resilience. Resilient supply chains are underpinned by forecasting, anticipating

TABLE 12.2 Example scope table for supply chain risks.

	Supplies considerations		**Product considerations**	
Description	**Included**	**Excluded**	**Included**	**Excluded**
People: People involved in management of the risk and people potentially impacted if risks are not managed				
Locations: Locations or areas where the risk exists or that could be impacted by the risk event if it occurs				
Equipment: Equipment and plant exposed to the risk and/or involved in managing the risk				
Activities: Activities include normal and abnormal tasks and functions exposed to, could impact or be impacted by risk				
Timeframes: To capture time considerations that risk needs to be managed	Short-term/ Acute: Long-term/ Chronic:	Short-term/ Acute: Long-term/ Chronic:	Short-term/ Acute: Long-term/ Chronic:	Short-term/ Acute: Long-term/ Chronic:
Scenarios: Known risk scenarios that need to be considered				
Other assumptions: Captures other assumptions and issues being considered				

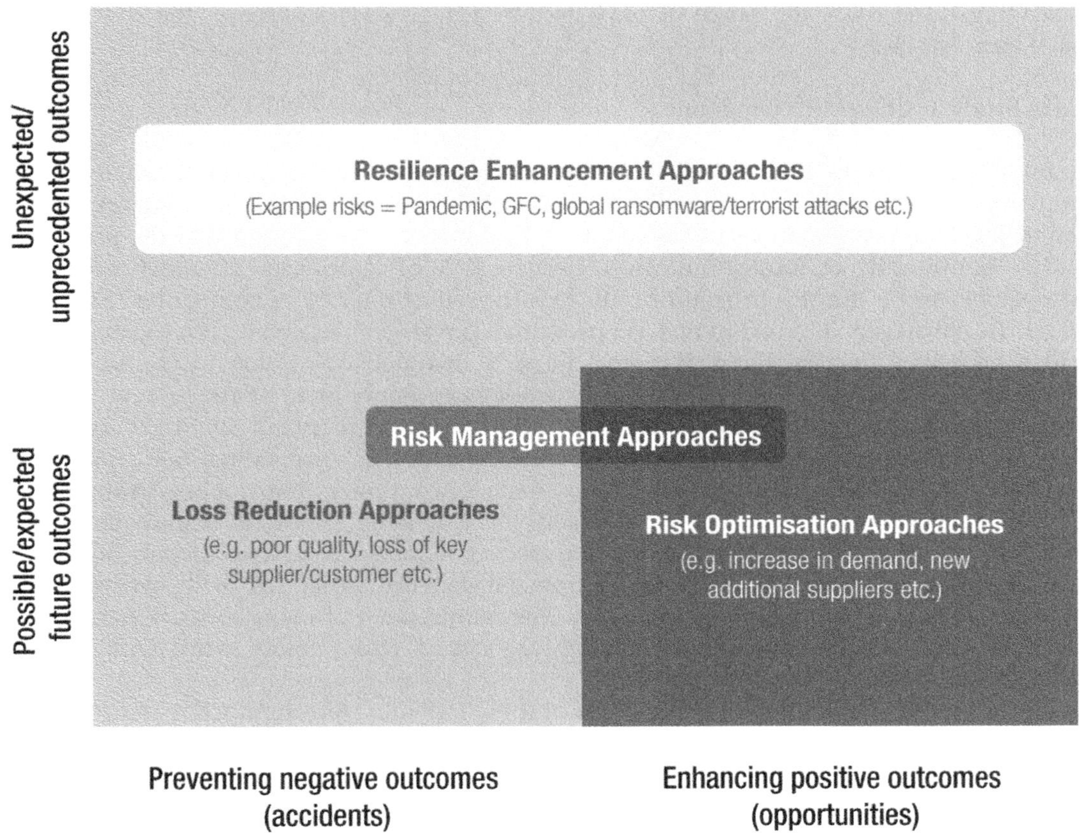

FIGURE 12.2 Example of supply chain risks on scope of risk and uncertainty management.

and quickly responding to future risks – both threats and opportunities (Fu, Chien, 2019). They understand and leverage surveillance data to create plans to control known, or respond to emergent, risks. Such responses include (Gartner, 2021; McKinsey, 2020):

- Having diversified options for suppliers and customers (e.g. a number of different suppliers and customers onshore as well as in different regions offshore).
- Having a range of options for where products are manufactured.
- Implementing agile production capacity.
- Ensuring inventory buffer capabilities that can absorb shocks.

12.4 Case study examples

In this section we present some supply chain risk case studies associated with the process industries. Through these case studies we hope to illustrate the importance of managing these risks, the range of impacts they can have and some of the lessons that have been learned.

Case Study 1: COVID-19 pandemic

In 2020 through to 2022 the COVID-19 pandemic impacted the world and had significant impact on processing companies and associated supply chains. In a survey of 200 senior-level supply chain executives in late 2020, Ernst & Young found that the pandemic had a 'significantly or mostly' negative effect on 72% of companies, no effect on 18% of companies, and a 'mostly or significantly' positive effect on 11% of companies (Harapko, 2021). In particular, 97% of 'industrial products companies' reported that the pandemic had a negative effect on them. Negative impacts included disruption to the workforce, where many employees were requested to work from home or to adapt to new requirements for physical spacing and enhanced PPE, and lower demand for many products such as restaurant supplies, footwear, apparel, accessories, furnishings and appliances (Wilson, 2020). These products often use raw materials supplied from process industries.

Conversely, there was increased demand for sanitising products, pandemic-related PPE, COVID tests and vaccines. Consumer grocery items such as toilet paper, flour, rice and canned goods were also exposed to unusual demand variations, with high demand experienced at the beginning of lockdowns. The requirement of many to work from home also increased the demand for technology associated with remote work and internet services.

The consequential supply chain impacts of the pandemic were significant and included (Statista, 2021):

- Delays in cross border and domestic land transportation
- Severe delays in acquiring critical supplies
- Inability to acquire critical supplies
- Delays in, and capacity problems with, air transportation
- Delays in sea transportation
- Absences in supply chain staff
- Insufficient stockpiles of goods
- Critical suppliers went into liquidation
- Significant increases in freight costs

As we write this chapter, the pandemic is still very much a global concern, and supply chain impacts are still ongoing. The future will provide many valuable insights and learnings for how to better prepare for, and manage, supply chains in the events of major global disruptive events. In the meantime, lessons learned so far from COVID-19 that relate to supply chain risks include:

- Planning must be based on vulnerability or impact rather than frequency. 'The pandemic underscored the imperative of manufacturers and supply chain partners to do more than plan for infrequent and 100-year events' (Witham, 2021).
- Just-in-time inventories philosophies should be revisited using risk-based approaches using knowledge gained from pandemic-related shortages to determine whether the

just-in-time benefits are worth the trade-off when faced with future disruptions (Witham, 2021).
- 'Re-imagining your supply chain model' to review what work is preferably done locally, regionally or globally (Harapko, 2021).
- Reviewing the supply chain footprint to ensure diversified sources of supply (Deconinck et al., 2020; Harapko, 2021).

Case Study 2: Pharmaceutical company's response to Hurricane Maria

A number of the world top pharmaceutical products and biologic drugs are manufactured in Puerto Rico. These facilities were impacted in a range of ways when Hurricane Maria, a devastating hurricane, hit Puerto Rico on September 20, 2017. A summary of the insights relating to supply chain risk management from Jarvis (2018), and McKinsey (2020) are described next.

The hurricane damaged and flooded buildings and crippled the islands power supply and communication systems. This affected the production of 17 major drugs that were sold to the United States. In the aftermath of the hurricane, companies prioritised looking for employees. Due to the number of employees, the spread of residents and the status of the communication systems, this took weeks. For example, Amgen and Johnson & Johnson took nearly a month to locate all of their employees. Some of these employees were without food, gas or electricity and could not stay in the damage affected areas. One of the biggest lessons learned was the need to re-establish society to ensure food, shelter and order after the chaos (Jarvis, 2018). Emergency supplies, generators and gas stoves had to be shipped in. Another key learning was that 'Among the issues laid bare by the storm was the widespread assumption that communications and operations would continue so long as power was on. Everyone planned for plants to run according to paper versions of the standard operating procedures' (Jarvis, 2018). The lack of internet impacted interconnected systems, thus preventing the start-up of operations and tracking of inventory (Jarvis, 2018). Purified gas, required to keep delicate cell lines alive, was also in short supply and had to be shipped in from Europe (Jarvis, 2018). According to a source 'one company risked losing $200 million in material as it got down to its last two days of oxygen' (Jarvis, 2018). Post Hurricane Maria, industry executives conceded that more severe natural disasters could occur due to climate change. This raised questions of how the drug companies should prepare to mitigate big natural disaster events to avoid disastrous drug shortages. The result was that the manufacturers identified several key areas where improvements could mitigate, but not eliminate risks. These included:

- Risk map important supplies and suppliers based on criticality of the item to operations, how easily the item can be substituted with an alternative and how easy the item and/or alternatives can be source from another supplier.
- Develop more robust power and communication with reliable backup systems that deliver in extended outages.
- Develop a 'plant in a box' that is 'able to run our quality systems, our maintenance system, our production system within the plant without connectivity' (Jarvis, 2018).

- Work with government to improve the coordination of relief efforts and infrastructure needed to enable effective relief activities.
- Have emergency supplies set aside in a location away from the affected area.
- Review the entire supply chain for critical drugs to ensure that countries like the United States do not suffer disastrous drug shortages as a result of future disruptive events.

Case Study 3: Global chip shortage 2020–2022

Digitalisation means that more companies are dependent on semiconductor chips. In 2020 and 2021, the demand for semiconductor chips outstripped supply. This was a consequence of several coincidental major disruptive events including COVID-19, a major China-US trade war, and the closures of several manufacturing facilities as a result of severe weather events and fires. The impact was felt by global supply chains (Shein, 2021). The supply shortage impacted 169 industries including those supplying and buying electronic safety systems (e.g. sensors, instruments, display panels), computers, screens, phones, cars and other chip-enabled goods (Shein, 2021).

The lessons learned and documented so far include:

- Shortages in major supplies cannot be fixed quickly if increased capacity is the only solution. Increased factory capacity takes time (Fig. 12.3) especially considering it takes 3 months to produce a chip (Hanbury & Hoecker, 2021).
- Shortages due to technology supply chain disruptions will continue to occur and continue to impact multiple industries. Companies need to develop strategies that (Hanbury & Hoecker, 2021; Sood, 2021):
  - Assess the risk of suppliers and components. This should include considerations like lead time to add capacity, location of suppliers and demographics and demand projection of customers.
  - Build adaptability and redundancy capacity for high-risk supplies to effectively manage inventory in line with customer demands.

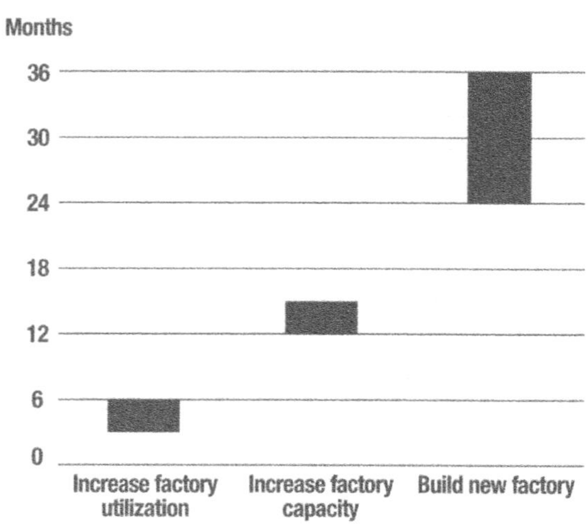

FIGURE 12.3 Time ranges required to increase capacity of microchip production. *Source: Data sourced from Hanbury, P., & Hoecker, A. (2021).* Two lessons the chip shortage taught us about supply chains. *September 20, 2021. Bain & Company. https://www.bain.com/insights/two-lessons-from-the-chip-shortage-supply-chains-tech-report-2021/.*

- Incorporate real-time, dynamic heat maps for high-risk supplies.
- Involve customers to help effectively manage inventory in line with customer demands.

Case Study 4: Energy crises

Energy in the form of gas, coal, and electricity is often used in the processing industries. The supply and therefore the price of energy can vary over time as shown in Fig. 12.4. When there is a significant gap between demand and supply this often causes significant increases in prices. For processing operations reliant on energy, this poses a significant risk that can threaten profits or in worst case scenarios shut processing plants down. Examples of reported cases of processing plants being shut down or suspend operations due to energy costs include:

- The Tata soda ash plant in Cheshire in the United Kingdom (Bounds & Powley, 2014).
- The BASF pigment plant in Paisley in the United Kingdom (Bounds & Powley, 2014).
- The New Zealand Tiwai Point aluminium smelter (Rio Tinto, 2020).
- Incitec Pivot Gibson Island fertilizer plant (Toscano, 2021).
- CF Industries two fertilizer plants in Billingham and Ince, United Kingdom (Elkin & Almeida, 2021).
- The Century Aluminium mill in Haweville, Kentucky, United States (Deaux & Malik, 2022).
- Belgium fertilizer producer Yara's Tertre, Wallonia operations (The Brussels Times, 2022).
- Aperam, stainless steel Genk site in Belgium (The Brussels Times, 2022).

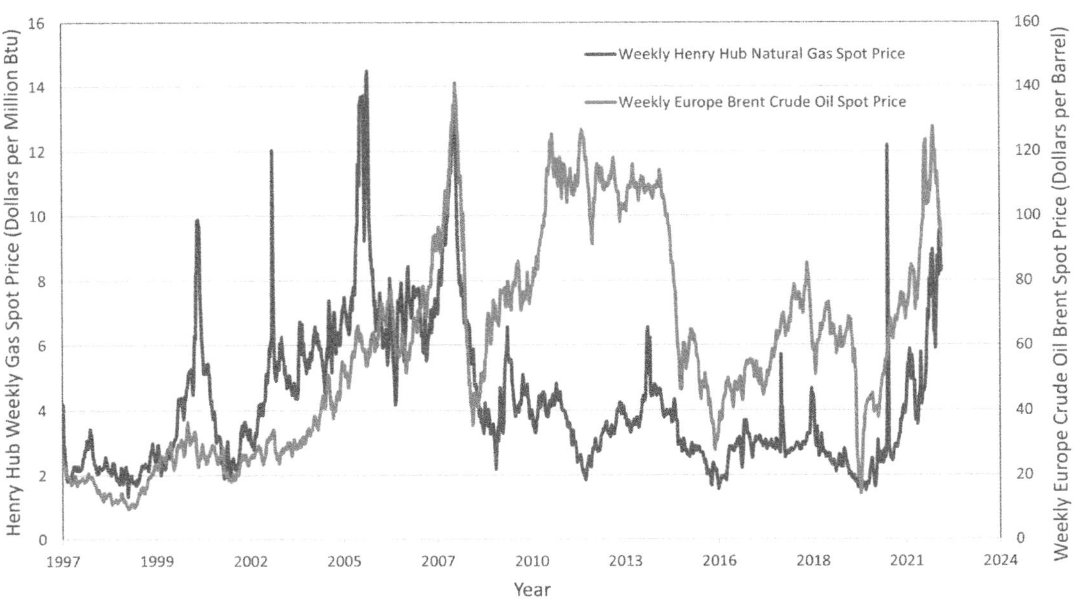

FIGURE 12.4 Weekly average spot price for gas and crude oil. *Source: Data sourced from http://www.eia.gov.*

Companies exposed to risks associated with supply and/or price of utilities can adopt a number of measures to prevent and mitigate such risks. These measures include:

- Having diversified supply options that may include their own source of renewable energy (e.g. solar and wind).
- Adopting energy-efficient technologies (Nichols, 2017). Choudhry et al. (2015) list examples of such technologies for different industries. Their list includes use of:
  - Advanced process control systems and membrane gas separation in oil refining and chemical processing.
  - Advanced thermomechanical pulping with heat recovery systems for pulp and paper industry.
  - Coke-dry quenching and endless-strip production in steelmaking.
  - Automated ventilation systems and high-pressure grinding for mining.
  - Combustion system improvements and high-efficiency grate coolers for the cement industry.

  Other examples include use of energy-efficient lights, appliances, plant and equipment.
- Avoiding unnecessary spikes in energy use (e.g. by staggering equipment start-ups) (Nichols, 2017).
- Reducing usage from grid during peak demand periods (Nichols, 2017). This could be done by reducing actual usage or by drawing from energy stored in batteries.
- Adopting energy-efficient practices including monitoring and addressing waste, maintaining equipment to ensure it maintains its efficiency, shutting down idle equipment and shutting down lighting and air conditioning systems when not in use (Nichols). A bowtie analysis can be used to illustrate the preventative and mitigation measures relevant to a specific company. Performing a control assurance management system analysis would also highlight what was required and who was responsible for implementing, monitoring and maintaining the prevention and mitigation measures. However, any such analysis should be conducted in a manner that is congruent with the changing paradigm of supply chain risk management. This is discussed next.

12.5 The changing paradigm of supply chain risk management

As we have seen in the preceding sections and case studies, supply chain risk management is still an evolving field of study and practice. A synopsis of how the dominant paradigm has changed was presented by Deloitte (2015), and is summarised in Table 12.3.

The findings and recommendations from the preceding case studies are consistent with the 'emerging approach'. Strategies for effectively managing supply chain risk should employ a broad focus and include third-party partnerships, focus on strategic agility, seek to identify and optimise opportunities as well as mitigating threats, and the senior executive team need to own the risk and be accountable for managing it.

TABLE 12.3 Changing supply chain risk paradigm.

Traditional approaches	Emerging approaches
Narrow focus on supply chain management	Ecosystem approach that leverages third parties, support services, business partners, strategic alliances and subsidiaries for marketing, sales and distribution
Cost saving is priority driver	Strategic agility, competitive advantage and access to special skills also considered as priority drivers
Threat is primary risk	Both threats and opportunities considered as part of risk analyses
Focus on technology to enable connectivity	Extending connectivity to include virtual networking and collaboration
Accountability held at operational levels	Accountability held at all levels including Executive and Board room levels
Limited analysis of stakeholder impact	Extended analysis based on premise that supply chains can have a significant impact on stakeholders

From Deloitte (2015). Third party governance and risk management – Turning risk into opportunity. Deloitte Touche Tohmatsu Limited. https://aicd.companydirectors.com.au/-/media/cd2/resources/advocacy/governance-leadership-centre/pdf/deloitte-uk-third-party-governance-risk-management.ashx.

12.6 Summary

Supply chain risk is a high priority for industry. This has been further emphasised by contemporary major disruptive events such as COVID-19 and Hurricane Maria. Processing companies are pivotal players in complex global supply chains and can be significantly impacted by supply chain disruptions. Priority supply chain risks include economic and political instability, quality and inventory of supplies, regulatory changes, cyber security, terrorism and a broad range of natural disasters.

Despite supply chain risks being complex and often globally significant, our approach for addressing them should be consistent with the standard risk management framework that features throughout this book. The challenge, and opportunity, lies in how to effectively apply the framework to identify, assess, analyse and treat the major disruptive events that impact supply chains. The case studies featured in this chapter have helped to elucidate contemporary learnings and strategies for managing supply chain risks.

By definition, supply chain risks are strategic risks, and the senior executive team need to own them and be accountable for managing them. However, as we have seen through contemporary case studies, supply chain risks can directly impact process operations, and many of the 'risk treatment controls' are also operational. It is thus very important that process engineers have an appreciation of the significance of supply chain risks, are able to incorporate them in their risk identification exercises, and can assess, analyse and effectively treat them.

Review questions

1. List 10 supply chain risks and classify them using the PESTLE framework.
2. List five significant impacts that industry may experience due to supply chain disruptions.

3. How has the supply chain risk paradigm changed, and what are the key aspects of emerging approaches?
4. List five key lessons learned from contemporary case studies of supply chain disruption
5. What is a resilient supply chain? State three characteristics of a resilient supply chain.

References

BDO. (2017a). *2017 BDO manufacturing riskfactor report*. BDO USA. https://www.bdo.com/getattachment/07362692-3d80-4640-ac0a-d310bc60b946/attachment.aspx?2017-Manufacturing-Riskfactor-Report-Brochure_WEB.pdf.

BDO. (2017b). *2017 BDO oil & gas riskfactor report*. BDO USA. https://www.bdo.com/getattachment/a1bf67be-1beb-42b1-8f0c-f3db2446c6ed/attachment.aspx?2017-Oil-Gas-Riskfactor-Report-Brochure-WEB.pdf.

Bounds, A., & Powley, T. (2014). *UK factories shut down to avoid high power costs*. Financial Times. February 19, 2014. https://www.ft.com/content/cdf768e0-98b4-11e3-8503-00144feab7de.

Choudhry, H., Lauritzen, M., Somers, K., & Van Niel, J. (2015). *Greening the future: New technologies that could transform how industry uses energy*. McKinsey & Company. https://www.mckinsey.com/~/media/mckinsey/business%20functions/operations/our%20insights/technologies%20that%20could%20transform%20how%20industries%20use%20energy/greening_the_future.pdf.

Deaux, J., & Malik, N. S. (2022). *The US Industrial Complex Is Starting to Buckle From High Power Costs*. Bloomberg Asia Edition. July 7, 2022. https://www.bloomberg.com/news/articles/2022-07-07/high-power-costs-force-us-industrials-factories-to-shut?leadSource = uverify%20wall.

Deconinck, K., Avery, E., & Jackson, L. A. (2020). Food supply chains and Covid-19: Impacts and policy lessons. *EuroChoices, 19*, 3.

Deloitte (2015). *Third party governance and risk Mmanagement – Turning risk into opportunity*. Deloitte Touche Tohmatsu Limited. https://aicd.companydirectors.com.au/-/media/cd2/resources/advocacy/governance-leadership-centre/pdf/deloitte-uk-third-party-governance-risk-management.ashx.

Dittmann, J. P. (2014). *Managing risk in the global supply chain*. A Report by the Supply Chain Management Faculty at The University of Tennessee.

Elkin, E., & Almeida, I. (2021). *Soaring energy prices in Europe are forcing U.K. factories to shut down*. Bloomberg News. September 16, 2021. https://financialpost.com/commodities/energy/oil-gas/europes-energy-crunch-is-forcing-u-k-factories-to-shut-down.

Fu, W., & Chien, C. (2019). UNISON data-driven intermittent demand forecast framework to empower supply chain resilience and an empirical study in electronics distribution. *Computers & Industrial Engineering, 135*, 940–949. Available from https://doi.org/10.1016/j.cie.2019.07.002.

Gartner. (2021). *Leadership Vision for 2022, Chief Supply Chain—Officer Leadership Vision 2022-3 Strategic Actions for Success*. https://www.gartner.com/smarterwithgartner/6-strategies-for-a-more-resilient-supply-chain.

Hanbury, P., & Hoecker, A. (2021). *Two lessons the chip shortage taught us about supply chains*. Bain & Company. September 20, 2021. https://www.bain.com/insights/two-lessons-from-the-chip-shortage-supply-chains-tech-report-2021/.

Harapko, S., (2021). *How COVID-19 impacted supply chains and what comes next. EY*. February 18, 2021. https://www.ey.com/en_au/supply-chain/how-covid-19-impacted-supply-chains-and-what-comes-next.

Jarvis, L. (2018). Hurricane Maria's lessons for the drug industry. *Chemical & Engineering News, 96*(Iss 37). Available from https://cen.acs.org/pharmaceuticals/biologics/Hurricane-Marias-lessons-drug-industry/96/i37.

McKinsey. (2020). *McKinsey Global Institute 2020 Risk, resilience, and rebalancing in global value chains*. McKinsey Global Institute. https://www.mckinsey.com/business-functions/operations/our-insights/risk-resilience-and-rebalancing-in-global-value-chains; https://www.safetyonline.com/doc/kaisers-gramercy-facility-back-on-line-follow-0001.

Nichols, M. (2017). *Five ways to reduce energy waste in manufacturing processes*. Plant Engineering. August 23, 2017. https://www.plantengineering.com/articles/five-ways-to-reduce-energy-waste-in-manufacturing-processes/.

Ponomarov, S. Y., & Holcomb, M. C. (2009). Understanding the concept of supply chain resilience. *International Journal of Logistics Management, 20*(1), 124–143.

Rio Tinto. (2020). *NZAS terminates electricity contract and plans to wind-down operations following strategic review*. Rio Tinto News Release. July 9, 2020 https://www.riotinto.com/en/news/releases/2020/NZAS-terminates-electricity-contract-and-plans-to-wind-down-operations-following-strategic-review.

Shein, E. (2021). *Global chip shortage: Everything you need to know*. TechRepublic. November 22, 2021. https://www.techrepublic.com/article/global-chip-shortage-cheat-sheet/.

Sood, I. (2021). Three key lessons from the chip shortage crisis. *SupplyChainBrain*, *3*. Available from https://www.supplychainbrain.com/blogs/1-think-tank/post/33204-three-key-lessons-from-the-chip-shortage-crisis.

Statista. (2021). How has COVID-19 negatively impacted your supply chain? Statista. https://www.statista.com/statistics/1182233/impact-covid-19-supply-chain/.

The Brussels Times. (2022) *Spiralling energy costs force Belgian companies to shut down*. The Brussels Times. August 27, 2022. https://www.brusselstimes.com/278843/spiralling-energy-costs-force-belgian-companies-to-shut-down.

Toscano, N. (2021). Incitec to shut down manufacturing plant in 2022 as gas prices bite. The Sydney Morning Herald. November 8, 2021. https://www.smh.com.au/business/companies/incitec-to-shut-down-manufacturing-plant-in-2022-as-gas-prices-bite-20211108-p596wn.html.

Wilson, G. (2020). COVID-19: Five priorities for retail supply chain. *SupplyChain*, *17*, 2020. Available from https://supplychaindigital.com/supply-chain-2/covid-19-five-priorities-retail-supply-chain.

Witham, J. (2021). Supply chain lessons learned from the Covid-19 pandemic. *Forbes*, *9*, 2021. Available from https://www.forbes.com/sites/forbestechcouncil/2021/09/09/supply-chain-lessons-learned-from-the-covid-19-pandemic/?sh = 79d6dd3324ca.

CHAPTER

13

Future risks and the future of risk management

OUTLINE

13.1 Introduction 272

13.2 Future process industry risks 273

13.3 The future of risk management 276

13.3.1 Digital twins 279

13.3.2 Virtual reality/augmented reality technologies 279

13.3.3 Automation and autonomy 280

13.3.4 Big data and artificial intelligence 280

13.3.5 Industrial internet of things 282

13.4 Professional engineering practice into the future 282

13.5 Summary 283

Review questions 283

References 284

Fundamentals of Risk Management for Process Industry Engineers
DOI: https://doi.org/10.1016/B978-0-12-820320-0.00013-1

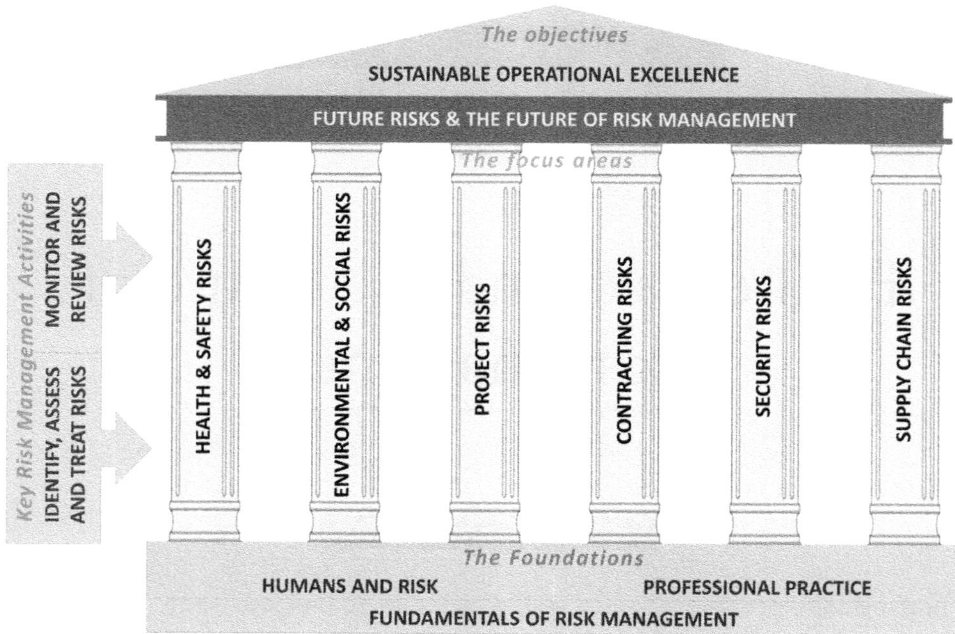

'You cannot improve your future if you are not willing to try something new and risk making mistakes and learning from them'. Robert Kiyosaki

13.1 Introduction

We started this book by proposing that risk management is one of the most critical tasks performed by process engineers, and that it is crucial for engineers to learn how to manage risks effectively. Having got this far in the book, we hope that you now feel far more aware of the vast array of risks that you will encounter and options for identifying and managing them. In this last chapter, we discuss future risks and the future of risk management.

In the future, the identification and management of risks is going to become more important, more integrated, broader in scope (spacially, timewise), more detailed, and more augmented with Industry 4.0 technologies. In the rest of this chapter we will consider emerging future risks, discuss the trends influencing these risks, and then also consider how risk management approaches may evolve as a consequence of, and in response to, the Industry 4.0 'revolution'. We will also identify several opportunities and threats that may emerge associated with future risk management approaches, especially as they pertain to Industry 4.0. We hope that this chapter will provide useful insights to better prepare you for the challenges associated with managing future risks in the workplace.

13.2 Future process industry risks

We saw in Chapter 1 that we live in a time where the future sustainability of the processing industries is increasingly being impacted by risks emanating from important global trends. These include:

1. Faster rate of change which is making the future more difficult to predict and gives business less time to respond (McKinsey & Company, 2021; Withers, Gupta, Curtis, & Larkins, 2015).
2. Globalisation that is increasing the interactions and interdependencies between businesses and means that risks and impacts from one region or business can quickly spread to others (Marsh & McLennan, 2015; Withers et al., 2015).
3. Increasing complexity of business processes, technologies and supply chains which increases the chance of hidden faults and unexpected outcomes (Marsh & McLennan, 2015).
4. Significant shifts in costs and demand of process industry supplies—both inputs and outputs. Examples include:
   a. increasing cost competitiveness associated with import alternatives and the corresponding loss of local suppliers and customers (Allianz, 2015; Aon, 2014; Mooney, 2014).
   b. energy transitions from fossil fuels to renewables, hydrogen and nuclear (McKinsey & Company, 2022)
   c. increasing demands for synthetically produced clothing, food, medical implants and medicines (EY, 2020).
5. Changing stakeholder norms and expectations (Allianz, 2015; Ernst & Young, 2015; EY, 2020; Poole & Sullivan, 2021) and regulatory requirements (Allianz, 2015; Aon, 2014; Poole & Sullivan, 2021).
6. Increasing pressure to deliver net zero carbon emissions targets (EY, 2020; Naughtin et al., 2022).
7. Transformational change within society and in industry delivered by Industry 4.0 technologies (Babier, 2017; Marr, 2022).
8. Increasing frequency and severity of climate change related impacts on weather, water supplies, food and the suitability of locations for dwellings and industry infrastructure (Manyika et al., 2020; Naughtin et al., 2022).

These trends will form the backdrop within which we will be managing risks into the future. Most of the risks mentioned in previous chapters have been current risks that are expected to continue into the future. In addition to these risks, there are potentially newly emerging risks. Listing these risks against a framework like the P-E-S-T-L-E framework as shown in Table 13.1 can help decision makers identify and document the range of risks that could impact their operations.

Different future risks can emerge or re-emerge at different rates and in different contexts. The identification of re-emerging and emerging risks should be done by reviewing relevant literature—both academic and grey literature. For re-emerging risks it can be done by looking at historical risks and incidents. Examples of such risks might include pandemics, black lung, natural disasters. For emerging risks, there are many publications that present perspectives on future risks. Examples include future risks reports by major consulting houses, major insurers and the World Economic Forum. The determination of

TABLE 13.1 Example of emerging risks that could impact process industries.

Category	Examples
P: Political factors	• Shift in energy sources from fossil fuels to renewables and hydrogen • Regulatory/legislative changes related to energy use, carbon and other emissions, process and personal safety, and sustainability goals
E: Economic factors	• Increasing expectations to contribute more to local and national economic development • Merger/acquisition/divestment/closure of operations • Changes in prices, demand and availability of supplies and products • Introduction and increase in taxes and charges relating to use of natural resources and production of emissions
S: Social factors	• Radical changes in stakeholder interest and influence on operations – including shareholders, executives, workers, contractors, community and governments • Attraction and retention of competent employees • Expectation of employees with respect to flexible work conditions and pay • Proactive management of psychosocial risks and provision of decent, psychologically safe work • Privacy and bias risks associated with advanced technologies (see next row) • Human-system interaction risks associated with introduction of new technologies • Increasing adoption of synthetically and renewable products including food, clothes, energy
T: Technology factors	• Digital twinning • AI software • Cognitive-augmented technologies to aid human decision making • Validity of 'big data' • Increasing use of automation and autonomous technologies • Remote supervision of unmanned operations
L: Legal factors	• Increased regulation and increased consequences of noncompliance • Privacy and ethical issues associated with advanced technologies and use of AI • Increasing focus and litigation associated with director and management obligations
E: Environmental factors	• Net zero carbon objectives • Decreasing consumption of resources such as energy and water and increased levels of reuse and recycling • Reduction in allowable emissions and waste • Management of hygiene and disease • Increase in extreme weather events

re-emerging and emerging risks should also include undertaking a futures forecasting or foresighting exercise.

Futures forecasting or foresighting is purposeful analysis conducted to:

- Examine prospective developments typically over 5–25 year timeframes in order to anticipate multiple, plausible futures (Smith & Saritas, 2011).
- 'Anticipate and understand the potential direction, rate, characteristics, and effects of technological change, especially invention, innovation, adoption, and use' (Firat, Woon, & Madnick, 2008, p. 1).
- Identify probable, plausible, possible, preferable futures as shown in Fig. 13.1.
- Highlight emerging opportunities, threats and uncertainties.

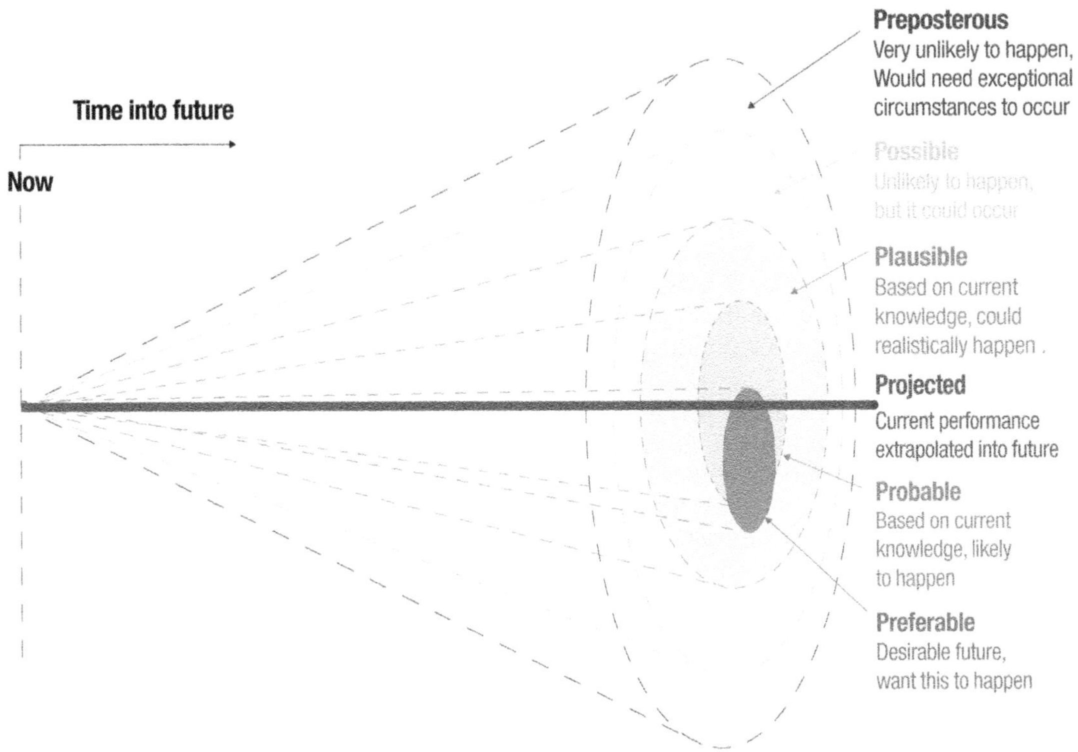

FIGURE 13.1 Examples of possible futures (Hancock & Bezold, 1994; Voros, 2019). *Source: Adapted from Hancock, T. and C. Bezold (1994). Possible futures, preferable futures. The Healthcare Forum Journal 37(2); Voros, J. (2019). Big History and Anticipation. Handbook of Anticipation: Theoretical and Applied Aspects of the Use of Future in Decision Making. R. Poli. Cham, Springer International Publishing: 425-464.*

- Understand the local, regional, national and global impacts (Sanderson et al., 2014).
- Help decision makers and planners improve readiness and obtain or sustain competitive advantage.

The foresighting process has been described by Cheah (2020) as comprising the following steps:

1. Scoping
2. Gathering intelligence (horizon scanning, literature review, data mining)
3. Imagining scenarios
4. Evaluating impact across the process
5. Setting priorities
6. Developing and implementing action plans

Imagining scenarios can be done by a broad range of methods which can be grouped as follows (Firat et al., 2008; Technology Futures Analysis Methods Working Group, 2004):

- Extrapolate current trends into the future (e.g. growth curve fitting, correlation analysis, trend prediction)

- Extract predictions from abstract or mathematical models (e.g. simulation models, adaptive system models, causal models)
- Think imaginatively to unlock insights of possible futures (e.g. creative workshops, expert option, Delphi technique, brainstorming)

Evaluation of impact can be done by expert opinion (e.g. via delphi or multi-criteria analysis techniques), backcasting, risk assessment, cost–benefit analysis, SWOT analysis, consequence modelling and simulation modelling. Developing and implementing action plans can be done by using backcasting as well as strategic or other planning processes.

It is important to note that it is best to use a number of methods when conducting futures foresighting as it will increase the depth, breadth and even clarity of insights gained (Aaltonen, 2009; Glenn & Gordon, 2009; Smith & Saritas, 2011; Technology Futures Analysis Methods Working Group, 2004).

In addition to the literature review and future foresighting exercises, it is also possible to obtain the views of a futurist. Futurists are people who specialise in thinking about possible futures and future possibilities. It seems to be a growing profession. One of the topics often covered by futurists is 'the future of work'. So it is worth reviewing commentary or seeking advice from good futurists to input into future foresighting exercises.

Once future risks have been identified, they will need to be managed. Just as risks will change in the future so will risk management approaches. The following section considers how risk management approaches may change into the future.

13.3 The future of risk management

At present, many risk management exercises conducted in the process industries are manual, spreadsheet-based, siloed, point in time processes. As we transition into the era of Industry 4.0, technology advancements now present us with the opportunity to create integrated, intelligent, real-time and predictive risk management systems that help humans make informed decisions and take effective actions to manage risks.

Contemporary navigation systems provide a nice analogy for us to consider as shown in Fig. 13.2. In a short period, street navigation systems transformed from paper-based, static, discrete and voluminous street maps and directories to the digitalised, dynamic, integrated, compact navigation systems that guide us with real-time data with prediction analysis. Why can't we do the same with our risk management systems?

Industry 4.0 is a term to refer to the fourth industrial revolution. The first industrial revolution was associated with steam power, the second with electricity, the third with digital computing and now the fourth industrial revolution is associated with cyber physical systems, autonomous systems, advanced data analytics and the internet of things.

The aspects of Industry 4.0 most relevant to the process industries are shown in Fig. 13.3. They are discussed in detail in the following subsections.

It is exciting to note that the risks associated with the introduction of these technologies do not only represent threats, but they also offer significant opportunities. Industry 4.0 technologies can be used through the lifecycle of a processing plant to improve design, hazard identification, mechanical integrity, reliability-based maintenance, operational efficiencies, operating procedures, and worker's competencies (Lee et al., 2020). Baur and Wee (2015) cite the example

FIGURE 13.2 Examples of technological advancements for navigation. *Source: Image 1: Road map - photo courtesy of Steve Buissinne sourced from https://pixabay.com/photos/map-vacation-travel-driving-2789052/. Image 2: Car dash navigator - photo courtesy of Lynda Sanchez sourced from https://pixabay.com/photos/car-gps-na-navigation-automotive-4073514/. Image 3: Stacks of paper - image courtesy of OpenClipart-Vectors sourced from https://pixabay.com/vectors/papers-stack-heap-documents-576385/. Image 4: Crane lifting photo courtesy of John R Perry sourced from https://pixabay.com/photos/oil-rig-refinery-industry-gas-fuel-514035/.*

of a goldmine that increased yield by 3.7% by use of more available real-time data, and they claim that by utilising advanced analytics, most companies can reduce production costs, and improve gross margin by 30%. 'A well thought-out and implemented Industry 4.0 strategy has the ability to greatly improve existing knowledge management practices and provide opportunities to capture and embed the experience or wisdom of experienced personnel within computer systems' (Lee et al., 2020, p. 18). However, these technologies can pose significant threats too. Implementing Industry 4.0 technology without a well-designed strategy can add more systems and complexity to a processing operations, divert resources away from other essential activities, and increase chance of errors in decision making—all of which can lead to catastrophic accidents (Lee et al., 2020). This is further compounded by the fact that, despite the potential and hype associated with Industry 4.0, there is no exemplar implemented within the process industries (Lee et al., 2020). So it will be important to ask questions like:

- What technology be accepted and used?
- How, when and why will the technology be used?

- Will users find it usable and useful?
- Will outputs be used correctly to inform not mislead decision makers?

It will also be important to have competent people interacting with the technology who understand its capabilities and limitations because at the end of the day, technology can provide truthful or false representations of reality as shown in Fig. 13.4. These are important considerations to keep in mind as we go through each of the technologies in detail in the next subsections.

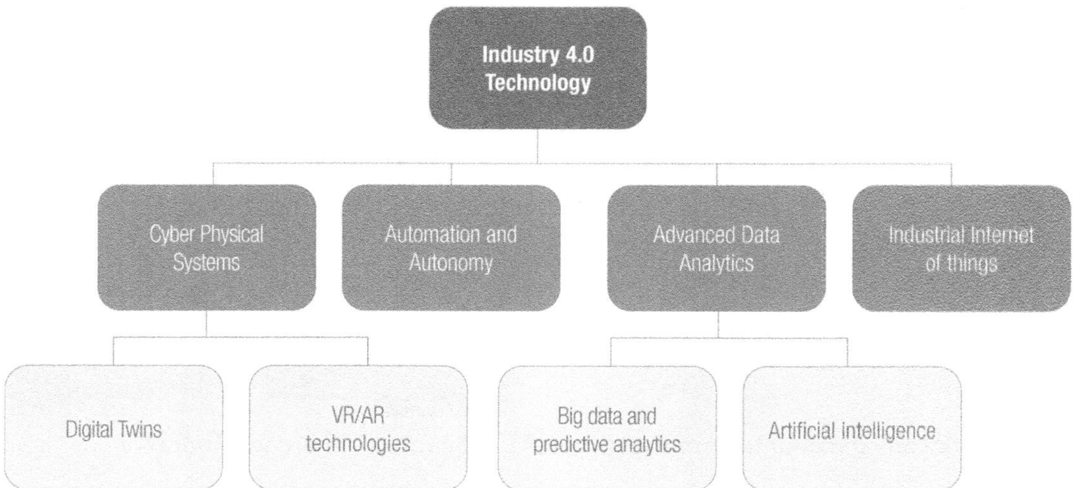

FIGURE 13.3 Industry 4.0 technologies most relevant to process industries.

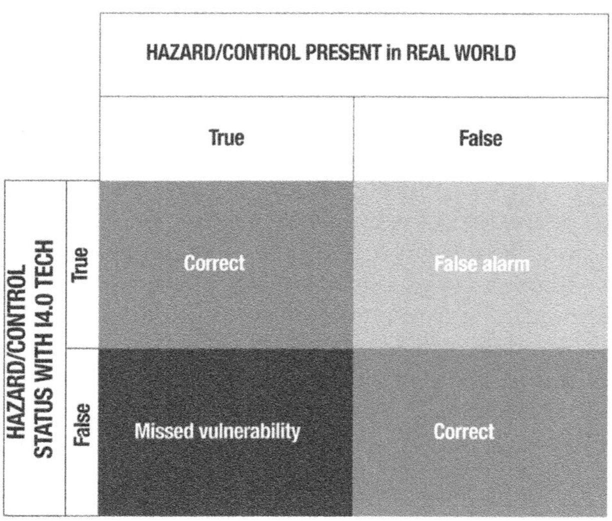

FIGURE 13.4 Possible outcomes from an overreliance of technology.

13.3.1 Digital twins

Digital twins are a digital model of the real world that can be fed with hypothetical, historical or real-time data. They are becoming increasingly used in the process industries for 'analysis of data and monitoring of systems to head off problems before they even occur, prevent downtime, develop new opportunities and even plan for the future' (Marr, 2017). Examples include testing process designs and control system design, alarm rationalisation, simulator-based training, performing risk assessments, trouble-shooting, management of change and scenario analysis. The benefits they offer in regard to risk management include providing objective information about consequences, reliability of human responses and effectiveness of controls (Lee et al., 2019). However, as George E. P. Box stated 'All models are wrong, but some are useful'. The same is true for digital twins and significant subject matter expertise is required to identify inaccuracy and even then it can be difficult to confirm how well a digital twin mimics an actual plant (Lee et al., 2022). Inaccuracies can stem from deliberate decisions not to model some sections and/or behaviours of a plant. They can also be as a result of human errors in setting up the model. In addition, if the model is not updated to reflect changes in the plant then this can lead to inaccuracies. Such inaccuracies can result in decision makers using wrong information to make decisions, operators being trained incorrectly and maintenance being done at wrong intervals all of which can lead to unnecessary downtime, incidents and accidents.

13.3.2 Virtual reality/augmented reality technologies

Virtual reality (VR) and augmented reality (AR) provide interfaces through which the user is presented with a combination of real-world and computer-generated information. VR involves immersing the user into a computer-generated 3D graphics model of an environment in a manner that gives the user an interactive experience (Cipresso et al., 2018). AR presents digital information overlaid onto visions of the real world through glasses, headset or mobile devices (PWC, 2019).

In the process industries, VR is utilised for design assessments, operability studies, training and education. A key benefit of VR is that it allows for the realistic re-creation of interactive experiences which would not be possible in the real world such as those associated with disasters (Cipresso et al., 2018). Examples include managing process plant upsets, practicing evacuation procedures, practicing high-risk and complex tasks. AR systems can be used for expert advisory systems, training and development, hazard identification, risk assessments, control monitoring, inspections, auditing and incident analysis. The benefits include enabling in-field personnel to access remote expertise, and enabling real-time access and interaction with 'live' information. As such AR systems can help provide maintainers, operators, investigators, assessors and auditors with information they need, in real-time, to detect and correctly address the situations they are confronted with.

Downsides include user acceptance, claustrophobia, motion sickness (PWC, 2019), image stability issues, impairment risks, wearability, maintenance and overall costs. Resources and expertise is required to ensure computer-generated information used in VR and AR systems is accurate and up-to-date otherwise it could misinform users.

Such misinformation could lead to people making erroneous decisions and responses that could result in adverse safety, environmental, production and/or other business outcomes.

13.3.3 Automation and autonomy

Automation and autonomy refer to plant and mobile equipment that perform work without human intervention. Automated equipment often refers to equipment at this set-up to repeatedly perform set, well-defined tasks. The operation of automated equipment is often supervised by a human located in a remote control room who is responsible for responding if the equipment stops working as required. Examples of automated equipment include automated minerals processing plants, assembly line manufacturing, goods packaging and warehousing processes. Autonomous equipment are equipment that are responsible for performing tasks and making decisions on how to perform the tasks. Examples of autonomous equipment include autonomous trains, cars, trucks and drones. The benefits of automation and autonomy include removing people from hazardous work areas and from doing harmful work. Examples include:

- Use of automated equipment has led to not-normally manned offshore oil and gas rigs.
- Use of inspection drones removes the need for people to go into confined spaces or to work at heights to do inspections.
- Use of autonomous mine haulage vehicles has removed people from working long shifts doing monotonous driving which exposed them to heavy vehicle accident risks.
- Use of robots on assembly lines and in warehousing has removed people from doing repetitive manual work and from having to be around hazardous work areas where there are a lot of mobile plant movements.

The potential downsides of using automated and autonomous technologies are that they may introduce hazardous human–automation interactions, require new types of expertise and resources to maintain functionality and capability, make employees redundant and, if not designed well, expose control room operators to high stress working environments.

13.3.4 Big data and artificial intelligence

The term 'big data' is used to describe large, diverse and complex datasets that can only be stored, understood and used with the help of special tools and methods (Cambridge dictionary). In the process industries, big data might comprise process measurements (e.g. temperature, pressure, flow, composition), alarm logs, maintenance information, conditional monitoring data, atmospheric/water/soil monitoring data, procedures and human resource information. Analysis of big data can provide information, knowledge and insights into relevant trends, presence of weak signals indicating safety problems and/or equipment degradation (Wang & Wang, 2021). Such analyses can provide real-time evidence-based insights to inform decision makers. Contemporary methods for analysing big data can range from the application of basic data analytics through to the use of

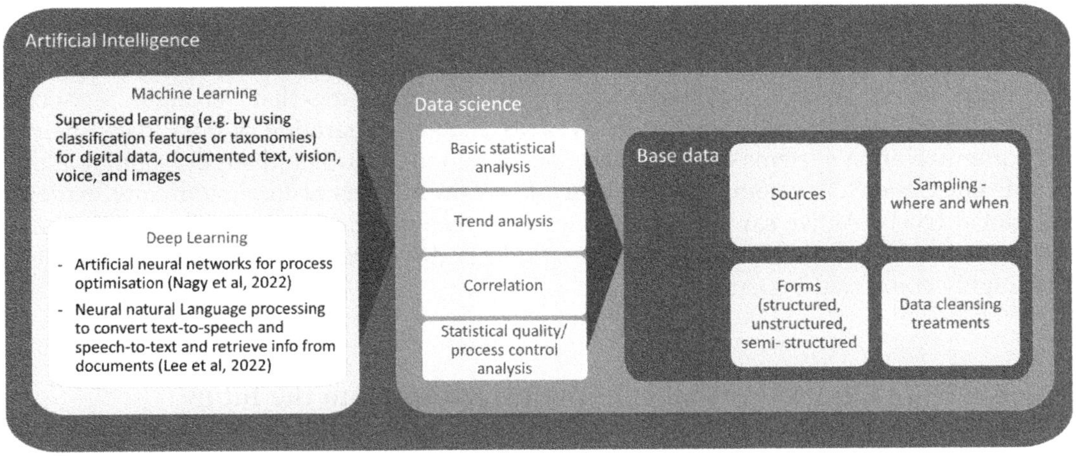

FIGURE 13.5 Data analysis approaches (Lee et al., 2022). *Source: Adatped from Lee, J, Cameron, I, & Hassall, M (2022). Information needs and challenges in future process safety.* Digital Chemical Engineering, 3, 1-15.

artificial intelligence (AI) using deep learning and/or machine learning approaches as shown in Fig. 13.5. AI 'encompasses wide-ranging and very diverse concepts like knowledge representation, reasoning, planning and manipulation' whereas 'the machine learning subdomain typically focuses on learning from data and pattern recognition' (Rybak & Hassall, 2021, p. 3). In machine learning the learning 'rules' or data features are established by humans (IBM Cloud Education, 2020). Deep learning is a subset of machine learning that automates the learning process by automatically extracting data features and determining what is important (IBM Cloud Education, 2020). Thus machine learning applications are trained by humans, deep learning applications train themselves (Fig. 13.5).

In the process industries, AI approaches can be used to analyse video, voice communication, digital data and written data to detect anomalies, outliers and inconsistencies, look for clusters that represent commonality, correlations and weak signals and make predictions. However, big data analysis can suffer from the garbage-in-garbage-out phenomenon whereby only quality insights will be derived from big data analytics if these analytics are performed on quality input data. Quality input data relies on: (1) Good understanding of the optimum location of sensors and sampling rates; and (2) Data cleaning that removes erroneous or unnecessary data and addresses data gaps without introducing biases (Lee et al., 2022; Rybak & Hassall, 2021). Another danger is the 'black box' nature of AI where the algorithms and processes used by the AI are not transparent, comprehensible or retraceable by humans and therefore the limitations, fairness and accuracy of the outputs is not fully understood (Haupt et al., 2021). Explainable AI has been proposed as a potential solution to help users of AI understand how and why an output was reached but further work is required to develop it (Haupt et al., 2021).

It is also important to note that even when quality input and analyses have been performed, big data analytics are still limited by the data and provide retrospective analysis. In performing such analyses, it is also important to consider privacy, fairness and bias.

13.3.5 Industrial internet of things

The industrial internet of things (IIoTs) refers to the use of the internet to connect sensors, instruments, devices, machines, equipment and plants together via digital networks both locally, nationally and internationally. This enables integration of data, data analysis and communication systems as well as the use of remote control and supervision. It allows for full integration of systems and analysis of system impacts at local, company, national and global levels. As we saw in the Chapter 11, cyber security is a significant risk associated with the IIoTs. Another potential downside is the increased consequences resulting from highly connected systems.

13.4 Professional engineering practice into the future

Throughout this book we have emphasised the importance of professional engineering practice in guiding how we manage risks. This will continue into the future, as we are challenged to manage the ever increasingly complex future risks and future risk management approaches discussed above. It will be critical to have competent people to identify and assess emerging risks and interacting with emerging technologies who understand the upside and downside risks for the process industries. In Chapter 2, we advised that professional engineers:

- should be able to demonstrate *prudent understanding of what should be done in a practical situation.*
- must make a lifelong commitment to your own development as a professional.
- must take responsibility for this.

In fulfilling these responsibilities, it is important to consider the future as well as the present. It is also critical to remember

> 'Those who do not remember the past are condemned to repeat it' George Santayana.

So as we gaze into the future to envision an exciting new world of process engineering, and to consider the raft of new risks and opportunities that will be posed, we should also reflect on one of the key messages in this book—that is, it is critical to learn from past experiences when identifying, assessing and treating risks. To put this another way, as we drive into the future, we must maintain sharp attention in the rearview mirror. Whilst we should not rely on the past as a good predictor of the future, it is also very important that we do not forget the lessons from the past.

As we have seen many times in this book, many process industry incidents are reoccurring events, based on known threats. If we are to learn anything from our history, it is that we can be certain that our future will comprise incidents that are similar to incidents that we have seen before. The Piper Alpha oil rig fire and explosion occurred in 1988 resulting in 167 fatalities and enormous asset damage. Twenty two years later, the Deepwater Horizon drilling rig suffered a blowout resulting in an explosion and fire that resulted in 11 fatalities, and enormous asset and environmental damage. In 1984 there was a disastrous release of a toxic gas from a pesticide plant in Bhopal, India, which resulted in appalling numbers of injuries and loss of life for nearby residents. Thirty six years later in Andhra Pradesh, India, a toxic gas release from a petrochemical plant killed 11 nearby residents. In 2006, workers were conducting hot works at the Bethune point wastewater treatment plant (Florida, USA). Sparks from the hot work ignited flammable methanol vapours from a nearby methanol storage tank. The tank exploded, killing two workers and severely injuring another. In 2017, at the PCA's DeRidder Paper Mill (Louisiana, USA), workers were conducting hot work above a storage tank that contained some hydrocarbons. Sparks from the hot work ignited hydrocarbon vapours and the tank exploded. Three contract workers were killed and seven others injured. The ignition of flammable gases and vapours from hot works continues to be a major cause of serious incidents throughout the process industries. As does the storage and management of hazardous compounds such as ammonium nitrate. Despite there being a litany of ammonium nitrate disasters over the last century, including two explosions in the last decade at storage facilities in West (USA) in 2013 and Tianjin (China) in 2015, resulting in 15 and 173 fatalities, respectively, in 2020 there was an enormous ammonium nitrate explosion in Beirut, resulting in unprecedented damages and 218 fatalities.

It is undeniable. We keep making the same mistakes and apparently failing to learn from the past. Recall that most incidents are associated with a failure to implement well-known controls for well-known risks. As we move into the future, we must expect this to continue to be true. It is thus incumbent upon us to learn from the past, and to use key lessons learnt from past incidents when identifying, assessing and treating risks. This must be a key aspect of all of our futures.

13.5 Summary

The main aim of this chapter was to discuss future risks and the future of risk management for the process industries. We hope that the chapter has provided useful information to help you better identify and prepare for the emerging threats and opportunities posed by the future.

The sustainability of the processing industries will require engineers to have the requisite competencies to utilise innovative technology, approaches and thinking, to develop more integrated, comprehensive, and broad reaching and effective identification and management of risks. But we must also remember to keep learning from the past, and to use key lessons learnt from past incidents when identifying, assessing and treating risks in the future.

Review questions

1. What is Industry 4.0?
2. What are the different potential futures that can be identified in foresighting exercises?
3. What are some of the threats and opportunities associated with using AI in the process industries?

4. Consider an industry of specific interest to you. List some emerging risks that could have a significant impact on that industry?
5. What are some of the ways you can continue to develop your knowledge and competencies in risk management and stay up-to-date with potential emerging risks and learnings from past events?

References

Aaltonen, M. (2009). *Evaluation and organization of futures research methodology-v3.0*. Washington DC: The Millenium Project.

Allianz. (2015). *Allianz risk barometer: Top business risks 2015*. Munich, Germany. ey.com.

Aon. (2014). *Aon's 2014 Australiasian Risk Survey*. Australia. http://www.aon.com.au/australia/thought-leadership/risk-survey.jsp.

Babier, F. (2017). *5 trends for the future of manufacturing*. World Economic Forum. June 22, 2017. Available from https://www.weforum.org/agenda/2017/06/what-s-going-on-with-manufacturing-b013f435-1746-4bce-ac75-05c642652d42.

Baur, C., & Wee, D. (2015). Manufacturing's next act. *McKinsey & Company, 1*, 2015. Available from https://www.mckinsey.com/business-functions/operations/our-insights/manufacturings-next-act.

Cheah, S. L.-Y. (2020). *Strategic foresight: Accelerating technological change*. De Gruyter.

Cipresso, P., Giglioli, I. A. C., Raya, M. A., & Riva, G. (2018). The past, present, and future of virtual and augmented reality research: A network and cluster analysis of the literature. *Frontiers in Psychology, 9*(2018). Available from https://doi.org/10.3389/fpsyg.2018.02086.

Ernst & Young. (2015). *Buniness risks facing mining and metals 2015–2016: Moving from the back seat to the driver's seat*. UK.

EY. (2020). *Are you reframing your future or is the future reframing you? Megatrends 2020 and beyond*. Available from https://assets.ey.com/content/dam/ey-sites/ey-com/en_gl/topics/megatrends/ey-megatrends-2020-report.pdf.

Firat, A. K., Woon, W. L., & Madnick, S. (2008). *Technological forecasting-A review* (Working paper CISL# 2008-15). Cambridge, MA: Composite Information Systems Laboratory (CISL), Sloan School of Management, Massachusetts Institute of Technology.

Glenn, J. C., & Gordon, T. J. (2009). Integration, comparisons, and frontiers of futures research methods. In T. J. Gordon, & J. C. Glenn (Eds.), *Futures research methodology version 3.0*. Washington, DC: The Millinnium Project.

Hancock, T., & Bezold, C. (1994). Possible futures, preferable futures. *The Healthcare Forum Journal, 37*(2), 23–29.

Haupt, S., Chapman, W., Adams, S., Kirkwood, C., Hosking, J., Robinson, N., Lerch, S., & Subramanian, A. (2021). Towards implementing artificial intelligence post-processing in weather and climate: proposed actions from the Oxford 2019 workshop. *Philosophical Transactions A. The Royal Society*. (Machine Learning for weather and climate modelling).

IBM Cloud Education. 2020. *Deep learning*. Available from https://www.ibm.com/cloud/learn/deep-learning.

Lee, J., Cameron, I., & Hassall, M. (2019). Improving process safety: What roles for digitalization and industry 4.0? *Process Safety and Environmental Protection, 132*, 325–339. Available from https://doi.org/10.1016/j.psep.2019.10.021.

Lee, J., Cameron, I., & Hassall, M. (2020). Managing process safety and operational risks with industry 4.0 technologies. In C. M. Hussain, & P. Di Sia (Eds.), *Handbook of smart materials, technologies, and devices: Applications of industry 4.0* (pp. 1–27). Cham: Springer International Publishing.

Lee, J., Cameron, I., & Hassall, M. (2022). Information needs and challenges in future process safety. *Digital Chemical Engineering, 3*, 1–15.

Marr, B. (2017). What is Digital Twin technology - and why is it so important? *Forbes*. Available from https://www.forbes.com/sites/bernardmarr/2017/03/06/what-is-digital-twin-technology-and-why-is-it-so-important/.

Marr, B. (2022). *The 10 biggest future trends in manufacturing*. Forbes. January 25, 2022. Available from https://www.forbes.com/sites/bernardmarr/2022/01/25/the-10-biggest-future-trends-in-manufacturing/?sh = 8cef60d4d56f.

Manyika, J., Smit, S., & Woetzel, J. (2020). *Climate risk and response: Physical hazards and socioeconomic impacts*. Available from https://www.mckinsey.com/capabilities/sustainability/our-insights/climate-risk-and-response-physical-hazards-and-socioeconomic-impacts.

Marsh & McLennan. (2015). *Global risks 2015*. Geneva, Switzerland.

McKinsey & Company. (2021). *The top trends in tech—executive summary download*. Available from https://www.mckinsey.com/~/media/McKinsey/Business%20Functions/McKinsey%20Digital/Our%20Insights/The%20top%20trends%20in%20tech%20final/Tech-Trends-Exec-Summary.

McKinsey & Company. (2022). *Global energy perspective 2022: Executive summary*. Available from https://www.mckinsey.com/~/media/McKinsey/Industries/Oil%20and%20Gas/Our%20Insights/Global%20Energy%20Perspective%202022/Global-Energy-Perspective-2022-Executive-Summary.pdf.

Mooney, S. (Ed.), (2014). Asia risk report: The top concerns for Asian risk manageers - 2015 edition. Sydney, Australia: Newsquest Specialist Media Ltd.

Naughtin, C., Hajkowicz, S., Schleiger, E., Bratanova, A., Cameron, A., Zamin, T., & Dutta, A. (2022). *Our future world: Global megatrends impacting the way we live over coming decades*. Brisbane, Australia. Available from https://www.csiro.au/en/research/technology-space/data/Our-Future-World.

Poole, V., & Sullivan, K.B. (2021). *Tectonic shifts: How ESG is changing business, moving markets, and driving regulation*. Available from https://www2.deloitte.com/us/en/insights/topics/strategy/esg-disclosure-regulation.html.

PWC. 2019. *Seeing is believing*. PWC. Available from https://www.pwc.com/gx/en/industries/technology/publications/economic-impact-of-vr-ar.html.

Rybak, N., & Hassall, M. (2021). Machine learning enhanced decision-making. In C. M. Hussain, & P. Di Sia (Eds.), *Handbook of smart materials, technologies, and devices*. Cham: Springer. Available from https://doi.org/10.1007/978-3-030-58675-1-20-1.

Sanderson, P., & Hassall, M. (2014). *Cognitive work analysis and defence futures*. DSTO research report 2013/1192983/1; MyIP 5349.

Smith, J. E., & Saritas, O. (2011). Science and technology foresight baker's dozen: A pocket primer of comparative and combined foresight methods. *Foresight (Los Angeles, Calif.)*, *13*(2), 79–96. Available from https://doi.org/10.1108/14636681111126265.

Technology Futures Analysis Methods Working Group. (2004). Technology futures analysis: Toward integration of the field and new methods. *Technological Forecasting and Social Change*, *71*(3), 287–303. Available from https://doi.org/10.1016/j.techfore.2003.11.004.

Voros, J. (2019). Big history and anticipation. In R. Poli (Ed.), *Handbook of Anticipation: Theoretical and Applied Aspects of the Use of Future in Decision Making* (pp. 425–464). Springer International Publishing.

Wang, B, & Wang, Y (2021). Big data in safety management: An overview. *Safety Science*, *143*.

Withers, G., Gupta, N., Curtis, L., & Larkins, N. (2015). *Securing Austalia's future: Australia's comparative advantage*. Melbourne, Australia.

APPENDIX

A

Process industry hazards

A.1 Introduction to workplace hazardous chemicals and conditions

For engineers working in the process industries, hazard identification often includes the identification of hazardous chemicals and operating conditions. In this appendix we will provide a brief introduction to hazardous chemicals and operating conditions, firstly introducing the global framework for the classification of hazardous chemicals and then a number of priority operating conditions that are frequently associated with adverse events in industry. For a rigorous and detailed discussion of hazardous chemicals and conditions in the process industries, see Kletz (2009).

A.2 Classifying workplace hazardous chemicals

The UN Globally Harmonized System of Classification and Labelling of Chemicals (GHS) has been developed to ensure consistency in the classification and labelling of chemicals. The latest edition (9th) was released in 2021 (UN, 2021). This provides an excellent framework for considering hazardous chemicals. The nine GHS classifications and labels representing physical, health and environmental hazards are shown in Table A1. All of these categories of materials are relevant to the process industries. An example of a chemical label is shown in Fig. A1.

In accordance with the GHS, all hazardous chemicals must have material safety data sheets which are a consistent source of information about hazards. These are a vital resource when identifying and assessing chemical hazards.

Some other common compounds may not be classified as hazardous according to the nine classifications in the GHS, yet these are still potentially hazardous in processing plants. These include:

- *Nitrogen*: Whilst inert, nitrogen can act to, and is used to, displace oxygen in enclosed spaces. As such, nitrogen poses a significant asphyxiation hazard. Nitrogen

TABLE A1 The nine GHS classifications and labels.

Explosive		Flammable		Oxidising	
Gases under pressure		Acute toxicity		Corrosive	
Health hazards		Chronic health hazards		Environmental	

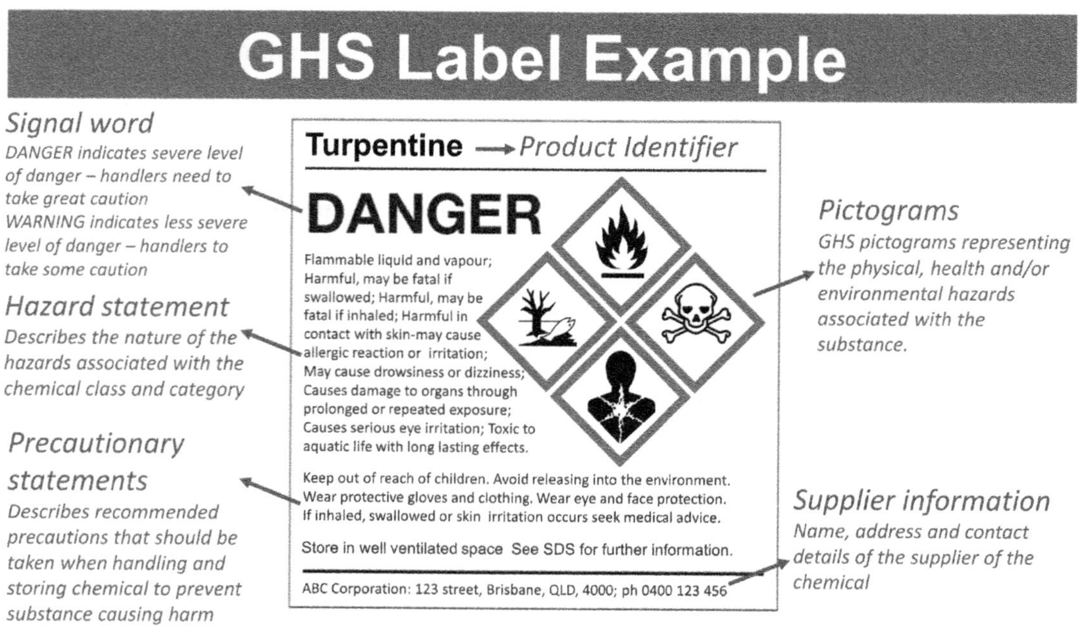

FIGURE A1 Example of a Globally Harmonized System label.

asphyxiation can happen with little warning due to the body's low oxygen sensing systems. Nitrogen is commonly used in industry to inert spaces, for preservation of material, fire suppression, or as a chemical reagent.

- *Carbon dioxide*: Whilst carbon dioxide is naturally present in air ($\sim 0.04\%$), and is not harmful at low concentrations, it can present a genuine threat to human health on processing plants in certain circumstances. At elevated concentrations, CO_2 can cause dizziness, headaches and confusion. Further, because CO_2 is heavier than air it can accumulate in lower areas and confined spaces, posing a serious risk of asphyxiation. It is thus critical that fuel burning equipment (e.g. generators) must not be placed near excavations or other confined spaces where work is being performed. CO_2 is classed as a "substantial hazard to health" under the UK Control of Substances Hazardous to Health Regulations 2002 (COSHH).
- *Steam*: Steam (water vapour) is a commonly used utility stream for heating in industry. Whilst commonly used, steam is hazardous in several ways:
  - It can cause vessel failures due to vacuums caused by condensing steam.
  - Steam leaks can often not be seen as some high pressure or superheated steams contain sufficient energy that the steam does not condense around the leak and cannot be seen.
  - Steam burns can be fatal.

A.3 Hazardous conditions

This section introduces a few conditions that are commonly encountered on processing sites, which can be hazardous, and have commonly featured in serious incidents. We are introducing them so you can identify them as hazards and address them in your risk management activities. Some of them, such as temperature and pressure, are routinely considered in activities such as HAZOP, but several others are not. You need to identify these risks when present.

A.3.1 Temperature

Temperature, whether high, low or changing can be a hazard. Low temperatures may result in blockages, burns, cold temperature embrittlement and changes in material properties. High temperatures may result in metal weakness, burns or changes to material properties. Rapid changes in temperature can be an indication of undesired reactions taking place or create new threats. The serious incident at Esso's gas plant at Longford in 1998 (Victoria, Australia) was, in part, a consequence of a fracture in a heat exchanger due to the metal becoming brittle because of unusually low temperatures (Hopkins, 2000).

A.3.2 Pressure

Trapped pressure or vacuum becomes a hazard when it is outside the design limits of the vessel or pipework it is contained in. Pressure is generated deliberately through pumps and compressors, but it can be generated accidently by mixing chemicals, heating material, cooling material, draining without adequate venting or moving material.

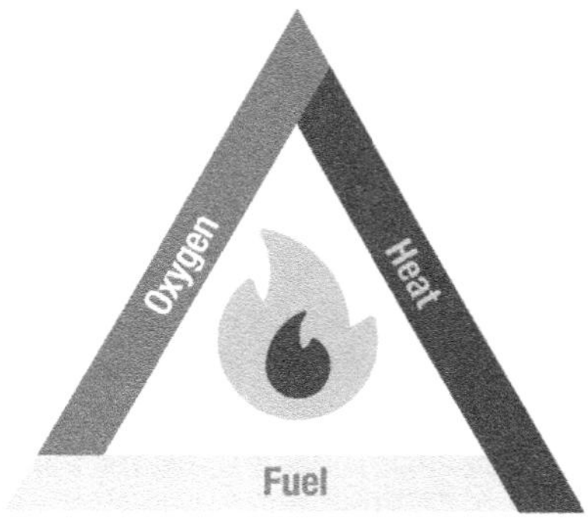

FIGURE A2 The fire triangle.

A.3.3 Flammable environments

The physical properties of flammable materials determine ease of ignition and possible severity of consequences should ignition occur. Relevant physical properties include flash point, lower explosive limit, upper explosive limit, autoignition temperature and density at atmospheric conditions. The well-known fire triangle shown in Fig. A2 is often used to illustrate that the three elements that a fuel needs to ignite are fuel, oxygen and heat.

However, whilst the fire triangle is theoretically correct, it is worth considering an alternative that was proposed by Trevor Kletz, a world leading pioneer of process safety (Kletz, 2009):

AIR + FUEL → BANG ! or AIR + FUEL → FIRE !

> If flammable gases or vapours are mixed with air in flammable concentrations, experience shows that sources of ignition are likely to turn up, as the amount of energy required to ignite a flammable mixture can be very small . . . as little as 0.2 mJ. (Kletz, 2009; p. 474).

This is very sage advice and strongly warns against relying solely on ignition controls to avoid explosions or fires.

A.3.4 Combustible dust

Combustible dust explosions occur when there is sufficient combustible dust in the air that an explosive mixture forms and is ignited. In addition to the dust, oxygen and ignition source, dust explosions also require confinement and dispersion of the dust as shown in Fig. A3.

Combustible dust explosions are unfortunately common and occur in a range of industries. Coal, sawdust, flour, sugar, cocoa, powdered milk, coffee, pharmaceuticals and metals can all form explosive suspensions in air. The US Chemical Safety and Hazard Investigation Board continues to highlight combustible dust safety as a critical issue (CSB, 2018a).

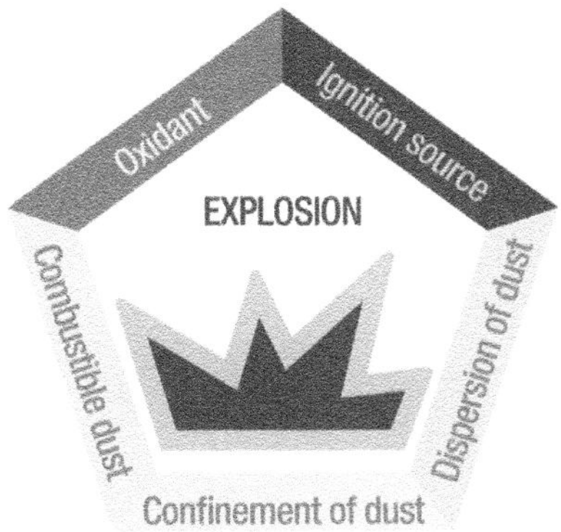

FIGURE A3 The five requirements for a dust explosion.

A.3.5 Conducting hot work

Hot work essentially refers to conducting work on site that may generate a source of ignition for flammable material. It includes activities such as burning, cutting, welding and grinding. Hot work is often conducted as part of maintenance around processing sites. The US Chemical Safety Board has reported that explosions and fires caused by hot work are among the most common incidents they investigate. The CSB reported 60 fatalities in the United States due to explosions and fires from hot work activities on tanks between 1990 and 2010 (CSB, 2010a). Unfortunately, they have investigated several further hot work incidents since 2010 (CSB, 2018b). As a consequence, the CSB has published some excellent (succinct and practicable) recommendations for conducting safe hot work practices (CSB, 2018a,b).

A.3.6 Confined spaces

Confined spaces are hazardous for people that enter them because they are not specifically designed spaces for people to work. They may contain a hazardous atmosphere which can result in injury or death. Many jurisdictions around the world have regulations for confined space entry. There are many, and similar, definitions for a confined space. The following definition from Safe Work Australia is a contemporary and specific example:

> A confined space means an enclosed or partially enclosed space that:
>
> - is not designed or intended primarily to be occupied by a person
> - is, or is designed or intended to be, at normal atmospheric pressure while any person is in the space; and
> - is or is likely to be a risk to health and safety from:
>   - an atmosphere that does not have a safe oxygen level, or
>   - contaminants, including airborne gases, vapours and dusts, that may cause injury from fire or explosion, or
>   - harmful concentrations of any airborne contaminants, or
>   - engulfment.
>
> (Safe Work Australia, 2020; p 5).

Entry to confined spaces must only be done under a properly controlled permit to work system after the prerequisite training has been completed. The hazards presented by a confined space entry must be re-evaluated if hot work or other work is performed in the space. The fire at the Xcel Energy Hydroelectric Plant (Colorado, United States) in 2007 is a sad and pertinent reminder of the risk of conducting work in a confined space. Ignition of flammable vapour resulted in a fire and the death of five contractors conducting maintenance work in the penstock (CSB, 2010b).

A.3.7 Start-up and shutdown

During start-ups and shutdowns, the operating conditions and flows of materials may be outside their normal range. This can affect the reliability of instruments used to control the process or plant. In some cases, safety instrumented systems may need to be temporarily overridden. There may also be unintended mixtures of material, such as air and hydrocarbon, and large temperature or pressure changes. A significant number of incidents occur during start-ups and shutdowns. Start-up and shutdown risks should be considered in the unit design.

A.3.8 Other hazardous tasks

A review of workplace fatalities around the world highlights that other common types of work activities that have resulted in worker deaths include the following:

- Driving vehicles or operating mobile plant: Many workers have been killed as a result of accidents with vehicles and mobile plant. These events include occupants being killed in crashes, pedestrians dying after being struck by a vehicle and accident victims perishing after vehicles have gone into water or they have been stranded and not rescued. Utilitising safe vehicles and traffic management systems as well as having trained, competent and compliant operators can help prevent and mitigate such accidents.
- Working near energised plant: Workers can be struck, entangled in, crushed by or electrocuted if they work near energised plant. Barriers should be erected to prevent people accessing live plant unless it is isolated.
- Working at height: Workers have been killed as a result of falling from elevated structures or into voids. Use of elevated work platforms and fall prevention and arresting systems should be used to prevent accidents when people need to work at height or near an edge.
- Airborne particulates: Breathing in coal dust, silica, diesel particulates, smoke and other forms of contaminated atmospheres have all been attributed to workers' disabilities and deaths due to occupational lung disease. Preventing the contaminates entering the atmosphere where workers are is the best means of addressing airborne particulate hazards followed by the use of personal protective equipment (PPE).

References

CSB. (2010a). *Seven key lessons to prevent worker deaths during hot work in and around tanks: Effective hazard assessment and use of combustible gas monitoring will save lives.* No 2009-01-SB. The U.S. Chemical Safety and Hazard Investigation Board.

CSB. (2010b). Investigation Report: Xcel Energy Hydroelectric Plant Penstock Fire. Report No. 2008-01-I-CO. August 2010. US Chemical Safety and Hazard Investigation Board.

CSB. (2018a). *CSB's drivers of critical chemical safety change: Combustible dust safety.* April 2018. The U.S. Chemical Safety and Hazard Investigation Board.

CSB. (2018b). *CSB's drivers of critical chemical safety change: Safe hot work practices.* April 2018. The U.S. Chemical Safety and Hazard Investigation Board.

Hopkins, A. (2000). *Lessons from Longford: The Esso gas plant explosion.* Sydney: CCM.

Kletz, T. (2009). *What went wrong? Case histories of process plant disasters and how they could have been avoided* (Fifth Edition). Gulf Professional Publishing.

Safe Work Australia. (2020). *Confined spaces: Code of practice.* Safe Work Australia. ISBN 978-0-642-33313-1.

UN. (2021). Globally Harmonized System of Classification and Labelling of Chemicals (GHS) (Ninth revised edition). New York and Geneva: United Nations.

APPENDIX

B

Tank filling: a process safety case study

B.1 Introduction

In this appendix, we will work through a petrochemical tank filling case study to demonstrate the use of a range of risk management tools discussed in Chapters 5–7. The information provided is for demonstration only. If applying these approaches to actual or proposed facilities, the work should be performed by appropriate competent professionals using all relevant information and the range of expertise required to fully identify and assess risks, controls and safety management activities.

Tank filling was selected as the case study topic because filling, storing, and emptying liquid using bulk storage tanks are common practices in the process industries. When the liquids involved are hazardous, these processes entail a number of process safety risks. Examples are as follows:

- The tanks can catastrophically fail. Examples include the tank failure that occurred at Ranger Uranium Mine in Australia in 2013 (Byrne, 2013) and the tank failure that occurred at Darling Cellars in South Africa in 2021 (Dreizen, 2021).
- Their contents can be ignited through hot work, lighting or other ignition sources, which can result in fires and explosions. For example hot work conducted on/near tanks have resulted in explosions at the City of Daytona, United States, where a wastewater treatment plant methanol tank exploded (U.S. Chemical Safety and Hazard Investigation Board, 2007), at the Avonmouth water recycling plant in the United Kingdom where a biosolids silo exploded (Aviram, 2021) and at the DeRidder Pulp and Paper Mill, Louisiana USA, where a 'foul condensate' tank exploded in 2017 (US Chemical Safety and Hazard Investigation Board, 2018).
- They can crack or corrode leading to leaks such as experienced by freedom industries when a tank leaked chemicals into the drinking water system of the town in 2014 (U.S.

Chemical Safety and Hazard Investigation Board, 2017b) and by ITC when leaking naphtha ignited and caused destruction of 12 tanks in Texas, United States (Marsh, 2020).
- There can be operational errors for example where incompatible chemicals are mixed in a tank such as what occurred at Greenbrook Water Treatment Plant in Canada in 2014 (Kang, 2014) and MGPI processing plant in Kansas, United States in 2016 (U.S. Chemical Safety and Hazard Investigation Board, 2017a).
- Numerous recurring accidents associated with overfilling of tanks happened in Buncefield, United Kingdom in 2005, which result in a huge fire and explosion (Waite, 2013), and in Bayamon Puerto Rico in 2009 (U.S. Chemical Safety and Hazard Investigation Board, 2015).

Therefore it is important to proactively assess and manage tank filling, storage and emptying risks. **This case study will focus on filling and overfilling of tanks** using the process, as shown in Fig. B1.

FIGURE B1 Case study hazard identification process.

B.2 Describing scope

The first step is to describe the scope. To aid in this step, it can be useful to collect site information from pictures as shown in Fig. B2, from engineering drawings, an example of

(Photo by Schwede66, CC BY-SA 4.0 <https://creativecommons.org/licenses/by-sa/4.0>, via Wikimedia Commons)

(Photo by Ted Auch, FracTracker Alliance, 2019. Aerial support provided by LightHawk. Retrieved from https://www.flickr.com/photos/fractracker/49802773158/in/photostream/)

FIGURE B2 Examples of bulk tank farm installations.

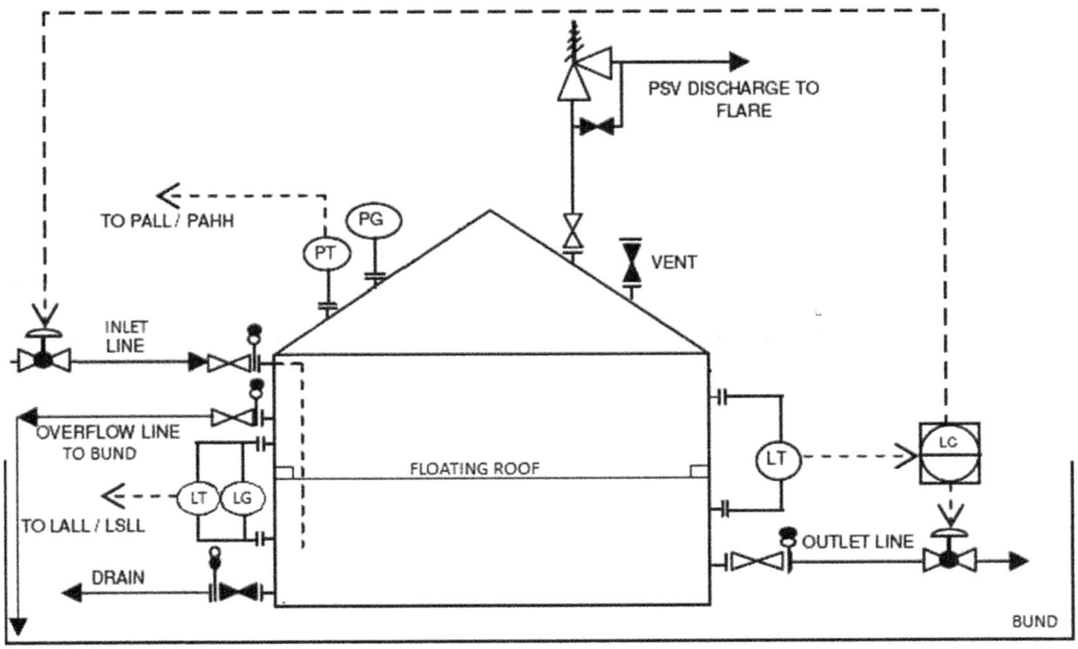

FIGURE B3 Example of an engineering drawing of storage tank. *Source: Modified from drawing retrieved from https://instrumentationtools.com/pid-guidelines-storage-tanks/.*

which is shown in Fig. B3, and other site documentation such as operating procedures and maintenance activities.

Using Figs. B2 and B3 as references, setting the scope should be done to clarify what is included and excluded from the risk assessment. As discussed in Chapter 2, a framework like PLEATESO can help ensure a range of factors to be considered. Presenting the scope in a table format can help users read, understand and assess the scope. Table B1 lists an example scope for the tank filling case study.

To ensure the scope and the subsequence risk assessment and risk treatment analyses capture relevant lessons learned from past events, it is worthwhile (and in some jurisdictions a requirement) to research and collate lessons to be learned from accidents and incidents that have occurred in similar processes and with similar systems. Examples of some overfilling events are tabulated in Table B2.

B.3 Identifying hazards, threats and consequences

As mentioned in Chapter 5, there are a number of tools for identifying hazards, threats and consequences. These include hazard identification (HAZID) or preliminary hazard analysis also known as PHA, hazard and operability studies also known as HAZOP and

TABLE B1 Example scope for tank filling case study.

Description	Included	Excluded
People. People involved in management of the risk and people potentially impacted if risks are not managed	Operators and maintainers involved in filling, emptying, monitoring and maintaining tank. Other people – both on- and offsite that could be in the plume and blast zone for vapour losses and liquid dispersion and radiant heat zone for liquid losses.	Unauthorised persons/trespassers.
Locations. Locations or areas where the risk exists or that could be impacted by the risk event if it occurs	Tank farm is located in Brisbane Australia. Risk assessment to consider tank bund areas, control room location and impact zones for gas dispersion, liquid losses, fire and explosion blasts should be specified, for example roads, residential areas and schools.	Outside impact zones associated with gas and liquid losses and potential fires and explosions.
Equipment. Equipment and plant exposed to the risk and associated with managing the risk	Tank, inlet piping from bund edge to tank, outlet piping from tank to bund edge, all valves, instruments, filters and other components fitted to piping and tank. Bund around tank and associated valves and drainage system. Field and control room control systems including sensors/instruments, controllers, communication lines, alarm systems, interfaces and programming.	Equipment used to perform maintenance and/or cleaning of tank.
Activities. Activities include operations, maintenance, startup, shutdown, normal and abnormal tasks and functions that are exposed to, could impact or be impacted by the risk	Control room operator in conjunction with two field operators, two mechanical and one electrical tradespeople manages the filling, emptying and maintenance of tanks and associated piping, fittings and instrumentation. Includes bund and its associated fittings.	Activities carried out on tank when out of service (e.g. cleaning, major refurbishment work).
Timeframes. To capture time considerations (e.g. the times of the day or days of year) plus other time considerations such as shift changes and how far into the future is being considered	Continuous – 24 hours a day, 365 days/year. Includes shift handover considerations.	Excludes major upgrades to tank, control system or site infrastructure. Excludes decommissioning.
Environment. Known weather and climate-related conditions (e.g. heatwave, tsunami, bush fire, earthquake and landslides)	Brisbane's climatic and environmental conditions include cyclones, electrical storms, floods, heat up to 45°C and high humidity.	Bushfires, earthquakes and tsunami are not considered.

(*Continued*)

TABLE B1 (Continued)

Description	Included	Excluded
Scenarios. Known risk scenarios that need to be considered	Loss of containment of liquid from tank, pipes, fittings or bund that could catch on fire or explode in a manner that could cause fatal or permanent disabling injuries or illnesses to people, environmental contamination and major asset losses.	Sabotage. Minor injury or asset damage scenarios.
Other assumptions. Captures other assumptions such as information on the hazardous chemicals and energies and environmental issues being considered	Tank is used to hold petrochemicals specifically unleaded petrol, which is extremely flammable and toxic.	

TABLE B2 Examples of tank overfilling incidents.

Date and location	Event	Causes and contributing factors	Consequences	Recommendations	References
2005 Buncefield, UK	Petrol transfer to tank resulted in tank overflowing	Operators were not monitoring filling, they relied on sensors, alarms and trips which failed.	Vapour cloud explosion which injured 43 people.	High-level alarm that alerts operators who shut off the feed. Independent level system interlocked with automatic shutdown system. Controls to prevent loss of containment escalating into a major offsite incident.	COMAH (2011)
2009 Puerto Rico	Petrol transfer from ship to tank resulting in overflow from the tank and the spilled flowed out of the bund through open drain valve and into the wastewater treatment pond where it was ignited resulting in vapour cloud explosion	Level gauge was stuck giving false reading. No high-level alarm or automatic high–high (HH) level shut off system was installed. Bund drain valve was left open – it was difficult for workers to differentiate between open and closed valves.	Vapour cloud explosion which destroyed 17 tanks and damaged 300 nearby buildings.	High-level alarm that alerts operators who shut off the feed. Independent level system interlocked with automatic shutdown system.	U.S. Chemical Safety and Hazard Investigation Board (2015)

(Continued)

TABLE B2 (Continued)

Date and location	Event	Causes and contributing factors	Consequences	Recommendations	References
2011 Reichstett, France	Tank-to-tank transfer of hydrocarbon resulted in overflow	Radar-level indicator showed wrong level and separate HH level system was inoperable.	Tank damage and loss of oil but no impact offsite.	Improve information on control room consoles. Replace HH system with more reliable one and ensure adequate testing	IMPEL - French Ministry for Sustainable Development (2013)
2014 Ohio, United States	Overfilling of tank during fuel transfer with spilt fuel running into the Ohio River	Operator failed to stop flow when tank was full. Valve left open on secondary containment which allowed oil to flow into river.	Environmental harm, Environment Protection Authority (EPA) fine.		Department of Justice U.S. Attorney's Office Southern District of Ohio (2016)
2018 NSW, Australia	Diesel twice overflowed from storage tank into creek	Alarm was not acknowledged to alarm flood issues and bund was not maintained so failed to contain fuel.	Environmental harm incurred fines from EPA.	Correct alarm system issues and maintain bund	NSW EPA (2020)

failure modes, effects and criticality analysis also known as FMECA. In this section, each of these techniques will be demonstrated using the tank filling case study in a manner that captures the learnings from past events.

B.3.1 Example of hazard identification and analysis (HAZID)

The HAZID process and spreadsheet used here are illustrated in Fig. B4. The risk matrix is shown in Fig. B5.

Details of the steps undertaken to perform the HAZID are as follows:

a. List the hazards associated with the processes in scope that have the potential to cause harm to human health and safety or have the potential to cause a process safety incident. The lists of hazards provided in Appendix A may be used as prompts.
For the tank filling, examples of hazards are:
- *chemical* – petrol which is flammable and explosive
- *pressure and thermal* – amount and pressure of petrol vapour changes with temperature
- *technology* – structural integrity of the tank, pipes, fittings and bund
- *human capability* – incorrect connection, monitoring and disconnection of tank

b. For each hazard, determine unwanted event scenarios (e.g. scenarios that could release and/or expose the hazard in a manner that could cause harm). Note: it is often better to write the unwanted event as a "description of the initial event or energy release that

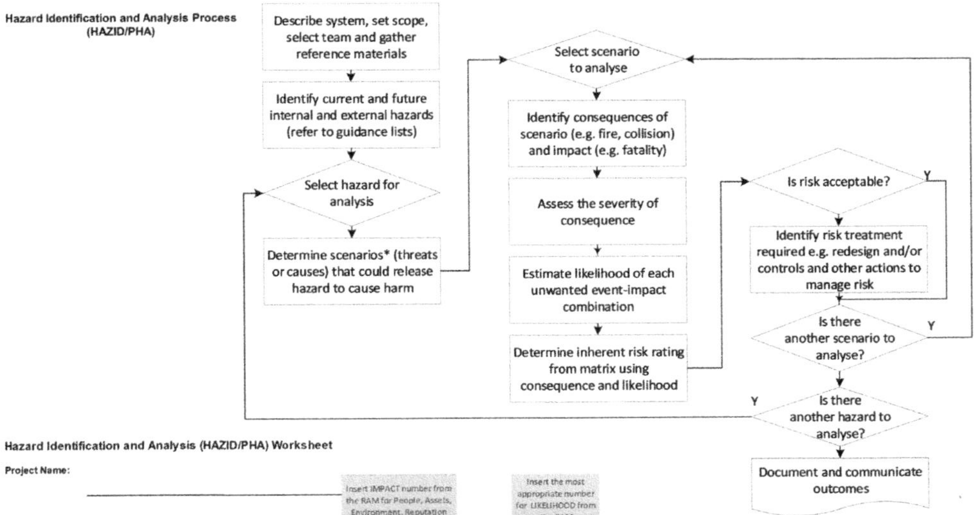

Hazard Identification and Analysis (HAZID/PHA) Worksheet

Project Name: ____________________

Insert IMPACT number from the RAM for People, Assets, Environment, Reputation

Insert the most appropriate number for LIKELIHOOD from the RAM

Description of Hazard	Description of unwanted event scenarios	Causes	Consequences	Potential Impact					Risk analysis		Risk evaluation and recommendations on risk treatment options (i.e. in terms of Inherently Safer Design and Defence in Depth/Hierarchy of Controls)	What controls are required to effectively manage the risk	What monitoring and review is required to ensure the risks and controls are effectively managed	Responsible person
				People	Assets	Environment	Reputation	Max Impact	Est Likehood	Overall Risk Rank				
										Yet to do				
										Yet to do				
										Yet to do				
										Yet to do				

FIGURE B4 HAZID process and worksheet used for tank filling case study. *HAZID*, Hazard identification.

creates an unsafe situation and that could potentially lead to a specific unwanted event. Typically an initiating event will be the initial loss of control of the hazard or a loss of awareness of the situation" (Hassall & Harris, 2017). Writing the unwanted event as the initial "loss of control" or "loss of containment" or unsafe situation helps to ensure that

– sufficient prevention controls have been identified to stop the threat causing the event.
– sufficient mitigation controls have been identified to ensure unwanted event does escalate to produce the adverse consequences.

For the tank filling, examples of unwanted event scenarios are:

– overfilling/overflowing the tank
– leaks from tank and/or piping
– overpressurisaton of tank/piping
– contamination of product
– ignition of fuel in tank
– loss of structural integrity of tank and/or fittings

c. For each unwanted event scenario, identify the different threats that could cause the release of the hazard.

RISK RANKING MATRIX					LIKELIHOOD				
IMPACT					Rare	Unlikely	Possible	Likely	Almost Certain
Impact	Injury and Disease (includes workers and community)	Business Cost (assets losses, fines, etc)	Environment	Social/Reputation	Event may occur under exceptional circumstances. (Estimated frequency between 1 every 10,000 years and 1 every 1000 years)	Event could occur at some time. (Estimated frequency between 1 every 1000 years and 1 every 100 years)	Event should occur at some time (May happen in plant life based on industry events. Est frequency between 1 every 100 years and 1 every 10 years)	Event will probably occur in most circumstrances (Probably occur once based on history of near misses. Est freq = 1 every 10 years or 1 per year)	Event expected to occur in most circumstances (e.g. does occur, definitely history of occurrence or estimated frequency = 1-10 times per year)
Catastrophic	Multiple fatality and/or significant irreversible exposure to a health risk that affects greater than 10 people	> $25 m	Permanent effects on the environment (e.g. air, water, soil, plants, animals). Potential loss of licence to operate. Prosecution of company and directors possible.	Very serious widespread/ irrepairable damage to significant relationships. Extremely adverse media internationally. Reputation impacted with majority of key stakeholders. Will not gain approval to operate.	Very High	Very High	Very High	Very High	Very High
Major	Single fatality, permanent disabling injury or exposure to a health risk that is permanently disabling and/or reduces life expectancy	$5m - $25 m	Major effects on environment (e.g. air, water, soil, plants, animals). Substantial cleanup costs. Personal & business prosecution possible	On-going serious social issues with significant relationships. Reputation impacted by adverse media and with significant number of key stakeholders. May lose licence to operate or not gain approval.	Medium	Significant	Significant	Very High	Very High
Moderate	Serious injury/exposure that require hospitalisation with moderate permanent effects (e.g. internal/head injuries, toxic inhalations).	$0.5m - $5m	Pollution with moderate effects on environment (e.g. air, water, soil, plants, animals). Required to report to authorities. Extensive cleanup may be required	Ongoing social issues. Attention from media. Heightened concern by local community. Criticism by local NGOs. Significant difficulties in gaining approvals. Reputation impacted with some key stakeholders.	Low	Medium	Significant	Significant	Very High
Minor	Significant injury. Medical treatment required from which recovery is likely (e.g. burns, broken bones, severe rashes/reactions, cuts).	$50k - $500k	Pollution with minor effects on enivronment. Required to inform authorities. May involve a cleanup. Visible discharge observed offsite.	Minor medium term impacts on local population / items of significance. Mostly repairable. Minor, adverse local public or media attention. Reputation is impacted with a small number of people.	Low	Medium	Medium	Medium	Significant
Insignificant	Minor injury not requiring medical treatment (e.g. cuts, mild skin irritations, no measurable physical effects).	< $50 k	Low Pollution = No observable effect to environment. No requirement to inform authorities. No observable discharges offsite.	Low-level social or cultural impacts. Public concern restricted to local complaints and individual concerns. No discernable impact on reputation.	Low	Low	Low	Medium	Medium

RISK RATING			
	High	Unacceptable Risk	Unacceptable – Operations do not to continue until risk is reduced. Risk treatment considerations should include inherently safe design and defense in depth analyses. Assign to Executive Team Member
	Significant	Tolerable Risk - Reduce risk to ALARP levels only stopping whenf risk reduction is impracticable or it's cost is grossly disproportionate to improvement gained)	Tolerable if actioned as a high priority and industry recognised risk controls measures are identified, implemented, monitored and reported on with respect to their ability to reduce risk. Assign senior manager responsible to action, monitor and review.
	Medium		Tolerable if actioned with industry recognised risk controls measures are identified, implemented, monitored and reported on with respect to their ability to reduce risk. Assign manager responsible to action, monitor and review.
	Low	Broadly Acceptable	Generally Acceptable - Manage with regular monitoring and review.

FIGURE B5 Risk matrix used to complete risk assessments for tank filling case study.

For the tank filling, examples of threats that could lead to a release of petrol are:

- incorrect operation
- incomplete or incorrect connections of fittings
- external heating (e.g. from hot work, adverse weather and external fire);
- component wear, failure or corrosion
- contaminants in feedstock

d. Identify the range of consequences – the initial and final outcome or impact – that might result from an uncontrolled/unplanned release of the hazard. Consequences could be adverse impacts on humans (e.g. fatality, injury, illness and disease), environment, production, assets, financial performance, reputation etc. These consequences should match the unwanted scenarios listed as in scope.

For the tank filling, examples of consequences in scope are:

- petrol spills escaping into and contaminating soil and/or water
- petrol pool fire leading to smoke inhalation, pollution, asset damage and reputational harm
- vapour cloud explosion causing multiple fatalities, major asset and production losses, environmental contamination and significant reputational issues
- contaminated product causing reputational damage and potentially financial and legal issues

e. Rank the cause–consequence scenarios using a risk matrix.

Make an assessment of likelihood to determine the risk ranking. The risk ranking exercise should refer to an organisation's risk matrix or risk appetite statement. *It is important to note that if the consequence is catastrophic then the highest level of action should be required regardless of the assessment of likelihood*. This is because:

- process safety events are often considered rare and the assessment of rare events has been proven to be inaccurate (as discussed in Chapter 5)
- even if an accurate likelihood assessment is made (e.g. 1 in 100 years), this does not mean that the event will not occur today, tomorrow or the next day so it needs to be actioned.

f. Determine risk treatment recommendations by first considering whether the risk is tolerable/to ALARP or intolerable and needs to be treated. Guidance on this assessment is often provided at the bottom of the risk matrix. The recommendations should also include whether there are inherently safer design options that should be implemented or investigated.

g. Identify if any prevention and mitigation controls and control management requirements that are needed to reduce the risk to ALARP standard.

h. Assign accountability for implementing the recommendations to an appropriate person. Guidance on the organisational level role that should be assigned the accountability can also be provided at the bottom of the risk matrix.

For the tank filling, an example of some outcomes from the HAZID or PHA process are shown in Fig. B6.

Options for performing further risk assessments include HAZOP and FMECA. It is often advisable to use more than one technique to check that the range of significant risks have been identified and are being actioned. HAZOP assesses deviations in operating parameters

Description of Hazard	Description of unwanted event scenarios	Causes	Consequences	Potential Impact: People	Assets	Environment	Reputation	Max Impact	Risk analysis: Est Likehood	Overall Risk Rank	Risk evaluation and recommendations on risk treatment options (i.e. in terms of Inherently Safer Design and Defence in Depth/Hierarchy of Controls)	What controls are required to effectively manage the risk	What monitoring and review is required to ensure the risks and risk controls are effectively managed	Responsible person
Petrol which is flammable and explosive	Overfilling tank	Inlet valve fails to close, outlet valves fails in closed position, blocked outlet pipe, faulty level sensors.	Liquid spills into bund and is ignited causing pool fire or vapourises to form vapour cloud that explodes causing multiple fatalities/ serious injuries, severe equipment damage and onsite environmental damage	5	5	3	3	5 Catastrophic	Likely	High	Unacceptable – Operations do not to continue until risk is reduced. Risk treatment considerations should include (1) inherently safe design options such as minimising inventories to reduce consequences, using non corrosive materials for tank construction and simplying filling process to reduce errors and (2) defence in depth controls which include adequate level control systems with sufficient SIL levels and external bunding to prevent external damage while containing any spills. FULL BOWTIE, CONSEQUENCE LOPA Analyses should be performed to ensure risk is reduced to ALARP.	Independent level protection systems that alerts/ alarms operator and if no response automatically shuts system down. Bunds installed that collect all liquid spills. Vapour detection sensors installed that detect all vapour losses. Fire/vapour suppression options should be investigated for bund fires and when vapour detected	Pre-start checks required on all tanks prior to filling and on a routine basis (e.g. daily) to ensure all controls are functioning as required and that levels and pressures are as expected and within acceptable limits.	Executive manager responsible for entire tank farm operations
Petrol which is flammable and explosive	Leak in feed line, leak from tank, leak from outlet piping	Corrosion, cracking, external damage, loosened flange		5	5	3	3	5 Catastrophic	Moderate	High				
External fire or extreme temperatures	Overpressurising tank	Excessive external heat, failed pressure relief/vent equipment	Vapour releases resulting in a vapour cloud explosion causing multiple fatalities/injuries, severe equipment damage and onsite environmental damage	5	5	3	3	5 Catastrophic	Unlikely	High	Unacceptable – Operations do not to continue until risk is reduced with risk treatment that removes external fire threats and safe designs with controls (such as sprinkers) to withstand exteme weather conditions. FULL BOWTIE, CONSEQUENCE LOPA Analyses should be performed to ensure risk is reduced to ALARP.	Automatic (or manual) deluge systems should be considered for tanks vulnerable to excessive heat from environment or other tank fires.	Daily checking that deluge systems are functional (fault free)	
Mixing of incompatible liquids	Petrol pumped into diesel tank	Connection error, misidentification of liquid and/or tank	Contaminated inventory that if undetected could be sold to customers causing damage to vehicles	2	2	1	4	4 Major	Moderate	Significant	Action as a high priority to identify risk control measures to reduce risk of putting petrol into diesel tank and for detecting when contamination has occurred.	Computer system interlocks that prevent petrol being pumped into diesel tanks.	Pre-unloading sampling to verify composition of contents to go into tank. Sampling to verify composition of tank contents prior to unloading tank.	Senior manager responsible for unloading operations
Falling objects	Overhead piping, fittings etc fall from tank	Corrosion, wear and tear, damage, incorrect fitment	Person injuried when struck by objects falling from height (leaks covered above)	4	2	1	1	4 Major	Rare	Medium	Action to reduce risk by regularly verifying integrity of tank and equipment	Only tanks certified as having requisite integrity of all components and fittings are used in operations	Routine and post-change integrity measurement and inspections on tanks	Manager responsible for tank integrity & maintenance
Lighting	Ignition of vapour in tank	Lightning strike on tank	Explosion in tank prod, destroying tank causing potential multiple fatalities/injuries, severe equipment damage and onsite environmental damage	5	5	3	3	5 Catastrophic	Rare	High	Unacceptable – Operations do not to continue until risk is reduced with risk treatment that removes external fire threats and safe designs with controls to withstand/mitigate exteme weather conditions. FULL BOWTIE, CONSEQUENCE LOPA Analyses should be performed to ensure risk is reduced to ALARP.	Tanks should be fitted with lightning rods or equivalent to protect against lighting strikes. Weather warning system that evacuates people to safe shelters when lightning in vicinity.	Routine checking and testing of lightning protection and weather warning and evacuation systems	Manager responsible for tank integrity & maintenance

FIGURE B6 SAMPLE/EXAMPLE ONLY of preliminary hazard analysis for tank filling.

(e.g. temperature, pressure, flow and level) and FMECA assesses failures in components (e.g. valves, pumps, pipes and sensors). These two techniques should produce complementary, but sometimes unique, outcomes depending on rigour and expertise involved. Both these approaches will be demonstrated in the next sections followed by a discussion of risk treatment.

B.3.2 Example of hazard and operability (HAZOP) study

The HAZOP process (as described in Chapter 5) is demonstrated here on the tank filling case study using the process and worksheet, as illustrated in Fig. B7. The same risk matrix (Fig. B5) is used.

Details of the steps undertaken to perform the HAZOP are as follows:

a. Select node for analysis.
For the tank filling, examples of hazards of nodes are:
  – tank inlet piping from upstream pump into tank
  – tank
  – tank outlet piping to downstream road tanker fill station

b. Identify parameters.
For the tank filling, examples of parameters include:
  – level
  – temperature
  – pressure
  – flow
  – operator response/task [insert descriptor]

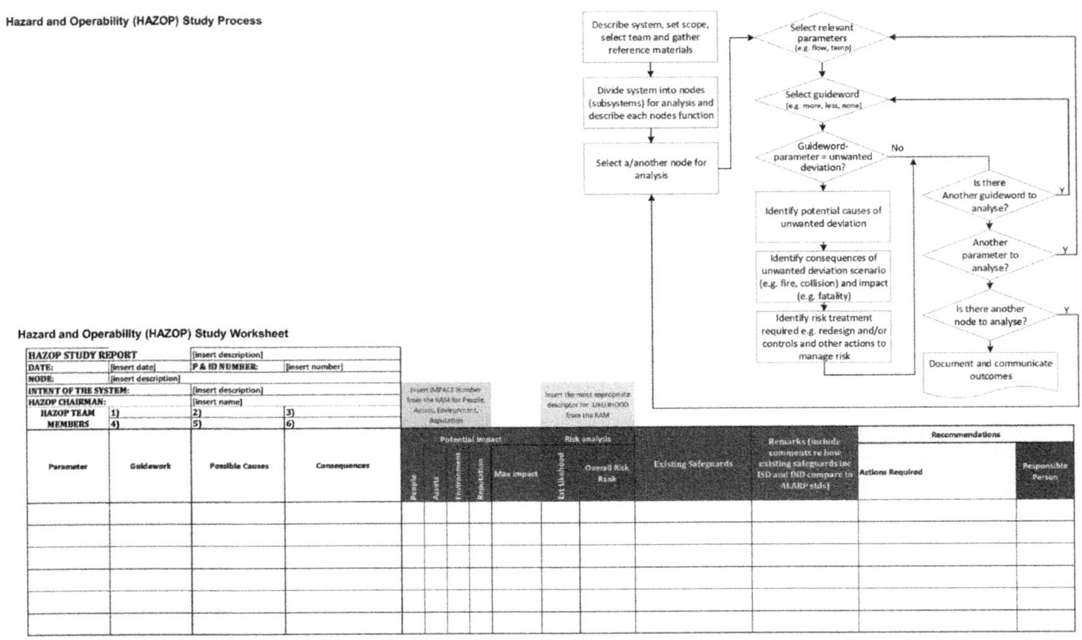

FIGURE B7 HAZOP process and worksheet used for tank filling case study. *HAZOP*, Hazard and operability.

c. Identify guidewords.
For the tank filling, examples of guidewords include:
- no
- more
- less
- as well as
- part of
- reverse
- other than
- sooner
- later

d. Use parameter-guideword combinations to identify consequences of significance. The consequences should detail the immediate event as well as the final impact on the objectives of value (e.g. safety of people, assets, environment and reputation).
For the tank filling, examples of parameters-guideword combinations and their significant consequences are as follows:
- More level: Tank overflows into bund and fuel catches on fire.
- More pressure: External heat causes overpressurisation of vapour in tank resulting in vapour release which explodes causing multiple fatalities, major asset and reputational damage.
- Reverse flow: Fuel flows out of tank back through inlet piping and spills at dock causing environmental harm.
- Other than: Operator chooses wrong tank and puts diesel into petrol tank.

e. Rank the cause–consequence scenarios using a risk matrix.

f. Make an assessment of likelihood to determine the risk ranking. The risk ranking exercise should refer to an organisation's risk matrix or risk appetite statement. *It is important to note that if the consequence is catastrophic then the highest level of action should be required regardless of the assessment of likelihood.* This is because:
- process safety events are often considered rare and the assessment of rare events has been proven to be inaccurate
- even if an accurate likelihood assessment is made (e.g. 1 in 100 years), it does not mean that the event will not occur today, tomorrow or the next day so it needs to be actioned

g. Identify any existing safeguards (i.e. prevention and/or mitigation controls) and any control support activities (e.g. monitoring and maintenance of controls).

h. Remark on whether current system has an inherently safe design and defence in-depth (DiD) layers of protections when compared with industry standards (for the risk profile).

i. Identify if any actions are required to address gaps in (1) existing safeguards and their required support activities and gaps, and (2) when compared against ALARP/industry standards.

j. Assign accountability for implementing the recommendations to an appropriate person. Guidance on the organisational level role that should be assigned the accountability can also be provided at the bottom of the risk matrix.

For the tank filling, an example assessment of risk done using a HAZOP sheet is shown in Fig. B8.

Parameter	Guidework	Possible Causes	Consequences	Potential Impact: People	Assets	Environment	Reputation	Risk analysis: Max Impact	Est Likelihood	Overall Risk Rank	Existing Safeguards	Remarks (include comments re how existing safeguards inc ISD and DiD compare to ALARP stds)	Recommendations: Actions Required	Responsible Person
Flow	No	Blocked inlet line, blocked filter, inlet valve failed in shut position, operation doesn't start filling process.	Tank doesn't fill, fuel is put elsewhere	1	1	1	1	1 Insignificant	Unlikely	Low	Regular cleaning and maintenance should minimise risk of blockages plus control room operator is required to monitor fill	A. Additional safe guards should be installed to ensure blocked lines do not result in lines bursting due to overpressurisation situations	1. Consider mass balance system that alerts operator of any deviations. Investigate if there are lines that could be overpressurised that require pressure detection and response systems installed	Process engineer
	No	Blocked outlet pipe	In inflow continues, tank overfills and fuel goes into a bund where it could catch on fire or form vapour cloud that explodes killing people, damaging assets and contaminating environment	5	4	3	3	5 Catastrophic	Moderate	High	Level control system, overflow pipe that puts spilling liquid into bund, emergency response.	B. Should consider hydrocarbon detection and vapour and fire suppression for bund	2. Initiate BOWTIE, QRA/LOPA study to identify and implement control suite that reduces risk to ALARP level	Site CEO
	More	Upstream pump malfunction	Tank fills too quickly and tank overfills as per above. Could also cause an increase in vapour that triggers pressure relief system which should go to flare	5	4	3	3	5 Catastrophic	Unlikely	High	Level control system, overflow pipe that puts spilling liquid into bund, emergency response. Pressure relief system that vents to flare when overpressure measured	C. If flare system failed, in certain situations vapour could be released, form a cloud and explode	3. Initiate BOWTIE, QRA/LOPA study to identify and implement control suite that reduces risk to ALARP level	Site CEO
	Reverse	Inlet valving left open	Tank drains out inlet piping which could leak into environment or flow into other tank causing contamination. If leaks to environment could catch on fire	2	2	4	4	4 Major	Moderate	Significant	None	D. Install one-way valve on inlet piping to prevent reverse flow	4. Install one-way valve on inlet piping to prevent reverse flow	Site Engineering Manager
Level	No	Tank leaked all fuel, level sensor failure	Loss of containment of fuel into bund which could catch on fire causing smoke inhalation, equipment damage, environmental pollution	2	4	3	3	4 Major	Moderate	Significant	Bund and emergency response	See B. above	Should be included as part of 2. above	Site CEO
	More	Level sensor failure, Operator fails to respond to high level alarm, HH level protection fails	Tank overflows and fuel goes into bund where it could catch on fire or form vapour cloud that explodes killing people, damaging assets and contaminating environment	5	4	3	3	5 Catastrophic	Moderate	High	Level control system comprising LAH, and HHL shutdown, overflow pipe that puts spilling liquid into bund, emergency response.	See B. above	Should be included as part of 2. above	Site CEO
Control room operator selects tank for fuel to be transferred into	No (tanks selected)	Operator absent/distracted	Fuel transfer not started	1	1	1	1	2	Moderate	Medium	Shift rostering to ensure sufficient and backup control room operators	None	Continue to monitor	Operations superintendent
	More (two tanks selected)	Permissible according to procedures	Can be difficult to monitor two separate tanks so overfilling more likely and it will spill into bund but volume will be less. Still could have pool fire in bund	4	4	2	2	4 Major	Unlikely	Significant	Level control system comprising LAH, and HHL shutdown, overflow pipe that puts spilling liquid into bund, emergency response.	See B. above	Should be included as part of 2. above	Site CEO
	Other Than (transferred to wrong tank)	Distracted/confused operator, misidentified fuel type	Fuel contamination e.g. petrol mixed into diesel which could damage customers' engines	1	4	1	4	4 Major	Likely	High	Having two operators confirm tank	E. Investigate options for mistake-proofing tank selection	5. Intiate Human Factors project to create an error-proof means of selecting tanks to receive fuel transfers	Site CEO

FIGURE B8 SAMPLE/EXAMPLE ONLY of HAZOP worksheet for tank filling. *HAZOP*, Hazard and operability.

B.3.3 Example failure modes effects and criticality analysis (FMECA)

Failure modes effects criticality analysis (FMECA) seeks to identify and assess impacts of equipment and/or component failures within a system. The FMECA process (as described in Chapter 5) is demonstrated here on the tank filling case study using the process and worksheet illustrated in Fig. B9. The same risk matrix (Fig. B5) is used.

Details of the steps undertaken for perform the FMECA for the tank filling are as follows:

a. Select a component for analysis.
Examples include:
- inlet piping
- inlet flow control valve
- tank
- pressure relief valve
- control room operator

b. Determine component function.
Examples of component functions include the following:
- Piping holds mass and permits flow
- Valves hold mass and regulate flow
- Operators select tank, connect pipes and stop flow

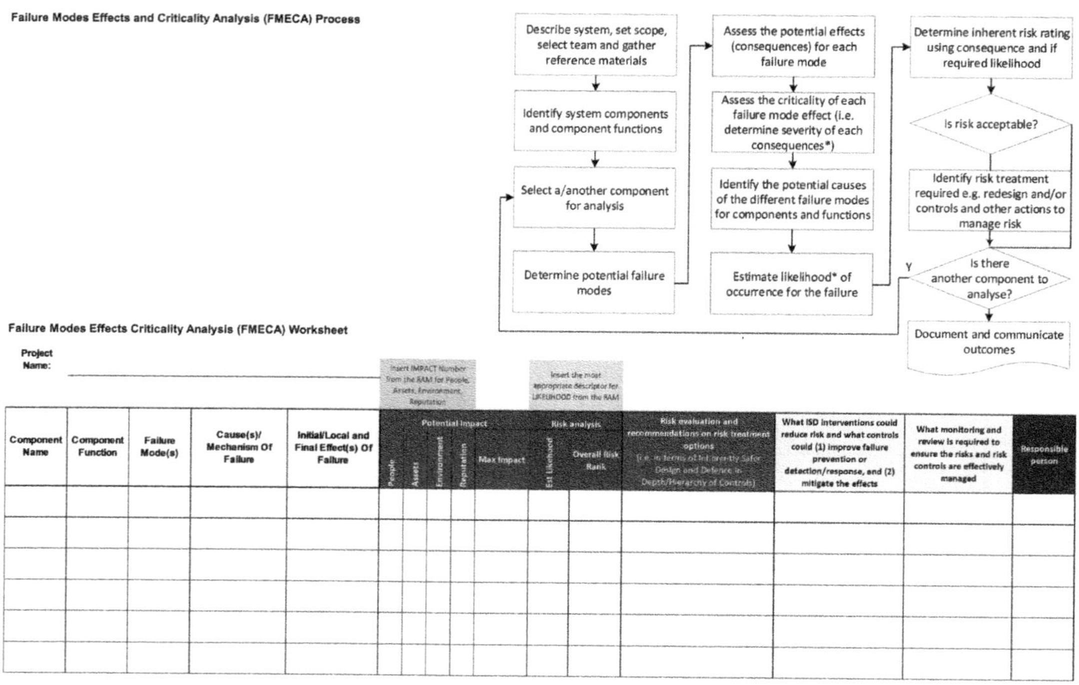

FIGURE B9 FMECA process and worksheet used for tank filling case study.

c. Determine potential failure modes.
Examples of failure modes are as follows:
- Piping releases mass, blocks flow
- Valves release mass, block flow and do not stop flow
- Operators incorrectly select tank, connect wrong pipes and fail to stop flow

d. Identify possible causes or mechanisms of the different failure modes.
Examples of threats that could lead to a release of petrol are as follows:
- Piping leaks (e.g. releases mass) could be caused by corrosion, external damage and incorrectly connected flanges.
- Valve failures could be caused by valve wear, corrosion and damage or failure of subcomponents.
- Operator error/mistake which could be caused by distraction, unavailability, operator attending other more urgent task, or operator judgement or input error.

e. Identify the range of consequences – the initial and final outcomes of the failure. Consequences could have adverse impacts on humans (e.g. fatality, injury, illness and disease), environment, production, assets, financial performance and reputation. These consequences should match the unwanted scenarios listed in scope.
Examples of consequences in scope are as follows:
- Release of petrol liquid that spills into bund. Liquid could ignite leading to pool fire that results in people suffering from smoke inhalation, environmental pollution, asset damage and reputational harm.
- Release of vapour or formation of vapour from spilt fuel that could form vapour cloud that explodes causing multiple fatalities, major asset and production losses, environmental contamination and significant reputational issues.
- Equipment failure/destruction resulting in projectiles or falling objects striking people causing serious injury.
- Contaminated product causing reputational damage and potentially financial and legal issues.

f. Rank the cause–consequence scenarios using a risk matrix.

g. Make an assessment of likelihood to determine the risk ranking. The risk ranking exercise should refer to an organisation's risk matrix or risk appetite statement.

h. Determine risk treatment recommendations by first considering whether the risk is tolerable/to ALARP or intolerable and needs to be treated. Guidance on this assessment is often provided at the bottom of the risk matrix.

i. Identify whether there are inherently safer design options that should be implemented or investigated and if any prevention and mitigation controls and control management requirements that are needed to reduce the risk to ALARP standard.

j. Assign accountability for implementing the recommendations to an appropriate person. Guidance on the organisational level role that should be assigned the accountability can also be provided at the bottom of the risk matrix.

For the tank filling, an example of FMECA risk assessment is shown in Fig. B10.

Component Name	Component Function	Failure Mode(s)	Cause(s)/Mechanism Of Failure	Initial/Local and Final Effect(s) Of Failure	Potential Impact					Risk analysis		Risk evaluation and recommendations on risk treatment options [i.e. in terms of inherently Safer Design and Defence in Depth/Hierarchy of Controls]	What ISD interventions could reduce risk and what controls could (1) improve failure prevention or detection/response, and (2) mitigate the effects	What monitoring and review is required to ensure the risks and risk controls are effectively managed	Responsible person
					People	Assets	Environment	Reputation	Max Impact	Est Likelihood	Overall Risk Rank				
Tank and pipework	Hold mass - fuel in liquid and/or vapour form	Release mass - leak	Corrosion, damage, incorrect fitment	Release of large amounts of fluid causes environmental damage, and/or catches on fire/explodes resulting in multiple injuries/ fatalities, significant asset damage and environmental harm	5	5	4	4	5 Catastrophic	Moderate	High	Unacceptable – Operations do not to continue until risk is reduced. FULL ADVANCED BOWTIE, CONSEQUENCE LOPA Analyses should be performed to ensure risk is reduced.	Risk treatment considerations should include (1) inherently safe design options such as corrosion resistant materials and isolating pipes/tank from damaging energies and (2) defense in depth controls which include adequate leak detection and response, vapour detection and response and bunding.	Daily checks that leak and vapour detection and response systems are fully functioning and that bunding has required integrity and bunds are empty with drain valves closed. Routine drills conducted to verify fire fighting and evacuation systems are complied with and effective	CEO of tank farm operations
	Permit fuel to flow/transfer	Block transfer/ flow of fuel	Sludge/ contaminants in fuel	No outflow could lead to overflow of liquid into bund resulting in pool fire that causes environmental and asset damage	3	4	4	3	4 Major	Unlikely	Significant	Tolerable if actioned as a high priority. BASIC BOWTIE analysis should be used to demonstrate industry recognised risk controls measures are identified, implemented, monitored and reported on.	Investigate inherently safe design for overflow system (e.g. overflow directed to empty tank) and controls that clean tank when sludge level is close to outlet/tank can be cleaned	Monitoring of sludge levels / sludge level detection system	Tank farm operations manager
Filter	Remove debris/ particles from feul	Captured contaminants block filter	Low/slow flow	Increase pressure causes line to spring a leak that could ignite causing jet or pool fire that damages equipment and might result in smoke inhalation	2	2	2	1	2 Minor	Likely	Medium	Tolerable if actioned with industry recognised design standards and risk controls measures are implemented.	Determine inherently safest combination of pressure vs pipe design. Controls should include pressure detection and response system for lines, leak detection and fire suppression	Daily checks that pressure and leak detection and response systems are fully functioning . Routine drills conducted to verify fire fighting and evacuation systems are effective	Engineering Superintendent
		Holed filter	Sludge/ contaminants in fuel	Contaminated feul in tank cannot be sold or if undetected goes to customer where it may cause blockages with their system	1	2	1	3	3 Moderate	Unlikely	Medium	Tolerable if actioned with industry recognised design standards and risk controls measures are implemented.	Fuel quality check and response system for all stored fuel	Check sampling is done and lab equipment is calibrated so quality checks are accurate	Maintenance Superintendent
Control Room Operator AND Inlet and outlet valves	Regulate flow of fuel into and out of tank	Outlet gate valve fails to open	Corrosion, damage, malfunction, operator fails to open valve when required	No outflow or failing to shut inflow could lead to overflow of liquid into bund resulting in pool fire that causes environmental and asset damage	3	4	4	3	4 Major	Moderate	Significant	Tolerable if actioned as a high priority. BASIC BOWTIE analysis should be used to demonstrate industry recognised risk controls measures are identified, implemented, monitored and reported on.	Investigate back up valve to shut inlet and inherently safe design for overflow system (e.g. overflow directed to empty tank)	Regular maintenance performed on all safety critical valves (as per manufacturers recommendations) and daily operational checks to verify they are fully functional	Tank farm maintenance manager
		Inlet valve fails to close	Wear and tear, damage, malfunction, operator fails to shut valve when required		3	4	4	3	4 Major	Moderate	Significant				
Pressure relief valve	Releases pressure of fuel vapour in tank if it is too high	Fails and opens when shouldn't	Spring failure, incorrect valve, incorrect fitment, corrosion, damage	Releases vapour potentially to atmosphere, resulting in some vapour igniting/exploding which could damage equipment	3	3	2	2	3 Moderate	Unlikely	Medium	Tolerable if actioned with industry recognised design standards and risk controls measures are implemented.	Ensure inherently safest pressure relief valve system installed along with vapour detection and response	Regular maintenance performed on all pressure relief systems (as per manufacturers specs) and routine checks to verify they are fully functional	Maintenance Superintendent
		Fails closed and doesn't open when it should	Corrosion, damage, malfunction	Tank overpressurises causing it to rupture leading to massive loss of containment which could lead to pool fire and/or vapour cloud explosion that injuries/kills multiple people, severely damages equipments and onsite environment	5	5	4	4	5 Catastrophic	Unlikely	High	Unacceptable – Operations do not to continue until risk is reduced. FULL BOWTIE, CONSEQUENCE LOPA Analyses should be performed to ensure risk is reduced.	Risk treatment considerations should include (1) inherently safe design for tank sizing, constuction and siting (e.g separation from other tanks) and (2) defense in depth controls which include adequate pressure detection and response, and pressure relief system.	Regular maintenance performed on all pressure detection and response system and pressure relief systems (as per manufacturers specs) and routine checks to verify they are fully functional	CEO of tank farm operations

FIGURE B10 SAMPLE/EXAMPLE ONLY of FMECA worksheet for tank filling.

B.3.4 Summary of risk assessment results using HAZID, HAZOP and FMECA

In summary,

- HAZID identifies risks associated with hazards using hazard sources as prompts.
- HAZOP identifies risks associated with deviations in critical process parameters.
- FMECA identifies risks associated with failures of equipment or components.

Each approach provides prompts to guide the risk assessment process. The completeness and accuracy of the process will depend on the people involved and the match between the techniques used and the potential risks associated with the system. For safety critical systems, it is often advisable to use more than one technique.

B.4 Risk treatment

If risk treatment is recommended for any hazards or hazardous scenarios identified in the HAZID, HAZOP or FMECA processes, then this usually involves the following steps:

– Determining if a safer design could be implemented using inherently safer design (ISD) principles for process safety events as shown in Fig. 7.6 and/or the hierarchy of risk treatment options as shown in Fig. 5.13.
  For the tank filling, applying the ISD principles involves the following steps:
  - Elimination: Eliminating the use of petrol which in this case is not an option.
  - Minimisation: Minimising the tank holding volumes and flow rates should be considered to reduce potential spill volumes. However, this assessment should consider other risks associated with having more smaller tanks.
  - Substitution: Substituting the flammable, explosive petrol with less volatile petrol which in this case is not an option.
  - Moderation: Current operations are already operating at atmospheric temperature and pressure. Refrigeration could be considered but this should only be done if risk reduction associated with cooler petrol would outweigh the cost and risk increases associated with refrigeration.
  - Simplification: Simplification of the piping, valve location and status, tank level detection and vapour sensing system should be implemented to ensure operators can easily detect current and project future status of the the system and whether it is in, or trending towards, a safe or unsafe state.

  Applying the hierarchy of risk treatment options to the tank filling involves the following steps:
– Eliminate hazard: Eliminating the use of petrol is not an option [as per ISD].
– Substitute hazard: Substituting the flammable, explosive petrol with less volatile petrol is not an option [as per ISD].
– Isolate hazard: Isolation should consider remote placement of tank in bunded and fire/blast protected area that is not near community, key assets or environment. Remote operations should also be considered so people do not have to go near the tank.

- Eliminate threats: Threats that can release the hazard include component damage, wear corrosion, incorrect operation etc. (as identified in HAZID).
- Implement prevention and mitigation controls: Using bowtie analysis and defence-in-depth DiD approaches to identify the controls needed to prevent the unsafe situation or should it occur mitigate the consequences to an acceptable level.

B.4.1 Defence-in-depth (DiD) analysis overview

If the risk treatment option for managing the hazard requires the implementation of controls, then the main approaches used in the process industries deploy the DiD principles, as shown in Fig. 7.8.

For the filling of a bulk fuel tank, application of DiD should consider inherently safe design options (as described in the previous section) followed by:

a. Basic process control system (BPCS) that receives information about the level of the liquid in the vessel and uses it to control inlet or outlet flows from the vessel. The aim of BPCS is to prevent unsafe operations by sustaining the tank within its normal operating levels.

b. Critical alarms with operator response describe layers of protection that comprise alarms to alert operators that the system is drifting to potentially unsafe operations that require operator response to bring the system back into its normal state. Examples of critical alarms requiring operator response for a tank include high and low-level alarms on vessels.

c. Safety instrumented systems (SIS) are automated control system responses to unsafe situations. These systems act independently of operators and the basic process control system. Examples of safety instrumented systems include very high (or high–high) or very low (or low–low) level measurement on a vessel that initiates action to open or close the outlet valves. The aim of safety instrumented systems is to prevent an unsafe operating state escalating into an unwanted event such as fuel overflowing.

d. Active protection devices and equipment that automatically activate in unsafe or abnormal situations in order to mitigate the consequences of accidents. Examples for tanks include pressure relief valves and automatic fire suppression systems.

e. Physical protection barriers also aim to mitigate or contain the consequences of accidents. Examples of physical protection barriers associated with tanks include overflow drains, bunds, blast walls/barriers and collection ponds.

f. Plant emergency response typically comprises fire fighting, first aid, medical treatment and environmental contamination response capabilities. The purpose of having plant emergency response is to mitigate accident consequences by having people available who can respond quickly and who are specifically trained to address the potential accident scenarios that could occur at a given plant.

g. Community emergency response is the external public emergency personnel who respond to emergency call numbers (e.g. 000 in Australia, 112 in Europe, 119 in Japan, 112 in India, 911 calls in North America and 999 in the United Kingdom).

B.4.2 Bowtie analysis

Bowtie analysis (as discussed in Chapter 5) can be used to visually show and analyse the adequacy of the DiD approach for a given system. The bowtie diagram should make it easy to ensure all threats and consequences are identified for a given unwanted event and that there are:

- prevention controls identified to address each threat so the likelihood of the unwanted event is reduced
- mitigation controls identified to address each consequence that could result from unwanted event so that the likelihood and severity of each consequence are reduced.

The development of the bowtie will be demonstrated using the approach described in Chapter 5. The first step is documenting the scope.

The next step is to determine the centre (knot) of the bowtie. This should represent the point at which the operation goes into an unsafe (no longer normal) state. It should not be the consequence (e.g. it should not be the ignition, fire or explosion). For safety bowties, this is very important as we want people to know whether they are operating in the safe or unsafe zone.

For the tank filling, an example of the knot of the bowtie is 'loss of containment of fuel from a tank and piping system'.

The next step is to determine the consequences that are in scope. Reviewing the event tree and brainstorming with stakeholders can help to determine the consequences that matter. Note consideration should be given to secondary consequences that matter. For example, if a person survived the fire and explosion, they could still be killed or permanently disabled due to secondary events such as asphyxiation or acute illnesses from breathing in fuel fire smoke.

Safety-related consequences that should be considered are:

- liquid spill that could catch fire causing a pool fire that injures people and damages assets and the environment.
- vapour release that could cause a toxic plume and/or form a vapour cloud that explodes both of which could result in fatalities or injuries to people onsite and injuries to people offsite.

The next step is to determine the threats or causes that could lead to a loss of containment of fuel. Note that these threats are inherent to the system and must not include failed controls. These are threats "that of themselves" can cause a release of fuel if the controls were not implemented. Threats usually fall under categories like work area attributes, equipment wear, tear or fail, inadvertent operation, interactions with other entities and environmental conditions.

Threats that should be considered are as follows:

- Inadequate structural integrity of tank and pipes due to poor material selection, corrosion and/or damage.
- Mechanical and/or electrical failure of safety critical components (but not failed controls). Examples include inlet and outlet valves.
- Overfilling/overflowing of the tank.
- Application of external heat sources such as fire from another tank, lightning and hot work on the tank

Now that we have a skeleton bowtie we can populate it with both prevention and mitigation controls. This can be done by referring to fault and event trees as well as brainstorming and relevant past incidents and industry guideline reports. Through this process, a number of artefacts will be considered – not all of which will be controls. The determination of which artefacts are controls can be done using the filter shown in Fig. B11 and Fig. 5.21. Note that the other artefacts are often important too as there are activities that ensure that controls remain effective or check if the control is implemented and how it is working. These activities are still documented as control supports and control verifications.

An example bowtie for the tank filling is shown in Fig. B12A and B. An example control support diagram (ConSupMap) showing critical controls with their associated control support and verification activities is shown in Fig. B13. An example control specification sheet is listed in Table B3.

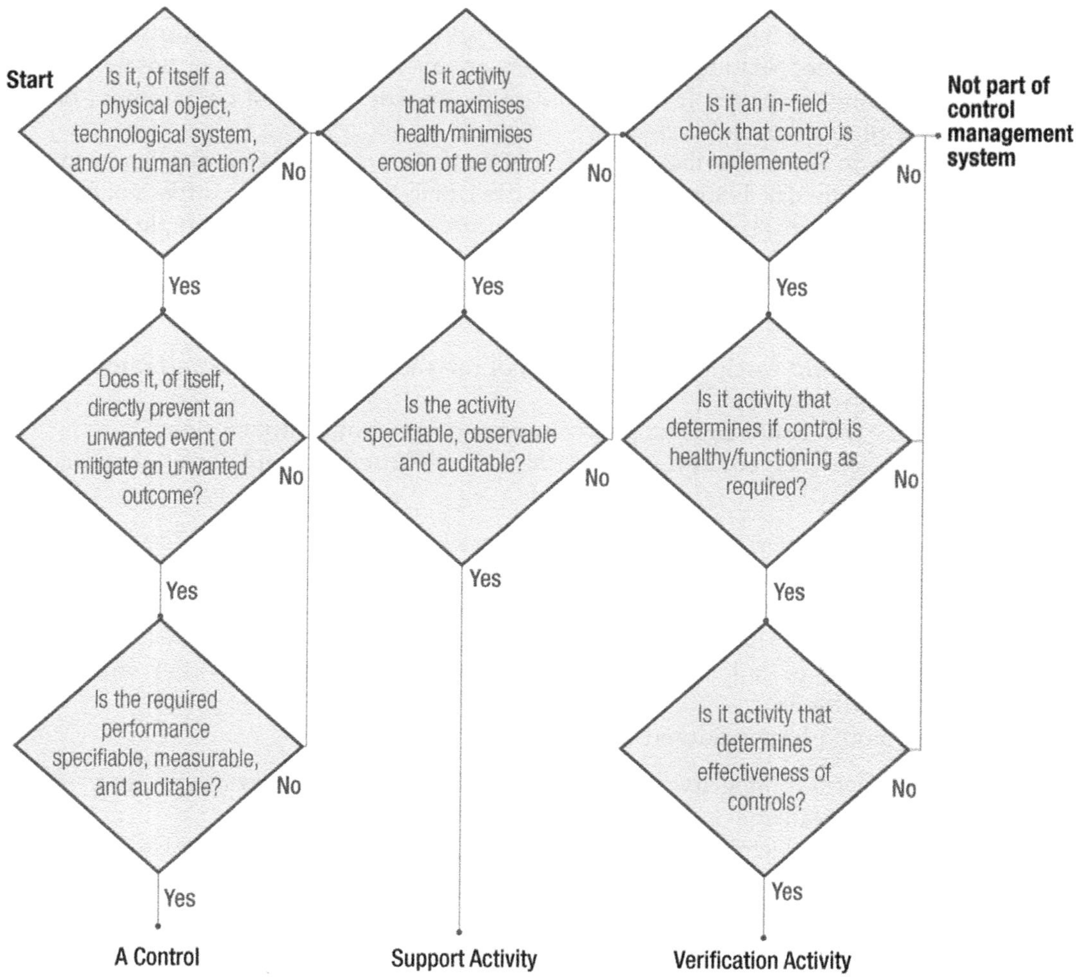

FIGURE B11 Filter to identify risk controls, and control supports and verification activities (Hassall & Harris, 2017).

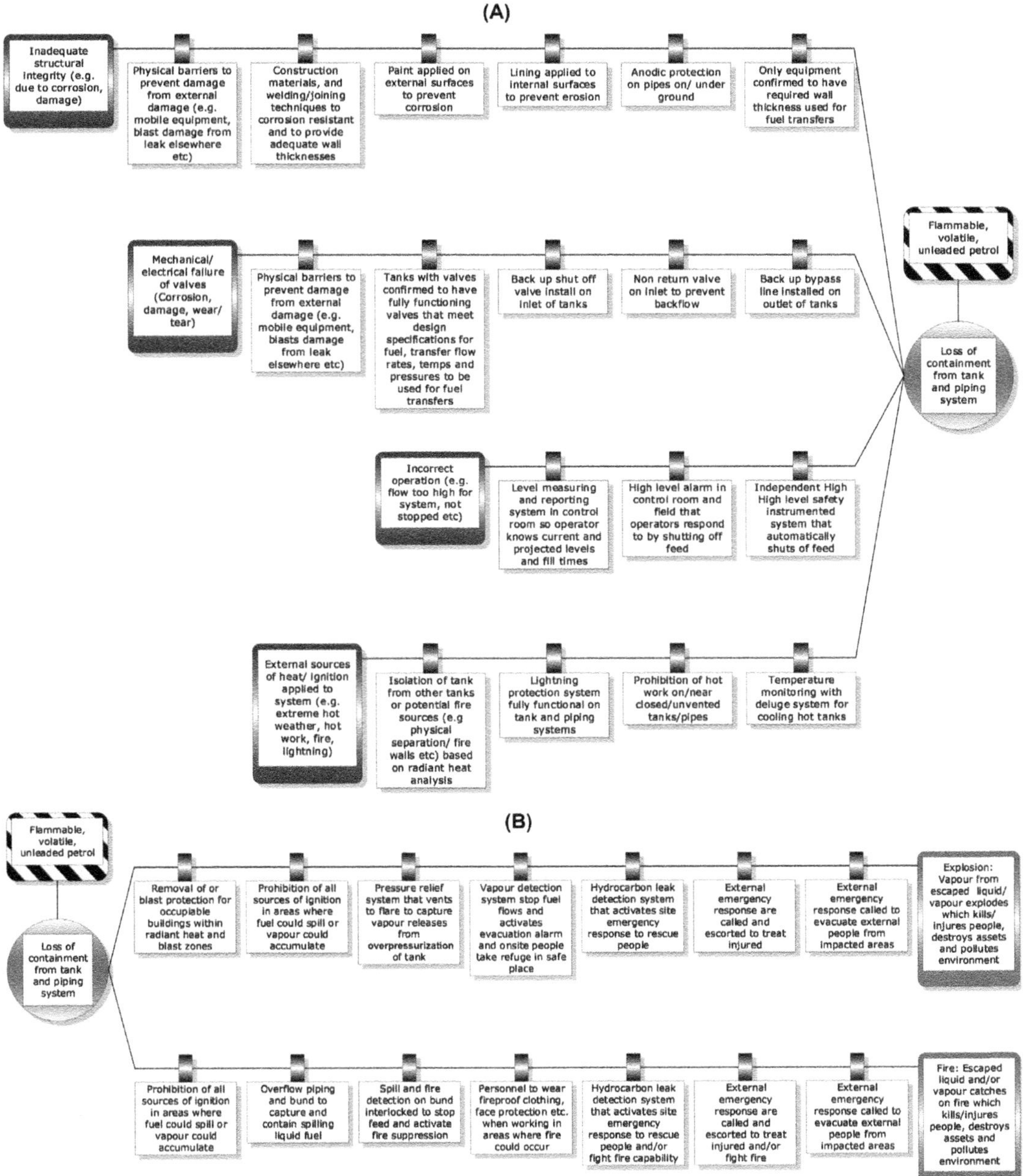

FIGURE B12 (A) Prevention (left) side of bowtie for loss of containment of fuel from tank. (B) Mitigation (right) side of bowtie for loss of containment of fuel from tank.

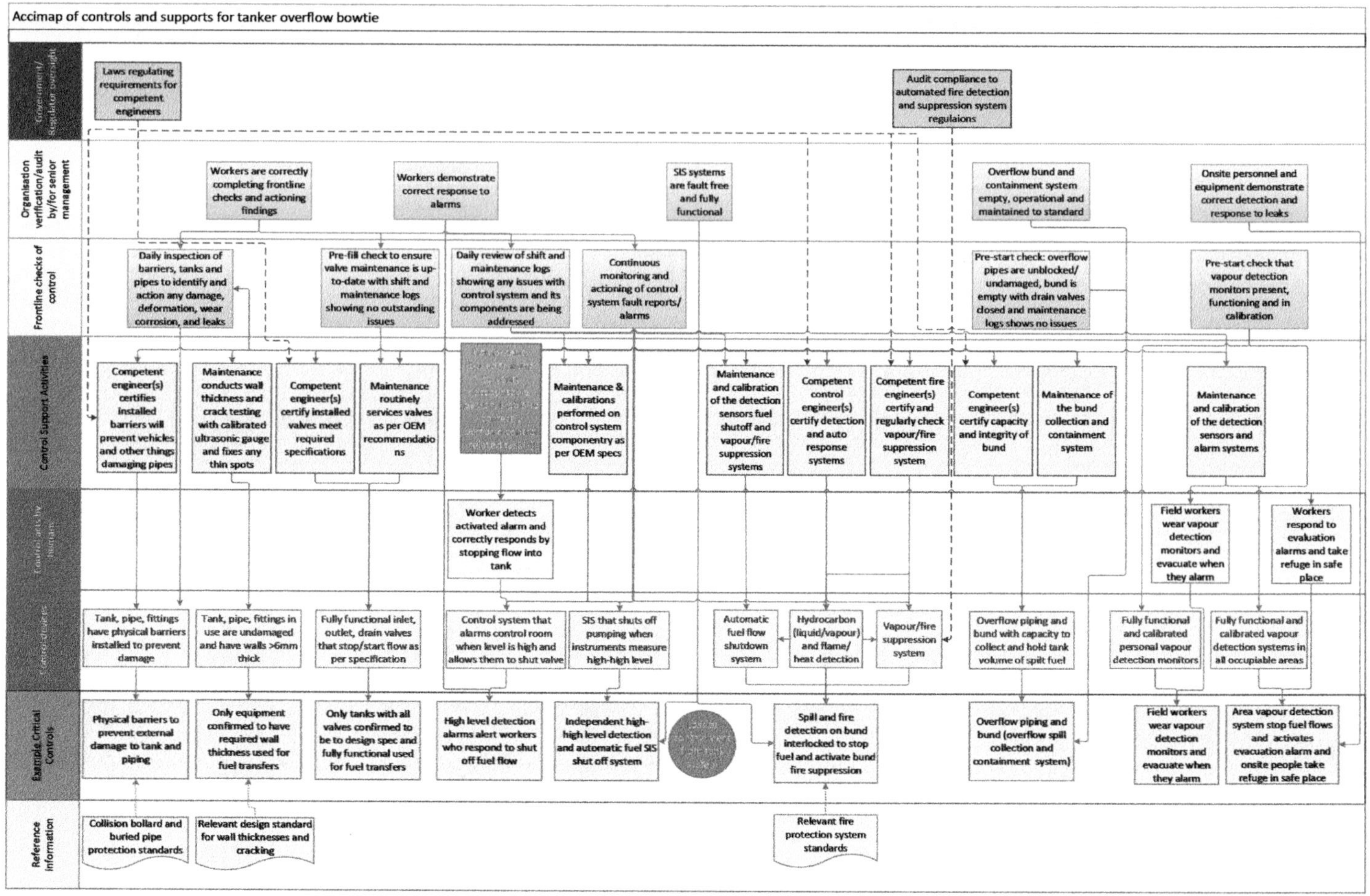

FIGURE B13 ConSupMap type presentation of key controls and control supports.

TABLE B3 Example control specification sheet.

1. What is the unwanted event?

Loss of containment of fuel from storage tank and piping system

2. Name of the control

Only equipment confirmed to have required wall thickness used for fuel transfer

3. What are the specific objectives of the control related to the unwanted event? Include a description of good/ideal control performance.

Objective:
To ensure fuel is contained by vessels with sufficient wall thickness to hold fuel at the range of flowrates, temperatures and pressures it might be transferred and stored at thereby preventing leaks from cracks, holes, worn joins etc)

Good/ideal performance:
Tank and all pipes and fittings to have minimum wall thickness as specified in [insert relevant standard] and tabulated below:

Pipe size (Nominal)	Minimum wall thickness
356mm to 405 mm	8 mm
203mm to 305 mm	6 mm
100mm to 152mm	5.5 mm
<100mm	4 mm

4. Support activity

Maintenance	Training	Other
- Wall thickness testing of tank, pipes and fittings every six months and replace/repair actual or predicted thin spots - Ultrasonic testing of wall to identify and correct any corrosion, cracking issues that could compromise wall thickness	Competency based training every 12 months for: - Operators doing daily checks - Maintainers doing thickness and ultrasonic testing - Maintainers doing wall repair/replacement	Mechanical engineer provides up-to-date version of pipe size vs wall thickness table for site

5. Control monitoring requirements

Worker daily visual inspection	Answer	Comments and actions
Is tank free from damage, corrosion, deformation and visual signs of leakage?	☐ Yes ☐ No ☐ Not applicable	
Is tank inlet piping free from damage, corrosion, deformation and visual signs of leakage?	☐ Yes ☐ No ☐ Not applicable	
Is tank outlet, overflow and drainage piping free from damage, corrosion, deformation and visual signs of leakage?	☐ Yes ☐ No ☐ Not applicable	
Manager Verification questions	**Answer**	**Comments and actions**
Has wall thickness of tank, all pipes and fitting been measured in last 6 months and any identified actual or predicted thin walled areas been actioned?	☐ Yes ☐ No ☐ Not applicable	
Has ultrasonic wall testing of tank, all pipes and fitting been measured in last 18 months and all identified issues been actioned?	☐ Yes ☐ No ☐ Not applicable	
Can workers explain what equipment needs to be inspected, how the inspection should be conducted and the actioning and reporting requirements	☐ Yes ☐ No ☐ Not applicable	
Are maintainers correctly using up-to-date version of pipe size vs thickness table when interpreting thickness results and doing repair/replacement work	☐ Yes ☐ No ☐ Not applicable	

The bowtie analysis can be used to subjectively assess the adequacy of a DiD approach. More objective approaches can also be used. These include fault tree analysis (FTA), event tree analysis (ETA), consequence modelling and layers of protection analysis (LOPA). These will be discussed and demonstrated in the following sections.

B.4.3 Fault tree and event tree analysis

Fault tree analysis (FTA) and event tree analysis (ETA) are techniques that can be used to visually show and analyse a DiD approach. An example of FTA is shown in Fig. B14. The fault tree was developed by applying deductive analysis and logic gates to determine specific causes or combination of causes – including equipment failure, human error and software failures – for undesirable events (Pilot, 2002). The process starts with identifying the top-level failure or fault condition then potential combination of the causes for the top-level failure or

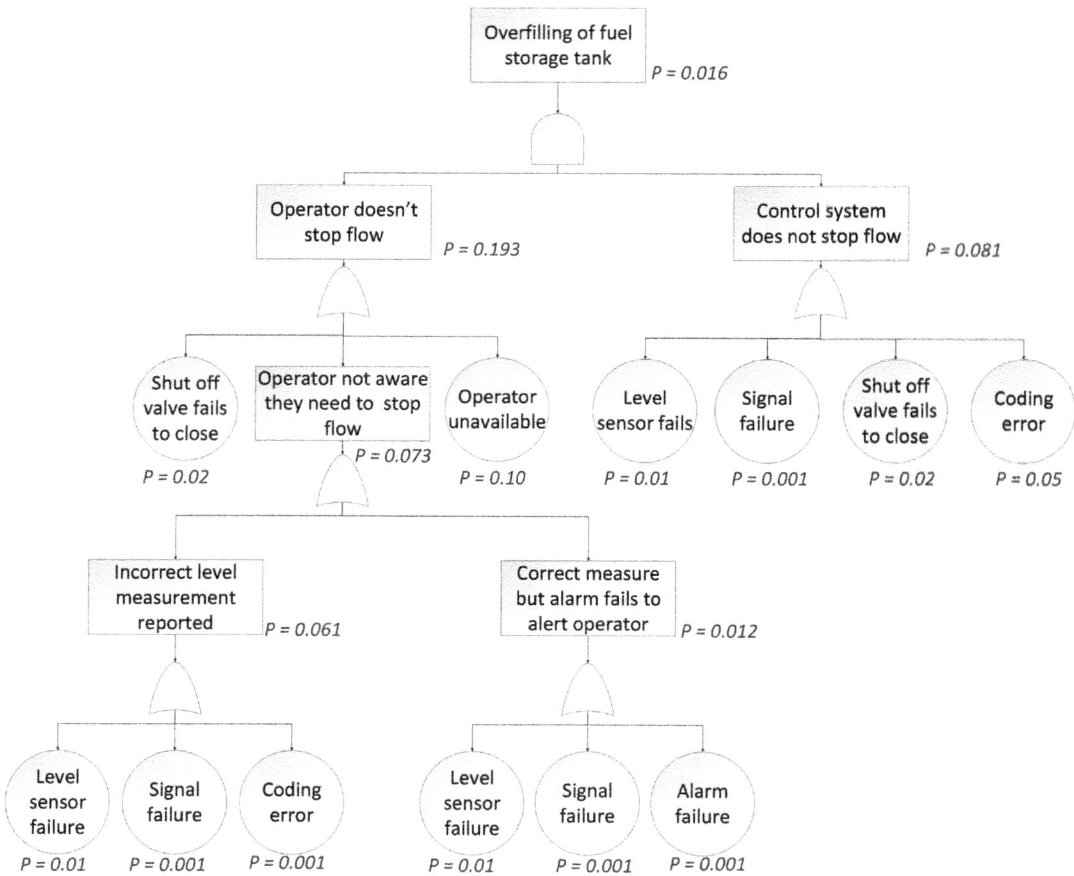

FIGURE B14 Fault tree analysis for overfilling a bulk storage tank (*blue numbers* show example and calculated probability of failure data).

fault. These combinations are shown using "AND" and "OR" logic gates. The reasons why these causes might occur are then identified and described on the next lower level of the diagram again with "AND" and "OR" logic gates. Breaking down each cause into subcauses with additional gates continues until a basic fault – equipment, human or software – has been identified. Thus FTA helps identify the potential causes and sequences of system failures. It can also include the quantitative probability of failure or reliability information to calculate the probability of occurrence of the top event. This is illustrated with blue text in Fig. B14. Note these numbers are examples only. The actual numbers used should be derived from historical records, test records, equipment manufacturers and relevant reliability databases.

The ETA explores the potential consequence pathways of an initial unsafe operating scenario taking into consideration the success and failure of responses that could occur along the event pathway. Probabilities for success and failure can be added to the tree to facilitate the determination of the probability of adverse versus non-adverse outcomes. An example of an event tree for overfilling a bulk fuel tank is shown in Fig. B15.

It is important to note that quantification of risk will only produce meaningful values in closed, unchanging systems where there is extensive and high-quality historical data. In most cases, however, quantified values of risk do not necessarily produce meaningful measures of risk but can be used in comparative analyses to assess which number and combination of controls deliver the most acceptable level of risk reduction.

The event tree and fault tree can provide insights into probabilities of failure and probability of consequences. The severity of consequences and effectiveness of controls can be assessed with consequence modelling and LOPA. Both of these are discussed in the following section.

B.4.4 Consequence modelling and layers of protection analysis (LOPA)

Once the risk has been assessed and controls have been identified, further analyses can be performed to estimate the severity of the consequences and the risk mitigation levels in a more quantitative/objective manner. Quantification of the consequences typically involves consequence modelling. Consequences of loss of containment events might include the release of a gas or vapour and/or the release of a liquid. On release, liquid can form pools and catch on fire or it can flash or evaporate to form vapour. In addition, vapour may also be directly released. If it is released under pressure, it could ignite creating a jet flame fire. Uncontained vapour can form plumes that travel downwind and/or form vapour clouds. Plumes or clouds can cause harm if they are toxic and people are exposed to them. If they are flammable, the plumes and clouds can ignite causing fires and explosions. Consequence models exist to determine discharge rates of released liquids and gases. They also exist for determining plume dispersion models, pool fires, jet fires, vapour cloud explosions, flash fires, and other forms of fire and explosions. There are also models for determining toxic gas effects, thermal effects and explosion effects. Refer to CCPS (2000) for more details.

As mentioned previously, for the tank filling, safety-related consequences that should be considered are as follows:

– Liquid spill that could cause a pool fire that injures people, and damages assets and the environment.

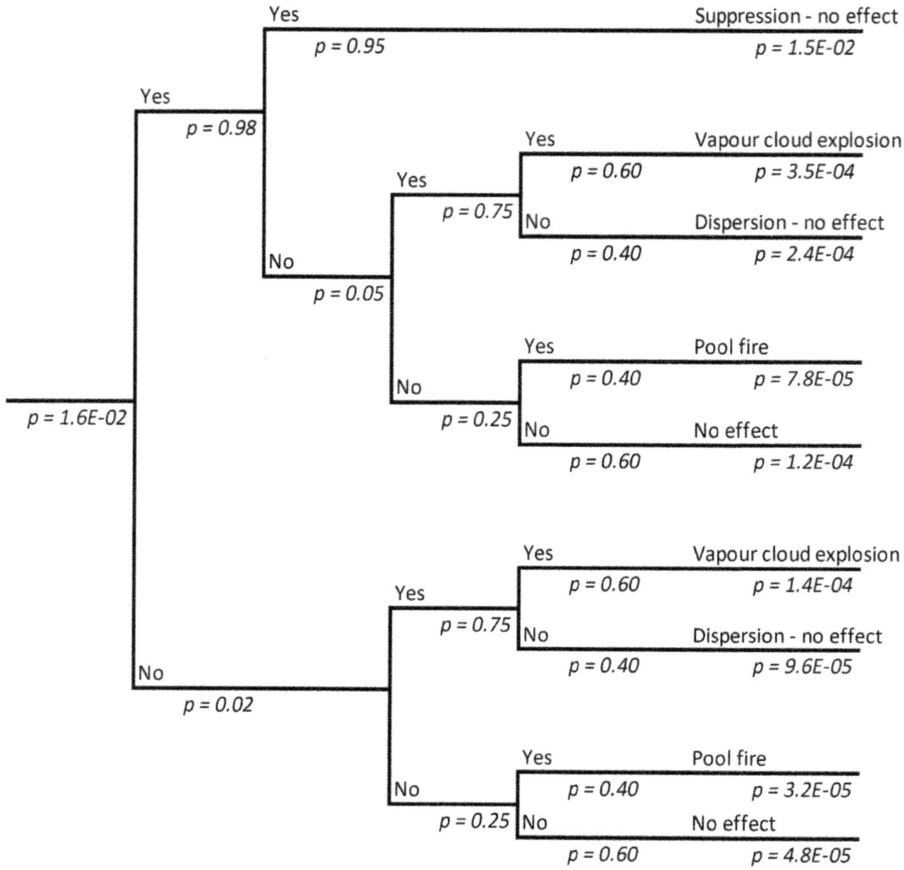

Probability of no consequences p = 1.5E-02
Probability of pool fire p = 1.1E-04
Probability of vapour cloud explosion p = 5.0E-04

FIGURE B15 Event tree analysis for overfilling a bulk petrol storage tank.

– Vapour release that could cause a toxic plume and/or form a vapour cloud that explodes both of which could result in fatalities or injuries to people onsite and injuries to people offsite.

Modelling the consequences of a pool fire in terms of injuries and fatalities can be done (using information from CCPS, 2000, example 2.30 and radiant heat impacts from Cameron and Raman, 2005, p. 263) to produce graphs as shown in Fig. B16. This information can then be used to determine the severity of a pool fire at various distances from the bund edge. The results can be superimposed on site maps that can be used to help determine the likelihood that people will be in the vicinity of the bund. The severity and assessments of likelihood can be used to more objectively determine the risk of injury and fatalities.

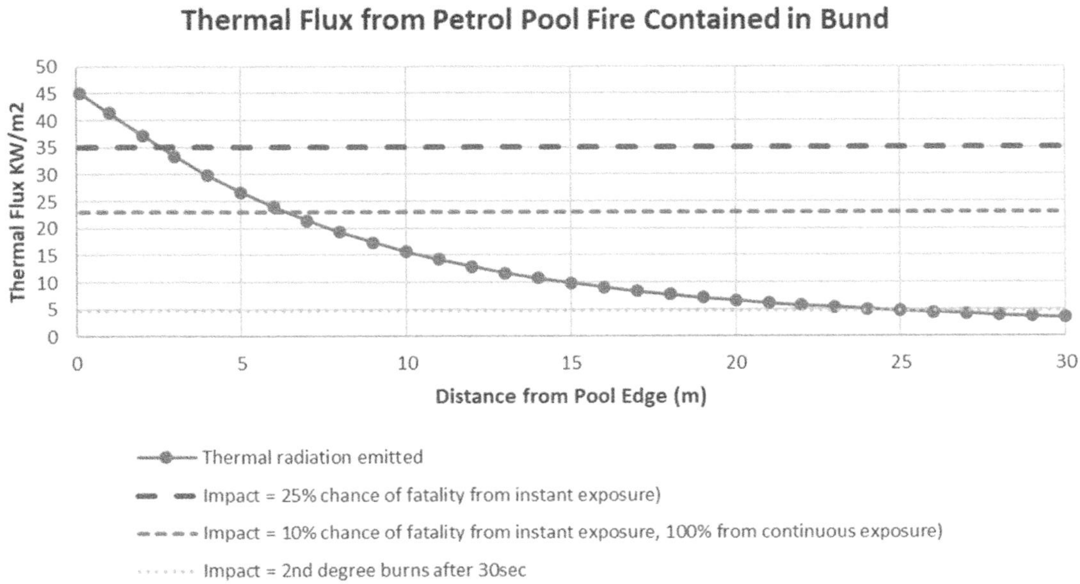

FIGURE B16 Model results of thermal radiation from pool fire.

In a similar fashion, modelling the release of vapour discharge and dispersion as well as vapour cloud explosions can also be done. Examples shown in Fig. B17 are for toxic exposure at level Acute Exposure Guidelines Level 1 (AEGL-1), which equates to reversible irritation and discomfort (these graphs were produced by following examples 2.13 and 2.16 from CCPS, 2000). AEGL-2 levels, which equate to irreversible, long lasting health effects, were not reached with this modelling analysis.

As shown above, consequence modelling can help to identify areas where adverse consequences might be experienced by people and assets that are exposed to these areas. Estimating the exposures of people as well as the probability of the event can then be used to estimate the inherent risk. Identifying the current controls and the probability that these controls might fail on demand can then be used to determine the mitigated risk. The mitigated risk number can be compared with company tolerability limits or industry ALARP standards to identify whether the controls are sufficient for this type of analysis. This assessment of the adequacy of the controls can be performed using a LOPA.

An example of LOPA sheet for a tank overfilling scenario is listed in Table B4.

B.4.5 Improving safety with human-centred design

It is important to note that critical controls often involve human actions (e.g. humans respond to high-level alarm). Control support activities also often involve human activities (e.g. maintenance and calibration of level sensors). Thus an analysis of human factors should be conducted to ensure work areas are designed and implemented in such a way that helps the users (1) detect and assess the situation in an accurate and timely manner and (2) respond in an accurate, efficient and safe manner (cf. chapter 3). A useful first step in performing a human factors

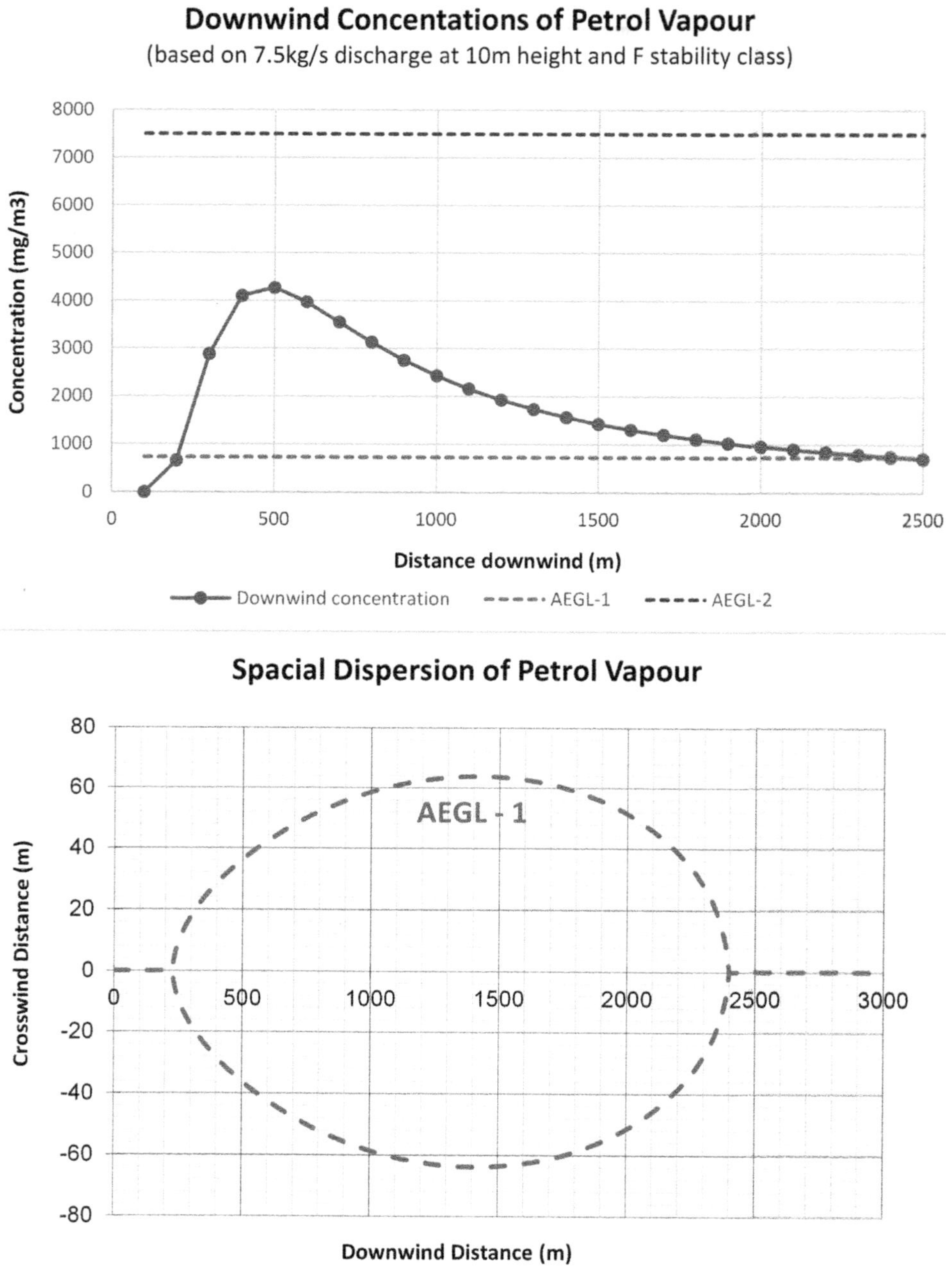

FIGURE B17 Example of a possible dispersion profile of petrol vapour escaping from tank.

TABLE B4 Example of layers of protection analysis sheet for overfilling tank.

ANALYSIS OF INDEPENDENT PROTECTION LAYERS (LOPA)				
Incident ID		Incident description		
Line or Equipment ID	Buffer Tank	Buffer tank overflows		
Document references				
INCIDENT ANALYSIS				
Factors	***Description***		***Probability***	***Frequency (/y)***
Consequence	Overflowing fuel pools in bund and is ignited causing a pool potentially harming people.			
Tolerability criteria	Maximum tolerability criteria for a fatality or permanent injury			**1E-5**
Initiating event	Level detection and alarm system (BPCS) failure			0.1[a]
Enabling event	None			
Potential modifiers	Probability of ignition		0.065[b]	
	Personnel exposure probability (assumes 50% occupancy rate in fatal thermal flux zone)		0.5	
	Fatality / injury probability		1.0	
Frequency of unmitigated incident / consequence				**3.25E-3**
IPL ANALYSIS				
IPL safeguard and its PFD	Bund to contain spill (this would contain the fire)		0.01[a]	
	Operator intervention on high level alarm – not independent of level control system failure = initiating event (therefor PFD = 1)		1.00	
	SIF to be added (see actions)		0.01[a]	
Total Probability of Failure on Demand (PFD) for IPLs			0.001	
Frequency of mitigated incident / consequence				***3.25 E-6***
Tolerability criterion met (yes/no)			Yes	If SIF added
ACTIONS				
Actions required	The actual frequency for potential fatalities due to overfilling tank and subsequent pool fire is high that tolerability limit unless and independent high level SIL 1 trip system comprising an independent sensor that remotely stops flow with a PFD of <=0.01 is installed.			
Further notes and justification				
Date [insert]	Author	[insert name]		
	Engineering review	[insert name]		

[a]CCPS. (2001). *Layer of protection analysis: Simplified process risk assessment*. American Institute of Chemical Engineers.
[b]Zhu et al. (2012). Study on ignition probability of flammable materials after leakage accidents. *Procedia Engineering 45*(2021), 435–441.

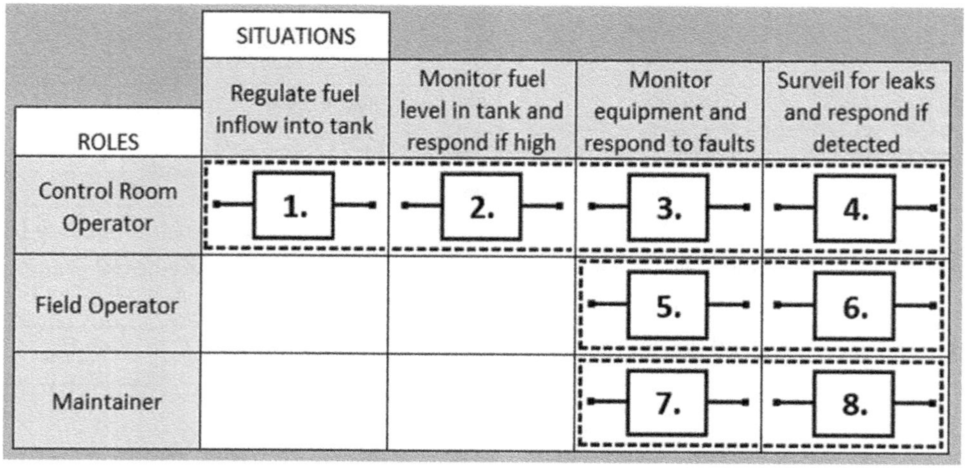

FIGURE B18 Contextual activity template identifying safety critical work performed by humans.

analysis is to document the safety critical control work that humans are expected to perform. This can be done using contextual activity diagrams (Naikar et al., 2006). Such analysis can be informed by the controls on the bowtie and the controls and control support activities performed by humans as shown on the ConSupMap (Fig. B13) or control specification sheets (Table B3). An example of a contextual activity template is shown in Fig. B18. It shows people or roles involved (as rows) and work activities or situations undertaken (as columns). Then, within the template, the tasks that are performed by each role associated with each work situation or activity are denoted with a number and box.

Human–system interaction type control tasks can also be described in more detail using controller flow diagrams (Systems Theoretic Process Analysis [STPA] Leveson and Thomas, 2018). The identification of these tasks can be informed by the ConSupMap (e.g. Fig. B13) and the abovementioned contextual activity template analysis. An example of a control flow diagram for the tank filling activity is shown in Fig. B19.

Once human control activity is described, then human factors analyses can be performed to identify design interventions that would help workers or users to accurately detect and assess the situation in a timely manner, with minimal cognitive and physical effort. This requires designs that very clearly and saliently:

- present the operational state of the system – whether it is currently in a safe or unsafe state and where that state is trending (i.e. is it worsening or improving).
- show the status of the layers of controls – whether they are present and fully functional.

SAfER analysis (as described in Chapter 3) can be performed to identify factors that help workers accurately detect and successfully respond to the situation. To help workers accurately detect a situation, an analysis should be conducted to determine what people, plant and work context factors need to be noticed and understood. To achieve this, the first part of the SAfER analysis, after documenting the scope and control diagram, involves performing the situation assessment, which involves identifying the factors that indicate whether the system's current and projected operating state is normal or abnormal

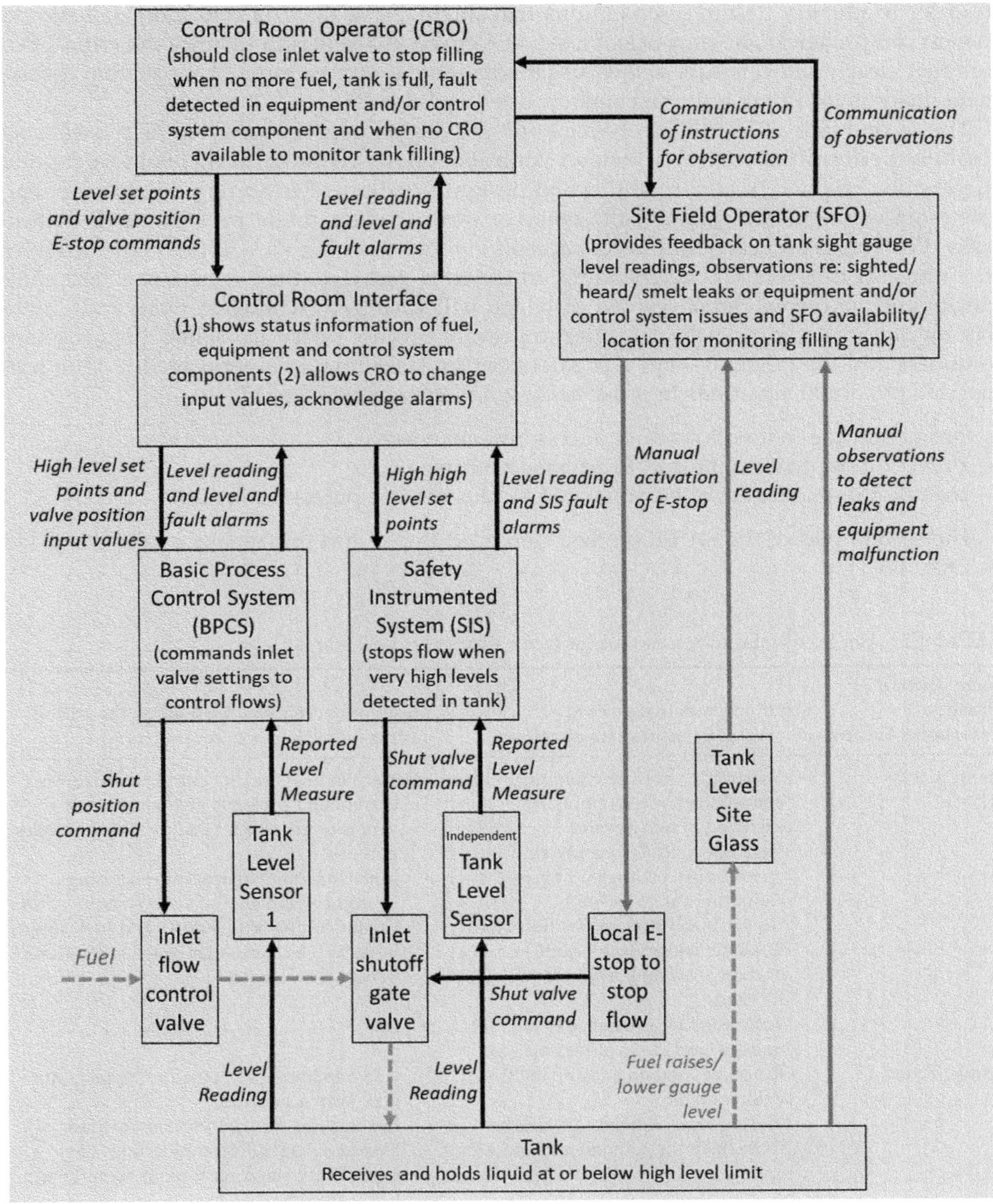

FIGURE B19 Human–system control diagram from managing tank fill levels.

in order to identify design interventions that might make these factors more salient and easy to comprehend. The purpose of a safety focused analysis is to identify the critical factors that clearly indicate safe versus unsafe operations. An example of a situation assessment analysis for filling bulk fuel tanks is listed in Table B5.

The second part of SAfER involves identifying design interventions that help users successfully perform the control actions or control supports required to maintain safety and successfully respond to abnormal (beyond design) situations. Performing this second analysis requires the identification of the range of ways workers might perform safety critical tasks. Workers try to carry out tasks successfully and in doing so will carry out activities in a range of different ways depending on the time pressure, their experience, task complexity and perceived risk levels (all of which will vary from worker to work and day to day for any given worker). So, in order to prevent process safety accidents, it is necessary to understand the range of ways that safety critical tasks might be performed in both normal and abnormal situations in order to design work contexts that:

- promote those that will produce successful outcomes
- prevent those that might lead to adverse outcomes
- tolerate those that cannot be prevented but no adverse outcomes result.

The second part of the SAfER method deployed to perform this type of analysis is listed in Table B6.

TABLE B5 Safety critical situation assessment factor analysis for tank filling.

Safety Critical Situation Assessment Factors	List the factors that need to be monitored to ensure safe operation	Design interventions for making the critical factors more salient to controllers
Plant/process factors	– Status of processing equipment pipes, connections, valves, pumps etc. involved in fuel transfer – Status of controls (e.g. physical components, coding and bypass, communication systems) – Contents and flow rate within pipes – Contents and level (absolute and rate of change) of tank included projected fill times – Reconciled mass balance to highlight unexplained losses/accumulations	Interface in field and in control room showing – current and projected changes in tanks versus target current and projected changes in levels – functional/fault alarms on processing equipment and control system components – Colour coding of fuel flows and fuel stored in tanks which are highlighted when these deviate from expected
People factors	– Real-time location and capabilities of field operator – Presence and vigilance of control room operator in monitoring transfer	– GPS monitoring of operator mapped onto tank farm schematic – Eye tracking for control room operator – Dedicated verbal communications (e.g. radios) between control room operator and field operator
Context factors	– Current and forecast weather – for lightning, gale force winds etc.	– Weather warning interlocked with control room operator alarm system and safety instrumented system (SIS) systems that alarm when weather forecast and shutdown when weather present

TABLE B6 SAfER analysis for filling bulk storage tanks with fuel.

Generic Strategy (prompt)	Actual strategies that might be used in domain	Reasons for using strategy	Implications of using strategy	Design to promote, prevent or tolerate strategy	What improvements would enhance the design (i.e. prevent/promote/tolerate the selection and execution of the strategy)
Avoidance: not done, defer or forget to do	Normal: not start filling	Control room operator not available, fault in system, no empty tank	Overfilling avoided, unloading delayed.	Promote	1. Adequate control room personnel to cover peak periods, shift changes and crib breaks.
	Normal: no field checks of level or spills/faults	Field operator not available; level, spills, faults not easily observable	No verification of level, undetected faults and leaks lead to excessive loss of fuel from system.	Tolerate	2. Hydrocarbon sensors to detect leaks in an around tank that automatically alert control room operators when concentrations are low or shut system down when concentrations are approaching lower explosive limits (LEL).
	Abnormal: not stop filling	Control room operator not monitoring filling, and/or is not alerted when a stop is required	Overfilling and loss of containment.	Prevent	3. Eye tracking/vigilance system that stops system if no one is monitoring. 4. Independent SIS that automatically shuts feed when high–high level reached.
	Abnormal: emergency response not available	Field operators and others not able and/or available to perform emergency response	Leaking fuel catches alight and/or forms vapour cloud that explodes. No evacuation, no first aid for injured.	Prevent	5. System that verifies there are adequate resources for emergency response otherwise fuel cannot be transferred. 6. [as per point 2] Hydrocarbon sensors to detect leaks and automatically shut system down when concentrations approach LEL.

(*Continued*)

TABLE B6 (Continued)

Generic Strategy (prompt)	Actual strategies that might be used in domain	Reasons for using strategy	Implications of using strategy	Design to promote, prevent or tolerate strategy	What improvements would enhance the design (i.e. prevent/promote/tolerate the selection and execution of the strategy)
Intuitive = automatic response, done without explicitly or deliberately using thought processes	Normal: Operators just assume everything is running ok (absence of chronic unease)	Site has not had a significant issue or incident	Overflow due to level sensor giving wrong reading or failure in control system/valve to correctly regulate flow or leak from system. Result could be fuel spill, pool fire, vapour cloud and explosion that could contaminate environment, damage assets, injure/kill people.	Prevent	7. Start of shift/day forcing function type checks which functionally test performance of critical prevention controls (a forcing function where filling cannot start unless system verifies check has been done and control is functional).
	Abnormal: High-high level shutoff relied upon to shut off fuel flows	Accepted normal practice. Operator not available to respond to high level alarm	Any malfunction in high-high level components would result in overflow. Result could be fuel spill, pool fire, vapour cloud and explosion that could contaminate environment, damage assets and injure/kill people.	Prevent	As per point 3
	Abnormal: worst case scenario for overflow is pool fire not vapour cloud explosion	Lack of awareness of events like Buncefield	Inadequate response to potential of vapour cloud explosion that could damage assets and injure/kill people.	Prevent	8. Area and personnel monitors used to detect hydrocarbon leaks that activate evacuation alarm and people can and do evacuate to explosion proof refuges which are installed and maintained in appropriate locations. This is regularly practiced as an emergency drill.

Arbitrary choice: guessed, scrambled haphazard or panicked response	Normal: Guessing which tank to put fuel in	Time pressure, issues with correct tank identification, inexperience	Mixing of fuels (e.g. petrol into diesel tank) could lead to issues with customer, filling full tank leading to overfilling Result could be fuel spill, pool fire, vapour cloud and explosion that could contaminate environment, damage assets, injure/kill people.	Prevent	9. Colour coding of fuel on control room consoles displays and forcing function to ensure cross-check is done by another person prior to allowing fuel to flow into tank.
	Abnormal: guessing whether alarm is true or false, especially in alarm flood situations	Time pressure, history of true/false alarms, inexperience, poor alarm information	Failure to acknowledge alarm could result in equipment damage and/or could lead to fuel spill, pool fire, vapour cloud and explosion that could contaminate environment, damage assets, injure/kill people.	Prevent	10. Ensure alarm system design utilises the leading alarm rationalisation and prioritisation approaches and is communicated using lead human-centred design approaches
Imitation strategies: copy how others do it or copy what has worked in the past	Normal: trusting level gauge and valve operation is working the same as it has in past	Risk normalisation, inability to test, time pressure, inexperience	Unable to stop flow could lead to overflow: fuel spill, pool fire, vapour cloud and explosion that could contaminate environment, damage assets and injure/kill people.	Tolerate	As per point 4
	Abnormal: assuming bund valves are closed	Difficulty in knowing when valve is open/closed, no checks done	Fuel spill to extend beyond bund leading to environmental damage, extensive fuel spill, pool fire, vapour cloud and explosion that could contaminate environment, damage assets and injure/kill people.	Prevent	11. Automate valve open/close and interlock filling with valve position so filling cannot occur when valve not confirmed closed.

(*Continued*)

TABLE B6 (Continued)

Generic Strategy (prompt)	Actual strategies that might be used in domain	Reasons for using strategy	Implications of using strategy	Design to promote, prevent or tolerate strategy	What improvements would enhance the design (i.e. prevent/promote/tolerate the selection and execution of the strategy)
	Abnormal: executing emergency stop as practiced	Regular training until it is habitual	Fuel flow stopped when required.	Promote	12. Emergency stops designed with user-centred design focus and regularly practiced.
	Abnormal: trusting emergency stop will function	Emergency stop function not tested – includes persons' willingness and ability to hit e-stop.	Emergency stop failure could result in equipment damage and/or could lead to fuel spill, pool fire, vapour cloud and explosion that could contaminate environment, damage assets and injure/kill people.	Prevent	As per point 12
Cue-based strategies: select Chosen Option using the Observed Info/Cues and Predict Consequences results	Normal: filling managed by monitoring tank levels using 3 independent level readings	Dedicated control room and field operators monitoring levels and the management of flow is easy for the control room operator to do	Prevents overfilling.	Promote	As per point 1 13. Interface design utilises leading control room console design practices to make level monitoring and management and easy, salient and efficient process. 14. Field level readings designed using human-centred design so operators can easily and accurate monitor level.
	Abnormal: Spills noticed by proactively monitoring tank, pipes and fittings	Dedicated field operator, with hydrocarbon leak detection monitors. Control room operator has cameras/mass balance to monitor	Minimises spill volumes reducing severity of any fire, vapour cloud and/or explosions.	Promote	15. Adequate field operators to cover peak periods, shift changes and crib breaks. 16. Control system performs real-time mass balance to detect and alarm on losses which is displayed to both control room and field operators.

					17. Camera feeds analysed by AI to alert operator on deviations from expected.
Compliance-based strategies: following procedures as they are written/practiced	Normal/abnormal: follow procedures	Inexperienced person learning system	Normal operations: if procedures accurate no spill.	Promote	18. User-centred design of procedures.
			Abnormal: delay in dealing with spill leading to worsening of spill, fire, vapour cloud and/or explosion.	Tolerate	19. Operators required to do 10-h training on training simulator to practice both normal and abnormal operations. 20. Regular emergency response drills for SFOs and emergency responders.
Analytical reasoning strategies: using analytical thinking to reason out the best way to perform task	Normal/abnormal: trial and error approach to managing operations	Normal: inexperienced/bored worker who does know/use procedure Abnormal: novel or beyond design situation	Potential for incorrect operation leading to damaged equipment and/or fuel spill, pool fire, vapour cloud and explosion that could contaminate environment, damage assets and injure/kill people.	Tolerate	As per points 19 and 20, which include novel and beyond design/novel situations.

Control action	Control Action NOT GIVEN	INCORRECT Control Action IS GIVEN	Control Action GIVEN AT WRONG TIME - TOO SOON/EARLY	Control Action GIVEN AT WRONG TIME - TOO LATE	Control Action GIVEN IN WRONG ORDER or FOR WRONG DURATION	Potential Consequence(s) and Significance (High priority - must address, Med priority - should address, Low priority - monitor for change, Negligable - No further action required)	Possible causes of unsafe control action	Assessment and recommendation for improving design (ISD) or controls (DiD) or control support activities
High level set points entered into control system	High level alarm and high-high level safety instrumented system will not work leading to overflow	Wrong set point numbers entered into system leading to an underfilling or overfilling of tank	No issue if entered before filling starts	Same as NOT GIVEN	If high level and high-high level numbers were mixed up, this would lead to unnecessary system shutdowns	Overfilling would result in fuel overflowing into bund where it could be ignited resulting in a pool fire or it could become vapour that forms a cloud which if ignited would create an explosion resulting in injuries/fatalities, equipment damage and potentially environmental pollution -> HIGH PRIORITY and must be addressed	Lack of knowledge, unavailability of personnel, input error (e.g. wrong tank, entry mistake etc), translation error by software code	(1) Only authorised person(s) to enter set points and set points peer review by an independent competent person
Control room operator monitors level in tank and adjusts valve position (% open or closed) to regulate flow	Not monitoring levels and not adjusting flows could result in fuel filling not being stopped when should leading to overflow	If valve is opened instead of closed this could result in overflow	If valve opened too soon it may result in filling a full tank leading to overflow. If closed too soon it may result in underfill	Delay in opening valve might delay ship unloading. Delay in closing valve could result in overflow	If opened instead of close = overflow. If closed instead of open = underfill.	Overfilling as above. Underfilling will result in delaying ship unloading which would increase shipping costs and potentially delay orders to customers.	Errors in level readings, control room operator distraction/ unavailability/ overload/ lack of knowledge, malfunction in code that transforms operator input into a valve position, valve malfunction	(2) Forcing function (e.g. eye tracking) that only allows filling to continue if operator vigilant. (3) Control room console interface design to ensure operator can accurately, easily and efficiently assess level situation and respond
Control room and field operators activates emergency stops when required	Emergency stop not activated when required leading to overfill	If emergency stop activated when not required, this could delay loading. If incorrect emergency stop hit (i.e wrong tank) could lead to overflow and equipment damage	If emergency stop activated when not required, this could delay loading	Emergency stop activated too late could result in overfill	If incorrect emergency stop hit (i.e wrong tank) could lead to overflow and equipment damage	As above for overfilling and delaying tanker loading. Equipment damage could also occurr if emergency stop not hit when operating equipment has fault/issue	Control room operator distraction/ unavailability/ overload/ lack of knowledge, malfunction in software/hardware that activates emergency stop	(4) Regular practice and testing of emergency stop processes
CRO monitors alarms and responds to active ones	Critical alarms not actioned leading to equipment damage and/or overflow	If alarms are cancelled without action or if response is incorrect this could cause equipment damage and/or overflow	Same as INCORRECT for cancelling alarms before/without investigating them	Delay in noticing/ responding to alarms could cause equipment damage and/or overflow	Responding to alarms in wrong order could result in equipment damage and/or overflow	As above for overfilling and equipment damage could also occurr if critical alarms are not responded to.	As above	(5) Alarm rationalisation and prioritisation done to industry leading standards (3) Control room console interface designed to ensure operator can accurately, easily and efficiently assess and respond to alarms
Safety instrumented system safely stops system when level is at high-high limit	Tank will keep filling, leading to overflow	As per NOT GIVEN	If High-High shutdown activated too early, this could lead to under filling	As per NOT GIVEN	If valve is shut before pumping stops could pressurise line leading to leaks	As per overfilling above plus overpressurising line could result in contained leaks of fuel that could catch on fire or form vapour that explodes resulting in injuries/fatalities, equipment damage and potentially environmental pollution	Errors in code, bypasses in system, malfunction of hardware	(6) Regular testing of the safety instrumented system functionality. Safety rating for safety instrumented system should match severity of consequences
Level sensors measure and report levels to operators and control system	No level reported then high level alarm and high-high level safety instrumented system will not work leading to overflow	Incorrect level measurement could result in under or over filling tank	As per INCORRECT	As per INCORRECT	As per INCORRECT	As per overfilling above	Faulty equipment, errors in translation from sensor to interface.	(7) Calibration checks between high level sensor, high-high level sensor and field readings with deviations resulting in alarm-response type actions

FIGURE B20 SAMPLE/EXAMPLE ONLY of STPA worksheet for filling a tank with fuel. *STPA,* Systems Theoretic Process Analysis.

Another form of analysis that can be that can be conducted is STPA. STPA uses the control actions from the control flow diagram (Fig. B19) and uses prompts associated with doing the control actions incorrectly to identify and assess risks related to incorrect operator actions. These prompts include control action not given, given incorrectly, given at the wrong time – either too soon or too late, and given in wrong order. The implications of doing the control action incorrectly are described along with the potential consequences and significance of any impacts. Possible causes are also identified. The consequences, significance and causes are then used to inform recommendations for improving the inherently safe design of the plant and the control system and its associated control supports. Fig. B20 illustrates the STPA for the tank filling example.

B.5 Summary

This case study has highlighted examples of techniques that can be used to assess safety risks in the process industries. Each technique can produce meaningful recommendations that focus on different aspects of the system. Using a number of techniques is often best to reveal the recommendations required to prevent process safety incidents. These recommendations should include the controls that prevent the event, controls that mitigate the event in the impacted areas, control support activities required to ensure controls are implemented, fully functional and effective at addressing the risk. The recomendations should also include human-centred design requirements to ensure that humans can successfully identify, assess and control both normal and abnormal situations and conduct control support activities as required.

References

Aviram, A. (2021). New details revealed about fatal Avonmouth explosion at Wessex Water plant. *The Bristool Cable*, August, 11, 2021. Retrieved from https://thebristolcable.org/2021/08/revealed-avonmouth-explosion-wessex-water-staff-allege-no-ignition-risk-assessment/.

Byrne, C. (2013). Photo shows scale of tank collapse at Ranger uranium mine. *NT News*, December, 11, 2013. Retrieved from https://www.ntnews.com.au/news/northern-territory/photo-shows-scale-of-tank-collapse-at-ranger-uranium-mine/news-story/eb6ea16557536b224da5a7395d7247c2.

Cameron, I. T., & Raman, R. (2005). *Process Systems Risk Management*. San Diego, CA: Academic Press.

CCPS. (2000). *Guidelines for chemical process quantitative risk analysis* (2nd ed.). Wiley-Interscience; American Institute of Chemical Engineers.

CCPS (2001). Layer of protection analysis: Simplified process risk assessment. American Institute of Chemical Engineers.

COMAH. (2011). *Buncefield: Why did it happen?* Retrieved from.

Department of Justice U.S. Attorney's Office Southern District of Ohio. (2016). *Duke Energy to Plead Guilty to Clean Water Act Crime and Pay $1M Fine for 2014 Oil Spill*. Retrieved from https://www.justice.gov/usao-sdoh/pr/duke-energy-plead-guilty-clean-water-act-crime-and-pay-1m-fine-2014-oil-spill.

Dreizen, C. (2021). Catastrophic tank collapse destroys 250,000 Liters of South African Wine. *Wine Spectator*, June 2, 2021. Retrieved from https://www.winespectator.com/articles/catastrophic-tank-collapse-destroys-250-000-liters-of-south-african-wine.

Hassall, M.E., & Harris, J. (2017). *Risk controls knowledge: Determining leading practice from case study analysis (ACARP report C25036)*. Retrieved from Australia.

IMPEL - French Ministry for Sustainable Development. (2013). *Overflow of a gasoline tank inside a refinery 22 October 2011*. Retrieved from https://www.aria.developpement-durable.gouv.fr/wp-content/files_mf/A41148_ips41148_003.pdf.

Kang, M. (2014). Kitchener treatment plant explosion due to chemical mix-up. *CBC News* April 2, 2014. Retrieved from https://www.cbc.ca/news/canada/kitchener-waterloo/kitchener-treatment-plant-explosion-due-to-chemical-mix-up-1.2596255.

Leveson, N.G. and Thomas, J.P. (2018). STPA handbook. Retrieved from https://psas.scripts.mit.edu/home/get_file.php?name=STPA_handbook.pdf

Marsh. (2020). *100 largest losses in the hydrocarbon industry 1974-2019*. Retrieved from.

Naikar, N., Moylan, A., & Pearce, B. (2006). Analysing activity in complex systems with cognitive work analysis: concepts, guidelines and case study for control task analysis. *Theoretical Issues in Ergonomics Science*, *7*(4), 371–394.

NSW EPA. (2020). *EPA fines AGL Macquarie $30,000 after diesel spill*. Retrieved from https://www.epa.nsw.gov.au/news/media-releases/2019/epamedia190305-epa-fines-agl-macquarie-$30000-after-diesel-spill.

Pilot, S. (2002). What is fault tree analysis?, American Society for Quality, Milwaukee.

U.S. Chemical Safety and Hazard Investigation Board. (2007). *Investigation report - Methanol tank explosion and fire - Bethune Point wastewater treatment plant, City of Daytona Beach, Florida, January 11, 2006*. Retrieved from https://www.csb.gov/assets/1/20/bethune_final_report.pdf?13742.

U.S. Chemical Safety and Hazard Investigation Board. (2015). *Final investigation report - Caribbean Petroleum tank terminal explosion and multiple tank fires* (2010.02.I.PR). Retrieved from https://www.csb.gov/caribbean-petroleum-refining-tank-explosion-and-fire/.

U.S. Chemical Safety and Hazard Investigation Board. (2017a). *Case Study - Key lessons for preventing inadvertent mixing during chemical unloading operations - Chemical reaction and release in Atchison, Kansas*. (2017-01-I-KS). Retrieved from https://www.csb.gov/dupont-la-porte-facility-toxic-chemical-release-/.

U.S. Chemical Safety and Hazard Investigation Board. (2017b). *Investigation report - Chemical spill contaminates public water supply in Charleston, West Virginia - Freedom Industries Inc. Charleston, WV - January 9, 2014*. (2014-01-I-WV). Retrieved from https://www.csb.gov/freedom-industries-chemical-release-/.

U.S. Chemical Safety and Hazard Investigation Board (2018). Non-condensable Gas System Explosion at PCA DeRidder Paper Mill. Report No. 2017-03-I-LA. April 2018. Retrieved from: https://www.csb.gov/packaging-corporation-of-america-hot-work-explosion-/

Waite, P. (2013). Recurring accidents: Overfilling vessels. *The Chemical Engineer*, *861*, 40–44.

Zhu., et al. (2012). Study on ignition probability of flammable materials after leakage accidents. *Procedia Engineering*, *45*(2021), 435441.

APPENDIX

C

Carbon storage case study

C.1 Introduction

In this appendix we will work through a case study to demonstrate the use of the social risk assessment and management tools introduced in Chapter 8. The case study focuses on the CarbonCap project, a proposed carbon capture and storage project. This case study is hypothetical, although based on real projects.

CarbonCap is assessing the potential of the offshore Gaia Basin to sequester CO_2 as a method of reducing greenhouse gas emissions. The International Energy Agency[1] asserts that CCS is necessary to achieve the deep emission cuts required in the Paris Agreement to balance anthropogenic (man-made) emissions with GHG sinks by the second half of this century. This is commonly known as "net zero emissions."

CarbonCap has conducted several remote exploration surveys to assess the suitability of the offshore Gaia Basin for storing CO_2. The most recent exploration activity was an offshore seismic survey in the area of Sunshine Beach. This generated negative community feedback due to noise volumes above the expected range. There are also other, less specific, community concerns about the project. CarbonCap intends to drill their first offshore well in 12-months time to further assess the geological suitability of the area. This well will assess the capacity of the sandstone to hold CO_2 within the pores as well as the integrity of the caprock that will trap the CO_2 in the target formation.

A preliminary stakeholder assessment is required, covering the full field development of the CO_2 transport and storage processes if the test drilling campaign is successful. The full field development will store 5 million tonnes of CO_2 per annum. The infrastructure required is displayed in Fig. C1 and will include the following:

1. Pipeline for transferring CO_2 from inland point sources. This will follow existing petroleum pipeline easements. The point sources will all be in the Rockstone industrial region, which comprises a cement manufacturing site, power station and a chemicals plant.
2. Coastal compression facility near Castello to boost CO_2 to supercritical conditions.

[1] https://www.iea.org/topics/carbon-capture-and-storage/

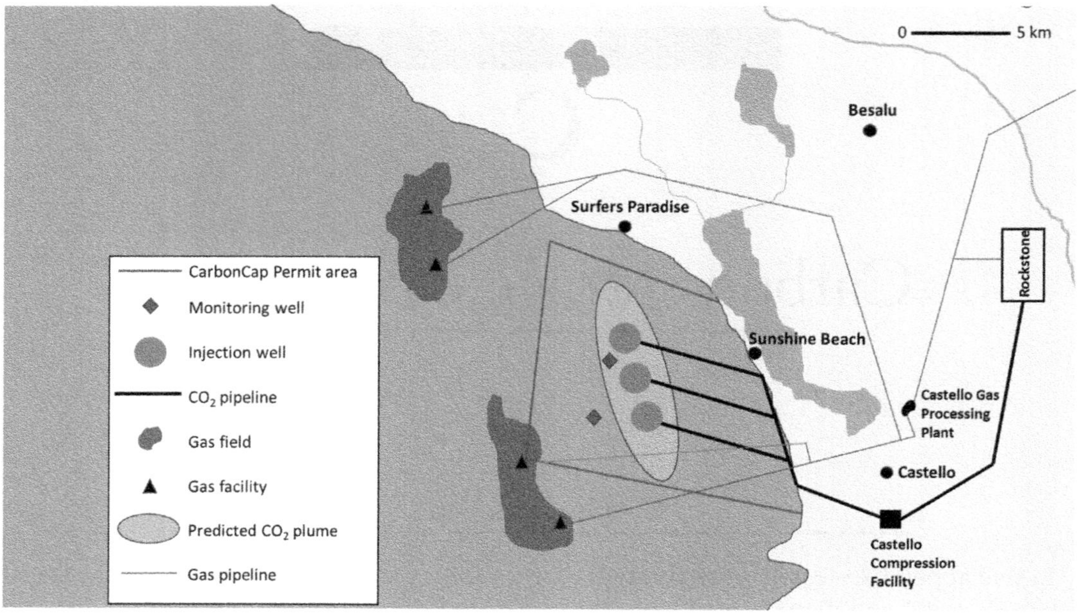

FIGURE C1 Proposed CarbonCap CCS project in the Gaia Basin.

3. Pipeline from compression site to three wellheads.
4. Three injection wells injecting supercritical CO_2 into a sandstone formation at 1500m below sea level.
5. Two monitoring wells to track changes in pressure and composition and calibrate these to numerical subsurface flow models.

The region has an industrial heritage. The Gaia basin is rich in hydrocarbon reserves, and oil and gas extraction has been occurring for several decades. There is a gas processing facility near Castello.

Sunshine Beach is a popular coastal town, comprising long stretches of sandy beaches and a marina for leisure craft and some fishing boats. It is also nearby to a popular natural wetland area and a set of inland lakes. The entrance to the lakes is 5 km from Sunshine Beach.

C.2 Your brief

1. Identify a list of stakeholders that will be impacted by or have potential influence on the project. Plot them on the power-interest grid shown in Fig. C2. In doing this, you may also like to consider the project from the perspective of each stakeholder which outlines their motives, attitudes and beliefs. In particular, focus on the social, health, environmental and cultural impacts of the project (i.e. "walk in their shoes").
2. Compile a risk register that identifies and classifies the risks (threats and opportunities) from the perspective of **three** selected stakeholders using Tables C1 and C2 as guides.

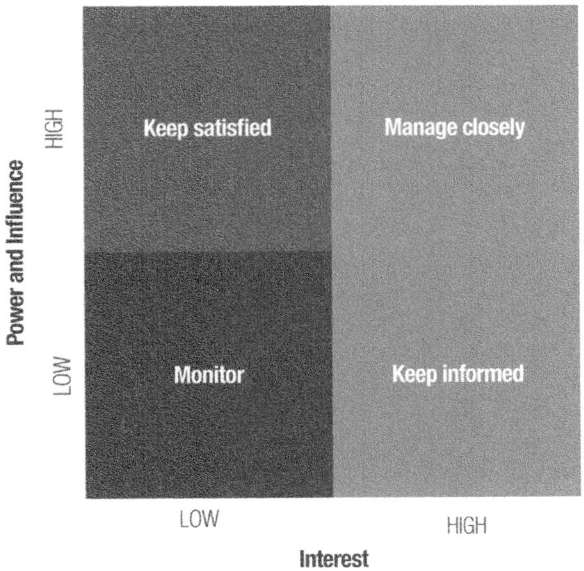

FIGURE C2 Power-interest grid.

3. Select **one** of the key impacts from the above and develop a draft Social Impact Management Plan using Table C3 as a guide.

TABLE C1 Stakeholder risk and opportunity assessment template.

Stakeholder Information			Potential Impacts (health/safety, environment, financial, societal)		
Name	Description	Classification (Fig C2)	Threats/Issues	Impact (Table C2)	Recommended Actions (Figs 8.6 & 8.7)
			Opportunities/Benefits	Impact (Table C2)	Recommended Actions (Figs 8.6 & 8.7)

TABLE C2 Social opportunity and threat risk assessment matrix.

		E – Rare Rarely occurs at either a specific stage of the project or more broadly	D - Unlikely Unlikely to occur at either a specific stage of the project or more broadly	C - Possible Possible to occur at either a specific stage of the project or more broadly	B - Likely Likely to occur either at a specific stage of the project or more broadly	A - Almost certain Very likely to occur either at a specific stage of the project or more broadly
5 – Serious/ Significant Impact	Irreversible serious impact and the stakeholder no capacity to adapt and cope with change. Significant opportunities from the project. Impact is expected to continue indefinitely.	E5	D5	C5	B5	A5
4 - Major Impact	Major long-term threat with recoverable impact. Stakeholder has limited capacity to adapt and cope with impact. Major opportunities emanating from the project. The impact is long-term.	E4	D4	C4	B4	A4
3 – Moderate Impact	Significant medium-term threat with recoverable impact. Stakeholder has some capacity to adapt and cope with impact. Significant opportunities emanating from the project. Probably a medium-term impact.	E3	D3	C3	B3	A3
2 – Minor Impact	Minor short-term threat with recoverable impact. Minor opportunities emanating from the project. Probably a short-term impact.	E2	D2	C2	B2	A2
1 – Negligible Impact	Local, small-scale, easily addressed threat with reversible impact. Very small-scale opportunities that can readily pursue and capitalise on.	E1	D1	C1	B1	A1

Low social impact or opportunity	Moderate social impact or opportunity	High social impact or opportunity	Extreme social impact or opportunity

TABLE C3 Social impact management plan template.

<table>
<tr><td colspan="2">Impact Category (e.g. air quality)</td><td></td></tr>
<tr><td colspan="2">Impact strategy (e.g. Minimise adverse impacts on air quality)</td><td></td></tr>
<tr><td colspan="2">Summary of Potential Threats/Opportunities that could affect Impact Category (e.g. construction activity, transport)</td><td>Impact Assessment</td></tr>
<tr><td colspan="2" rowspan="3">Threats

Opportunities</td><td>Anticipated social impact/opportunity without strategy:

Anticipated social impact/opportunity after strategy:</td></tr>
<tr><td>Key Stakeholders</td></tr>
<tr><td></td></tr>
<tr><td>Action Plans</td><td>Timeframes</td><td>Monitoring and Stakeholder Engagement Activities</td></tr>
<tr><td></td><td></td><td></td></tr>
<tr><td>Accountabilities:</td><td colspan="2">e.g. General Manager, Community Engagement</td></tr>
</table>

C.3 A worked answer

1. Large projects like CarbonCap usually have a long list of interested stakeholders. They may include the following:

- Regional government
- Environmental Protection Agency. Whilst being a part of the regional government, this is a significant enough for stakeholder to consider on its own
- Federal government
- Local community
- Broader community of interested citizens in the country
- Local industrial companies at Rockstone (the emitters of CO_2)
- Media
- Environmental activist groups
- Industry representative groups
- Fossil fuel lobby groups
- Renewable energy lobby groups
- National Fisheries Association

These stakeholders are mapped on the power-interest grid as shown in Fig. C3. Do you agree with this mapping?

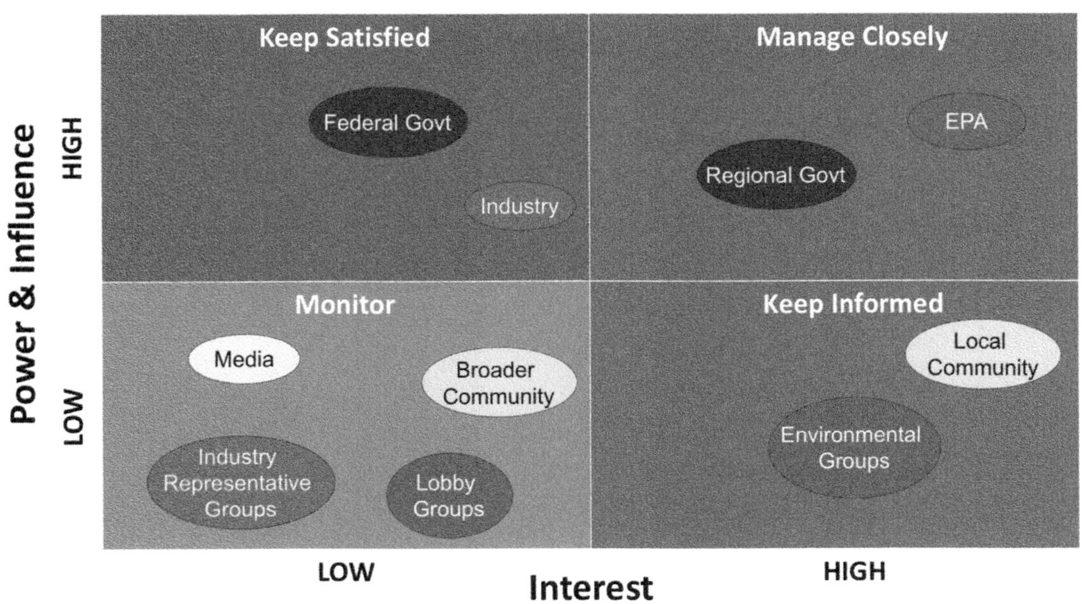

FIGURE C3 Example of CarbonCap stakeholders shown on power-interest grid.

2. The following stakeholder risk register shows threats and opportunities for three of the stakeholders shown above—that is the regional government, local community and industry (Table C4).
   Note that the register is referring to the threats and opportunities posed to the stakeholder by the project. This is not an assessment of the impacts to the project by this specific stakeholder group. Do you agree with the threats and opportunities highlighted? Can you think of other threats and opportunities to add for these stakeholders? Do you agree with the assessment for each risk? The recommended actions are the actions that CarbonCap can take to either mitigate the threat or enhance the opportunity.

TABLE C4 Stakeholder risk and opportunity assessment template

Stakeholder Information			Potential Impacts (health/safety, environment, financial, societal)		Recommended Actions
Name	Description	Classification	Threats/Issues	Impact	
Regional Government	The Regional Government is involved in several capacities. In addition to regulating and providing approvals for the project through the Dept of Environment, the Dept of Development has also provided grants for some early stage development work.	Manage Closely	Community concerns (weak or no social licence) about the project may affect support for the Government.	C3	**Reduce:** Engage with the public to inform them about the benefits of the project and to better understand the risks around CCS.
			The local community does not perceive that it is getting much benefit from the project.	B2	**Reduce:** CarbonCap to ensure that local community benefits are achieved through priority recruitment and contracting of locals and local businesses. Work with Govt to identify opportunities for enhanced spending on local community projects.
			Community feels that their concerns about the project are not heard by Govt.	B2	**Reduce and/or repair:** Engage proactively with the public to ascertain concerns. Inform Govt of concerns.
			The Govt will be held accountable if an unwanted event occurs (e.g. CO_2 leakage into the surrounding marine environment).	C4	**Avoid:** Communicate risk management processes with Govt and clearly illustrate how serious unwanted events are being mitigated with appropriate controls. Use bowtie's to illustrate this for highlighted unwanted events.
			Opportunities/Benefits	Impact	
			Continued use of low grade coal that is mined in the region. Thus maintaining royalty payments to the Govt.	A3	**Project Enhancement:** Enhance off-gas scrubber from the power station, thus improving overall emissions, as well as capturing the CO_2.
			Development of new industries to support the construction and operation of the project.	C3	**Local area enhancement:** Prioritise selection of local contractors and manufacturers.
			Skilled job opportunities for the local community.	C1	**Local area enhancement:** Prioritise recruitment from the local community.
			The project will assist in reducing CO_2 emissions and help the Govt to meet its stated emissions targets.	B4	**Wider area enhancement:** Report to Govt in regard to Regional benefits and contribution to CO2 emission reductions.
Local Community	Local people that live and work in the vicinity of the project at Castello, Sunshine Beach and Surfers Paradise.	Keep Informed	Negative impacts due to the project construction (e.g. dust, noise and traffic delays).	A2	**Avoid and/or reduce:** Avoid impacts from construction wherever reasonably practicable. Otherwise, reduce the impacts.
			Disruption to local businesses during construction.	B3	**Reduce:** Work with local businesses to understand their needs and to minimise negative impacts.
			Concerns about CO_2 storage in the Gaia Basin and the impacts of uncontrolled releases.	C3	**Reduce:** Engage with the public to inform them about the project and to better understand the risks around CCS.
			Decreased value of property due to negative impacts of the project.	C4	**Avoid:** Project designed to have little impact on residential property. Construction to occur within existing easements, and there is already significant industry in this region. Thus this project should have little impact.
			Improved infrastructure resulting from increased investment in the region because of the project.	B3	**Local area enhancement:** Ensure that project related infrastructure expenditure is targeting valuable public infrastructure.
			Job opportunities.	C2	**Local area enhancement:** Give priority to local applicants.
			Improvements to the local economy because of the project expenditures in the local region.	B2	**Local area enhancement:** Ensure priority selection of local suppliers and contractors.
Industrial companies at Rockstone Industrial park	The cement manufacturer, power station and chemicals plants that are based at the Rockstone industrial complex. These companies are the CO_2 emitters.	Keep Satisfied	Increased costs of operation as a result of the costs of the carbon capture and storage.	A4	**Avoid and/or Reduce:** This is a complex issue and depends on Govt policy. In the event of carbon taxing, then this project could offer cost savings. If not, then there will be inevitable increased costs. Work with Govt and Industry to find an optimum point.
			Damage to company reputation as a result of being associated with an unsuccessful and/or unpopular project.	C3	**Avoid:** Create an Industry Advisory Group for the project. Communicate risk management processes with Industry and clearly illustrate how serious unwanted events are being mitigated with appropriate controls.
			Having an increased negative profile as a 'high carbon polluter'.	C2	**Avoid:** Create an Industry Advisory Group for the project. Include Industry in developing and implementing the risk management strategy for managing local and broader community risks.
			Enhanced social licence to operate as a result of a successful project and thus reduced CO_2 emissions.	B3	**Wider area enhancement:** Dissemination of the CO_2 emission reductions to the wider community as a result of the project. Engage an independent third party to monitor and report on the actual CO2 reduction benefits.
			Reduced carbon emissions and thus assisting the company to move towards their emissions targets.	B3	**Project enhancement:** Ensure that project reporting (data) enables Industry to determine their carbon emissions reductions appropriately.
			Prolonged site lifespan as a result of a successful CCS project.	B4	**Project enhancement:** This is a significant benefit for Industry. Ensure that the project monitoring processes are also including an assessment of this benefit for the Industry. Share this data with Industry.

3. The draft Social Impact Management Plan for "Quality of life issues for local residents" is shown in Table C5. The plan focuses on minimising the adverse effects and maximising the beneficial opportunities the CarbonCap project delivers to the local residents.

TABLE C5 Social impact management plan template.

<table>
<tr><td>Impact Category (e.g. air quality)</td><td colspan="2">Quality of life issues for local residents</td></tr>
<tr><td>Impact strategy (e.g. Minimise adverse impacts on air quality)</td><td colspan="2">Minimise adverse effects and maximise beneficial opportunities for the local residents</td></tr>
<tr><td colspan="2">Summary of Potential Threats/Opportunities that could affect Impact Category (e.g. construction activity, transport)</td><td>Impact Assessment</td></tr>
<tr><td colspan="2" rowspan="3">Threats
• Negative impacts due to the project construction (eg dust, noise, traffic delays).
• Disruption to local businesses during construction.
• Decreased value of property due to negative impacts of the project.
• Community feels that their concerns about the project are not heard by Govt.

Opportunities
• Development of new industries to support the construction and operation of the project.
• Skilled job opportunities for the local community.
• Improved infrastructure resulting from increased investment in the region because of the project.
• Improvements to the local economy because of the project expenditures in the local region.</td><td>Anticipated social impact/opportunity without strategy:
• Discontentment and reduced licence to operate
• Negative media attention
• Damaging relationship with the Regional Govt
• Reputational damage

Anticipated social impact/opportunity after strategy:
• Enhanced quality of life for the local residents
• Enhanced licence to operate
• Community advocacy of the project
• Positive media attention</td></tr>
<tr><td>Key Stakeholders</td></tr>
<tr><td>Local community; Regional Government; Lobby groups; Broader community; Media; Training organisations</td></tr>
<tr><td>Action Plans</td><td>Timeframes</td><td>Monitoring and Stakeholder Engagement Activities</td></tr>
<tr><td>1. Targeted recruitment: Prioritise recruitment from the local community where possible. Plan targeted advertising and record key recruitment statistics.</td><td>Plan approved – 3 months.</td><td>1. Monitor wording and targeting of recruitment adverts; Monitor and report the recruitment statistics to the Key Stakeholders.</td></tr>
<tr><td>2. Training Programs: Work with local training organisations (vocational and tertiary) to develop a pipeline of suitably skilled locals. Provide Carbon Cap funding support to develop new programs and new scholarships for local residents.</td><td>Plan and budget approved – 6 months.</td><td>2. Monitor and report on the skills gap analysis and chosen training programs.</td></tr>
<tr><td>3. Preferential Supplier Program: Prioritise suppliers and contractors from the region. Provide assistance to local contractors to enable them to satisfy contractor qualification.</td><td>Plan approved – 3 months.</td><td>3. Monitor and report on number and type of contractors engaged. Report to Key Stakeholders.</td></tr>
<tr><td>4. Key infrastructure program: Work with Government to ensure suitable Govt support for needed local infrastructure.</td><td>GM Community Engagement to commence discussions with Minister asap.
Plan for on-going engagement approved – 3 months.</td><td>4. Monitor local infrastructure projects resulting from this project. Report to Key Stakeholders on project progress, expenditure, timeliness, and benefits for the local community.</td></tr>
<tr><td>5. Local Community Consultation Group: Initiate this Group to provide an enhanced communication channel with the local community, local businesses and Government.</td><td>Draft Terms of Reference approved – 1 month.
Call for nominations for membership from the community – 3 months.
LCCG established – 6 months.</td><td>5. Monitor and report on LCCG meetings. Report on Actions.</td></tr>
<tr><td>6. Avoid and/or minimise impacts from the Construction Project: The Project should be designed to avoid impact on the local community – eg using existing easements and working in existing industrial areas; selecting transport routes that have least local impact. The local community should be informed in advance of any potential impacts.</td><td>Commence reporting – 3 months.</td><td>6. Monitor key project statistics and report to Key Stakeholders and the Local Community Consultation Group. Key metrics include truck movements, construction zones, noise and air quality measurements.</td></tr>
<tr><td>Accountabilities:</td><td colspan="2">General Manager, Community Engagement</td></tr>
</table>

Index

Note: Page numbers followed by "*f*" and "*t*" refer to figures and tables, respectively.

A

Accident analysis mapping (Accimap), 32, 47, 106–107, 148–150
  missing control assurance elements, 149*f*
  strengths and weaknesses, 150*t*
  technique, 134
Accimap. *See* Accident analysis mapping (Accimap)
Accountabilities of professional practice, 78–80
  comparison of professional standards and legal requirements, 81*f*
  meaning of obligation, accountability and responsibility, 81*t*
ACM. *See* Association for Computing Machinery (ACM)
Active protection devices, 176, 312
Adaptations, 234
Adverse events, 129
Age of Human Factors, 46
AI. *See* Artificial intelligence (AI)
Air, 292
  conditioning systems, 266
Airborne particulates, 292
ALARP. *See* As Low As Reasonably Practicable (ALARP)
Alcohol testing, 237
American Institute of Chemical Engineers Code of Ethics, 75
Analytical reasoning strategy, 59*t*, 66*t*
AR. *See* Augmented reality (AR)
Arafura Resources Nolans Project, 210
Arbitrary-choice strategy, 59*t*, 66*t*
Arresting controls, 118*t*
Artefacts, 108
Artificial intelligence (AI), 280–281
As Low As Reasonably Practicable (ALARP), 23–25, 99
  risk levels mapped against ALARP standards, 28*f*
Association for Computing Machinery (ACM), 82
Attributes, 50–51
  of risk, 50
Augmented reality (AR), 279–280
Authority system, 80–81
Automated equipment, 280
Automation, 62, 280
Autonomous devices, 47–48
Autonomous systems, 276
Autonomy, 280
Avoidance strategy, 59*t*, 66*t*

B

Backup power generators, 246–247
Baseline studies, 200
BASF pigment plant, 265
Basic process control system (BPCS), 312
Bhopal accident (1986), 119
Bhopal Union Carbide plant, 163–164
BHP, 203
Big data, 280–281
Biologic drugs, 263
Biological hazards, 245
Bowtie analysis (BTA), 31, 105–108, 112, 134, 145–148, 147*f*, 266, 312–318
  advanced bowtie with associated CAMS table, 106*f*
  ConSupMap type presentation of key controls and control supports, 316*f*
  control support diagram which illustrates Accimap representation of CAMS, 107*f*
  event tree analysis for overfilling bulk petrol storage tank, 320*f*
  example control specification sheet, 317*t*
  fault tree analysis for overfilling bulk storage tank, 318*f*
  filter to identify risk controls, and control supports and verification activities, 314*f*
Bowtie diagram, 105
BPCS. *See* Basic process control system (BPCS)
BTA. *See* Bowtie analysis (BTA)
Buffer capabilities, 261
Bulk tank farm installations, examples of, 296*f*
Buncefield petrol overflow, 6–7
Business-critical processes and reports, 230

C

CAMS. *See* Control Assurance Management System (CAMS)
Carbon dioxide, 289

Carbon storage case study
  example of CarbonCap stakeholders shown on power-interest grid, 341*f*
  social impact management plan template, 340*t*, 344
  social opportunity and threat risk assessment matrix, 339*t*
  stakeholder risk and opportunity assessment template, 338*t*
CarbonCap, 335
  project, 210
  stakeholders on power-interest grid, 341*f*
Case studies
  approaches to risk management, 35–36
    availability of safety PPE, 36
    process plant operation, 35
  hazard identification process, 296*f*
CAST. *See* Causal Analysis using System Theory (CAST)
Catastrophic failures, 29–31
Causal Analysis using System Theory (CAST), 49
CCPS. *See* Centre for Chemical Process Safety (CCPS)
Cement operations, 231–232
Centre for Chemical Process Safety (CCPS), 183
  comparison between CCPS and OSHA PSMS information, 184*t*
Challenger space shuttle disaster, 32
Checking process, 118–119
Chemical hazard, 300
Chernobyl and Fukushima, 47
Civil aviation systems, 45
Client's safety management system, 235
Clostridium
  C. botulinum, 5
  C. sporogenes, 5–6
Coal bed methane project, 15
Coal seam gas, 15
Code of Ethics for Engineers, 75
Cognitive ergonomics, 62
Cognitive ergonomists, 62
Cognitive Reliability Analysis Method (CREAM), 31–32, 46
COMAH. *See* Competent Authority for the Control of Major Accident Hazards (COMAH)
Combustible dust, 290
Combustible dust explosions, 290
Combustion system, 266
Commercial fishing, 232
Communication
  and consultation, 26
  systems, 263–264
Community emergency response, 176, 312
Companies, 47–48
Competence of professional engineer, 75–78
Competent Authority for the Control of Major Accident Hazards (COMAH), 6–7
Complex industrial systems, 45
Complex nonlinear systems, 49–50
Compliance, 199, 201
  strategy, 59*t*, 66*t*
  systems, 79
Component function, 308
Computerisation, 62
Computerised models, 248–249
Computerised technology, 62
Confined spaces, 291–292
Consequence diagram, 29–30
Consequence modelling, 318–319, 321
ConSupMap, 314
  type presentation of key controls and control supports, 316*f*
Contemporary navigation systems, 276
Contemporary processing plants, 244
Contextual activity template identifying safety critical work performed by humans, 324*f*
Contract
  agreements, 233–235
  classification of contractors, 230–233
  contract agreements, 233–235
  contracting in process industries, 229–230
  contractor selection and management, 236–237
  Donaldson enterprises fireworks explosion Hawaii, 239
  Grosvenor coal mine gas explosion, 238–239
  management of contracted work, 239
  managing contractor risks, 237
Contractor
  management, 236–237
  risks
    categories of, 229*f*
    classification of contractors, 230–233
    contract agreements, 233–235
    contracting in process industries, 229–230
    contractor selection and management, 236–237
    Donaldson enterprises fireworks explosion Hawaii, 239
    factors that can distinguish employees from, 232*t*
    Grosvenor coal mine gas explosion, 238–239
    injury and fatality, 234–235
    management of contracted work, 239
    managing contractor risks, 237
Control
  failure modes analysis, 117–118
  defining, 116*f*
  management systems, 21
  specification sheet, 314
  systems, 176, 244
Control analysis, 114–118, 120*t*
Control Assurance Management, 25–26

Control Assurance Management System (CAMS), 105–106
  analysis, 266
Cooling system power supply, 246–247
Core components of foundations of book, 14
COVID-19 pandemic, 247, 260, 262
CREAM. *See* Cognitive Reliability Analysis Method (CREAM)
Cue-based strategy, 59*t*, 66*t*
Cyber security, 244, 259, 282
  in process industries, 248–253
Cyber-physical systems, 47–48, 244
Cyclones, 245

D

Data analysis approaches, 280–281, 281*f*
Decision gate, 201
Decision-making
  bias, 61
  strategies, 64
    range of possible, 59*t*
Deep learning, 280–281
Deepwater Horizon drilling rig, 282–283
  deepwater horizon fire, oil spill and cleanup by burning spilt oil, 7*f*
  explosion, 7–8
Deepwater horizon fire, oil spill and cleanup by burning spilt oil, 7*f*
Defence model, 253
Defence-in-depth (DiD), 172, 174–176, 306
  analysis, 312
  approaches, 312
  layers of protection, 175*f*
  used to protect castle residence, 175*f*
Delivery of replacement, 235
Demobilisation, 229
Design interventions, 62–64
DiD. *See* Defence-in-depth (DiD)
Dieselgate, 82
Digital control systems, 248–249
Digital twins, 49–50, 279
Digitalisation, 264
Disasters, 46–47
Domestic land transportation, 262
Donaldson enterprises fireworks explosion Hawaii, 239
Drones, 47–48, 246
Duty of care, 80
Dynamic models, 49–50

E

Earthquakes, 245–247
EIS. *See* Environmental Impact Statement (EIS)
Electricity, 263–264
Electronic safety systems, 264
Electronic systems, 248
Eliminate hazard, 311
Eliminate threats, 312
Elimination, 311
Emerging risks, 273–274
  process industries, 274*t*
Employees, 229–230
Employment, 238
Energy, 265–266
Engineering, 75
  accountabilities, 80–81
  decision scenarios, 14–15
  work, 228
Engineers, 11, 72
Engineers Australia, 77–78
Environmental and Social Impact Assessment (ESIA), 199–200
Environmental and Social Management System (ESMS), 199–200
Environmental Authority, 199
Environmental Impact Statement (EIS), 199
Environmental Licence, 201
Environmental Protection Agency, 341
Environmental risks, 195–198
  impact assessment, 200–202
  project lifecycle, 198–200
  project stages and lifecycle processes, 198*f*
  social licence to operate, 202–203
  tools to identify, assess, treat and monitor, 204–210
Environmental values, 197
Epidemics, 247
Equipment malfunctions, 29–31
Error, types of, 46*f*
ESIA. *See* Environmental and Social Impact Assessment (ESIA)
ESMS. *See* Environmental and Social Management System (ESMS)
Essential Factors Analysis, 134
ETA. *See* Event tree analysis (ETA)
Ethics of professional engineer, 74–75
  ethical dilemma grid, 76*f*
Event timeline. *See* Timeline
Event tree analysis (ETA), 29–30, 318–319
  for overfilling bulk petrol storage tank, 320*f*
Events, 97–99
  investigations, 127–128
    perform, 128–130
    reasons for, 130*f*
  unwanted event characterisation, 108
  unwanted event identification, 108–109
Expert consultants, hiring, 230
Explainable AI, 281
Extended risk management process for process industries, 24*f*
Exxon Valdez oil spill, 31–32

F

Failure Mode Effects Analysis (FMEA), 29, 44–45, 117–118, 297–300, 308
Failure modes effects criticality analysis (FMECA), 92, 297–300, 303–305, 308–311
  process and worksheet used for tank filling case study, 308*f*
  risk assessment results using, 309, 311
  sample/example only of FMECA worksheet for tank filling, 310*f*
Fault tree, 318–319
Fault tree analysis (FTA), 29, 318
  for overfilling bulk storage tank, 318*f*
Fermentation process, 14–15
Fertilizer plant, 246
FID. *See* Final investment decision (FID)
Final investment decision (FID), 217
Financial performance, 9*f*
Fire triangle, 290
Fishbone diagram, 134, 138–141
5 whys analysis, 133, 135–138
Flammable environments, 290
Flexible culture, 179
FMEA. *See* Failure Mode Effects Analysis (FMEA)
FMECA. *See* Failure modes effects and controls/criticality analysis (FMECA)
Fonterra (dairy companies), 5
Fonterra whey protein crisis, 5–6
Forcing functions, 62
Foresighting, 274–275
Formal articulation, 233
Fossil fuel industry, 203
FRAM. *See* Functional Resonance Analysis Method (FRAM)
FTA. *See* Fault tree analysis (FTA)
Fukushima Nuclear Power Plant disaster, 47
Functional Resonance Analysis Method (FRAM), 32–33, 49, 134
Future of risk management, 272, 276–282
  automation and autonomy, 280
  big data and artificial intelligence, 280–281
  digital twins, 279
  IIoTs, 282
  virtual reality/augmented reality technologies, 279–280
Future process industry risks, 273–276
Futures forecasting, 274–275
Futurists, 276

G

Gaia basin, 336
Gas leak detection, 114
Gas processing plant, 247
Gateway competence, 77
GHS. *See* UN Globally Harmonized System of Classification and Labelling of Chemicals (GHS)
Global changes, 237
Globalisation, 4
Golden Safety Rules, 62
Greenbrook Water Treatment Plant, 296
Grosvenor coal mine gas explosion, 238–239

H

Hazard
  analysis techniques, 90–92, 253
  hazard-focused risk identification tools, 92–94
Hazard and Operability Studies (HAZOPs), 31, 44–45, 92, 289, 297–300, 303–305, 311
  example, 305–307
  risk assessment results using, 311
Hazard Identification (HAZID), 90–92, 297–300, 303, 311
  example, 300–305
  risk assessment results using, 311
  techniques, 90–92
Hazardous chemicals, 287
Hazardous industries, history of poor technology and safety performance in, 30*f*
HAZID. *See* Hazard Identification (HAZID)
HAZOPs. *See* Hazard and Operability Studies (HAZOPs)
Health, 162, 232
  management system, 181
  risks, 163–166
    classifications of, 163*f*
    industry-related health risks, 164*t*
    personal and process safety trends over time, 166*f*
    relationship between process safety and personal safety, 165*f*
    silicosis, 164*f*
HEART. *See* Human Error Assessment and Reduction Technique (HEART)
Heat exchanger temperature control system, 174–175
HFACS. *See* Human Factors Analysis and Classification System (HFACS)
High hazard industries, 58–59
Highly Reliable Organisations (HROs), 56, 179
Hot work conduction, 291
HROs. *See* Highly Reliable Organisations (HROs)
Human Error Assessment and Reduction Technique (HEART), 31–32, 46
Human Factors Analysis and Classification System (HFACS), 32, 47, 134, 142–144
  analysis for Texas City Refinery explosion, 146*t*
  framework, 142*t*
  item examples, 143*t*

strengths and weaknesses, 146*t*
Humans, 31–32, 46
action, 120*t*
behaviour, 32, 42
capability, 300
changing role of human operator in industry, 43–50
changing nature of industrial work, 50
global, collaborative, cognitive, 47–50
history of industrial work, 44*f*
localised, direct and physical control, 43–45
regional, centralised and more cognitive control, 45–47
types of error, 46*f*
control activity, 324
decision-making process, 56–60
response strategies, 58–60
situation assessment, 56–58
error, 45
approach, 56
HAZOP, 31–32
risk assessment approaches, 31–32
factors, 42, 61
checklists, 31–32
techniques, 46
work, 61
human decision-making process, 56–60
components of human performance, 57*f*
designing for humans, 57*f*
situation assessment, 56–58
human-centred approach, 61
human–system control diagram from managing tank fill levels, 325*f*
human–system interaction type control tasks, 324
improving human performance, 61–65
example of designing for improved human performance, 64–65
example of life saving rules, 63*f*
improving safety with human-centred design, 321–333
contextual activity template identifying safety critical work performed by humans, 324*f*
human–system control diagram from managing tank fill levels, 325*f*
safety critical situation assessment factor analysis for tank filling, 326*t*
sample/example only of STPA worksheet for filling tank with fuel, 332*f*
interaction, 94–95
organisational decision-making, 43*f*
role of, 50–52
risk analysers, 52
risk controllers, 52
risk perceivers, 51
Hydrocarbon, 292
Hydrochloric acid (HCl), 14–15, 245

I

IChemE Safety Centre Guidance on Lead Process Safety Metrics, 180–181
ICMM. *See* International Council of Mining and Metals (ICMM)
IEEE. *See* Institute of Electrical and Electronics Engineers (IEEE)
IIoTs. *See* Industrial internet of things (IIoTs)
Imagining scenarios, 275–276
Imitation strategy, 59*t*, 66*t*
Impact
assessments, 95, 200–202
of consequences, 112
mitigation, 201
of occupational health, 164
safety, 89–90
Incident investigation techniques and application considerations, 132–154, 133*t*
accident analysis mapping, 148–150
bowtie analysis, 145–148
fishbone diagram, 138–141
5 whys analysis, 135–138
human factors analysis and classification system, 142–144
strategies analysis for enhancing resilience, 150–154
timeline, 134–135
user beware, 134
Independence of controls, 117
Industrial control transitions, 48
Industrial internet of things (IIoTs), 248–249, 282
Industrial revolution, 29–31
Industrial work, changing nature of, 50
Industry
changing role of human operator in, 43–50
industry 4.0, 276
aspects, 276
technologies, 276–278
operational risk management in, 28–33
projects, 217
Informal techniques, 90
Information, 235
Information system, 249
Information technology (IT), 244
Informed culture, 179
Inherently safe design (ISD), 173–174, 173*f*
options for road tanker filling fuel depot, 174*t*
principles, 311
Institute of Electrical and Electronics Engineers (IEEE), 82
Insurance information, 236

Interconnected industrial networks, 48
International Association of Oil & Gas Producers website (IOGP website), 95
International Council of Mining and Metals (ICMM), 118–119
International Energy Agency, 335
International Finance Corporation, 208–210
International standard
  AS/NZS, 22
  for risk management, 3
International Standard Occupational (ISO), 181
  ISO31000, 90
    extended version of ISO31000 risk management process, 87*f*
    framework, 86, 103–104
    ISO 31000:2018, 9–10, 22–23
    standard, 114
    standard for risk management, 52
  overview of ISO45001, 181–183
Internet-connected technology, 47–48
Intervention effectiveness, 64
Intuitive strategy, 59*t*, 66*t*
Investigation, 128
IOGP website. *See* International Association of Oil & Gas Producers website (IOGP website)
ISD. *See* Inherently safe design (ISD)
Ishikawa diagram. *See* Fishbone diagram
ISO. *See* International Standard Occupational (ISO)
Isolate hazard, 311
IT. *See* Information technology (IT)

J
JHA. *See* Job Hazard Analysis (JHA)
Job Hazard Analysis (JHA), 90–92
Just culture, 179

L
Laws, 73–74, 81–82
  of jurisdiction, 235
Layers of protection analysis (LOPA), 318–321
  sheet for tank overfilling scenario, 321
Learning
  culture, 179
  integration back into business, 154
Legacy issues, 198–199
Legal obligations, 78–80
Legal regimes, 79
Life cycle of projects, 217–219
Life Saving Rules, 62
Likelihood categories used in risk ranking matrix, examples of, 27*f*
Liquid, 15
LOPA. *See* Layers of protection analysis (LOPA)
Loss reduction approach, 34
LTIR, 180

M
Machine learning, 280–281
Macroergonomics, 62
Management
  of contracted work, 239
  of contractor risks, 237
  system, 229
Manufactured stone, 164
Medium-scale projects, 215
Membrane gas separation, 266
Methane gas, 197
Methyl isocyanate (MIC), 163–164
MIC. *See* Methyl isocyanate (MIC)
Military technologies, 45
Minimisation, 311
Ministry for Primary Industries in New Zealand, 5
Mitigating controls, 118*t*
Mobilisation, 229, 236
Moderation, 311
Modern industries, 47–48
Molten metal explosions, 246
Monitoring, 201
  and reviewing risks, 126
    defence model for monitoring and reviewing risk management activities, 128*f*
    incident investigation techniques and application considerations, 132–154
    integration of learning back into business, 154
    prospective and retrospective risk monitoring and review analysis, 127*f*
    purpose and theory behind investigations, 130–132

N
Nanotechnologies, 47–48
National Society of Professional Engineers, 75
  Code of Ethics, 75
Natural disasters, 259
Natural resources, 231–232
Net zero emissions, 335
New Zealand Government, 6
Nitric acid, 245
Nitrogen, 287–289
Nonpathogenic bacterium, 5–6
Nuclear power plants, 45, 246–247, 252
  accidents, 47

O
Obligations of professional practice, 78–82
Occupational health, 164

Occupational health and safety (OH&S), 95–97
Occupational Health and Safety Administration (OSHA), 183
  comparison between CCPS and OSHA PSMS information, 184*t*
Occupational safety risks, 172
OH&S. *See* Occupational health and safety (OH&S)
Oil
  and gas industry, 47–48
  refining, 45
Operational risks, 13–14. *See also* Social risks
  history of industrial performance, 31*f*
  history of poor technology and safety performance in hazardous industries, 30*f*
  management in industry, 28–33
  scope of risk and uncertainty management, 33*f*
Operational technology (OT), 244
Operator strategies for road tanker filling, designing improvements in response to possible, 66*t*
Option-based strategy, 59*t*
Organic peroxide products, 246–247
Organisation ergonomics. *See* Macroergonomics
Organisational culture and structure, 128
Organisational decision-making, 43*f*
Organisational factors, 32, 42, 46, 50–51
Organisational resilience, SCARPaR framework for, 177*f*
Organisational systems, 32
Organisational weaknesses, 46–47
Organisations, 33–34, 50–51, 179
OSHA. *See* Occupational Health and Safety Administration (OSHA)
OT. *See* Operational technology (OT)
Overfilling bulk petrol storage tank, event tree analysis for, 320*f*
Overfilling bulk storage tank, fault tree analysis for, 318*f*
Overfilling tank, example of layers of protection analysis sheet for, 323*t*
Overfilling vessels, 64–65
  designing improvements in response to possible operator strategies for road tanker filling, 66*t*

P

Pandemic outbreak, 13
Pandemic-related PPE, 262
Parameters-guideword combinations, 306
Paris Agreement, 335
People, Locations, Equipment, Activities, Timeframe and Scenarios (PLEATS), 110–111
  framework, 111
People, Locations, Equipment and plant, Activities, Timeframes, Environment, Scenarios assumptions (PLEATESO), 23, 297
Performance, 214
  of professional engineer, 78
  Warren Centre PPIR Protocol for Professional Performance, 79*t*
Permanent recruitment process, 229–230
Personal safety, 164–165
  personal and process safety trends over time, 166*f*
  relationship between process safety and personal safety, 165*f*
  risks, 172
PESTLE analysis. *See* Political, Economic, Social, Technology, Legal and Environmental impacts (PESTLE analysis)
Petrochemicals, 45
  tank filling, 295
PHA. *See* Preliminary Hazard Analysis (PHA); Preliminary hazard identification
Phronesis, 10
Physical control, 120*t*
Physical ergonomics, 61–62
Physical protection barriers, 176, 312
Physical security, 245
  in process industries, 245–248
Piper Alpha oil rig fire and explosion, 282–283
Planning, 262
Plant breakdowns, 29–31
Plant emergency response, 176, 312
Plant failure analysis techniques, 92
Plausible futures, 274
PLEATESO. *See* People, Locations, Equipment and plant, Activities, Timeframes, Environment, Scenarios assumptions (PLEATESO)
PLEATS. *See* People, Locations, Equipment, Activities, Timeframe and Scenarios (PLEATS)
Political, Economic, Social, Technology, Legal and Environmental impacts (PESTLE analysis), 92, 258, 273, 274*t*
Positive social impact, 194
Possible futures, 274, 275*f*
Potential failure modes, 309
Power supply systems, 263–264
Power-interest grid, 205
Praxis, 10
Preferable futures, 274
Preliminary Hazard Analysis (PHA), 90–92
Preliminary hazard identification, 297–300
Preliminary stakeholder analysis and management, 204–207, 204*f*, 205*f*
Pressure hazard, 300
Prevention controls, 105

Primary approvals, 199
Probable futures, 274
Process control system, 174–176
Process deviation analysis techniques, 92
Process engineering professional practice concept, 72
Process Hazard Analysis, 90–92
Process industries, 2, 22–23, 43, 90–92, 95, 108–109, 177, 262
  classifying workplace hazardous chemicals, 287–289
  contracting in, 229–230
  cyber security in, 248–253
  hazardous conditions, 289–292
    combustible dust, 290
    conducting hot work, 291
    confined spaces, 291–292
    flammable environments, 290
    hazardous tasks, 292
    pressure, 289
    start-up and shutdown, 292
    temperature, 289
  physical security in, 245–248
  process, 165–166
    accidents, 167*t*
  supply chain risks, 258–259
  workplace hazardous chemicals and conditions, 287
Process plant, 162
  operation, 35
Process safety, 164–165
  consideration for filling bulk storage tanks, 183–186
  effectiveness, 180–181
    lead process safety metrics, 181*t*
  personal and process safety trends over time, 166*f*
  relationship between process safety and personal safety, 165*f*
  risks, 172
Process safety management systems (PSMSs), 181
  comparison between CCPS and OSHA PSMS information, 184*t*
  overview of, 183
Process systems, 42
Productivity issues, 237
Professional engineers, 72, 78–79
  three dimensions of professional engineering, 74*f*
Professional Engineers Act (1990), 73–74
Professional Engineers Act (2002), 73–74
Professional engineering practice, 282–283
Professional Engineers Registration Act (2019), 73–74
Professional judgement, 20–21
Professional practice, 72
  case study, 82
  obligations, accountabilities and responsibilities, 78–82
    legal and regulatory obligations, 78–80
  professional engineer, 73–78
    competence, 75–78
    ethics, 74–75
    performance, 78
    three dimensions of professional engineering, 74*f*
  three-tiered system of knowledge, 73*f*
Professional standards and legal requirements, comparison of, 81*f*
Projects, 214–217
  life cycle, 217–219
    phases, value addition and cost of change, 219*f*
    phases associated with larger process industry projects, 218*f*
  relationship between project context and decision and consultation processes, 216*f*
  risks, 215*f*, 220
    assessment processes, 223*f*
    examples, 220*t*
    managing, 221–224
    register, 221–222, 223*f*
PSMSs. *See* Process safety management systems (PSMSs)

Q

QRA. *See* Quantitative risk assessment (QRA)
Quality systems, 263
Quantitative risk assessment (QRA), 95

R

Ransomware infiltration, 251
Raw materials, 262
RCA. *See* Root Cause Analysis (RCA)
Re-emerging risks, 273–274
Re-imagining your supply chain model, 263
Regulatory obligations, 78–80
Regulatory regimes, 79
Renewable energy, 266
Reoccurring events, 129
Reporting culture, 179
Residual risk, 103–104
  determined by effectiveness of controls, 104*f*
Resilience engineering, 49, 177–178
  SCARPaR framework for organisational resilience, 177*f*
Response strategies, 58–60
  range of possible decision-making strategies, 59*t*
Responsibilities of professional practice, 81–82
Risk, 2–3, 21, 100, 126, 194, 272
  case studies, 5–9
    Buncefield petrol overflow and vapour cloud explosion, 6–7, 6*f*
    deepwater horizon drilling rig explosion, 7–8
    Fonterra whey protein crisis, 5–6

risk maturity and financial performance link
between risk maturity and financial performance, 9*f*
synopsis, 8–9
changing role of human operator in industry, 43–50
changing nature of industrial work, 50
global, collaborative, cognitive, 47–50
history of industrial work, 44*f*
localised, direct and physical control, 43–45
regional, centralised and more cognitive control, 45–47
types of error, 46*f*
communication, 53–56
downside and upside, 35
engineering decision scenarios, 14–15
scenario, 14–15
guiding principles, 3–4
sample of protests from around world that relate to process, 5*f*
technology adoption, 4*f*
language, 21
management, 9–11
influences and impacts on/of industrial risks, 11*f*
three-tiered system of knowledge, 10*f*
matrices, 95, 97, 308
maps, 95–97
maturity, 9*f*
organisational decision-making, 43*f*
risk-based approach, 14
types of risks that should be considered by engineers, 11–14
top 10 risk concerns for industry, 12*t*
Risk analysers, 50–52
factors characterise high and low organisation risk appetites, 53*f*
Risk analysis, 21, 23–25, 95–99
example risk matrix, 96*f*
risk matrix, 99*f*
threats and opportunities matrix, 98*f*
Risk appetite, 52
Risk assessment, 90–102, 248, 260
process, 101
results using HAZID, HAZOP and FMECA, 311
risk analysis, 95–99
risk evaluation, 99–102
risk identification, 90–95
techniques, 29–31
Risk communication, 53–56
evolution of, 55*f*
methods of, 55*f*
model, 54*f*
risk management process, 54*f*
Risk communicator, 50
Risk controllers, 50–52
Risk controls, 114, 214
activities by organisational level, 115*f*
focused approach, 114
management approaches, 32
Risk evaluation, 21, 99–102
example risk register, 102*f*
risk tolerability and ALARP, 100*f*
selection of ALARP option for water treatment technology, 101*f*
Risk identification, 21, 23–25, 90–95
approaches, 92
approaches to risk identification, 91*f*
example of hazard, threat and consequence, 91*f*
generic method of different risk identification techniques in process industries, 93*f*
SWOT analysis framework, 94*f*
Risk management, 2, 20, 58, 72, 103–121, 126, 196, 214, 252–253, 272
approaches, 262–263
bowtie analysis, 105–108
case studies illustrating two approaches to risk management, 35–36
availability of safety PPE, 36
process plant operation, 35
control analysis, 114–118
layered bowtie for vehicle interactions, 119*f*
layers and types of control by function and response time, 117*f*
management of controls, 118–121, 120*f*
risk control activities by organisational level, 115*f*
types of controls, 118*t*
determining the scope of analysis, 109–111
framework, 26
identifying the possible consequences that could result from unwanted event, 112–113
example of consequences for unwanted vehicle interaction, 113*f*
examples of consequences result unwanted event occurs, 113*t*
identifying the threats that can cause unwanted event, 111
examples of threats for unwanted vehicle interaction, 112*f*
examples of threats lead to unwanted events, 111*t*
operational risk management in industry, 28–33
history of industrial performance, 31*f*
history of poor technology and safety performance in hazardous industries, 30*f*
scope of risk and uncertainty management, 33*f*
process, 20–27, 23*f*, 86, 88, 88*f*
example of risk ranking matrix, 26*f*
example scope table, 25*t*

Risk management (*Continued*)
examples of impact categories used in risk ranking matrix, 27*f*
examples of likelihood categories used in risk ranking matrix, 27*f*
examples of risk-based standards and industry guidelines, 22*t*
extended risk management process for process industries, 24*f*
risk levels mapped against ALARP standards, 28*f*
programme, 162
responsiblities for, 28
strategic *vs.* operational risk, 29*t*
risk language, 21
two approaches to modern risk management, 33–35, 34*t*
unwanted event characterisation, 108
unwanted event identification, 108–109
descriptions of unwanted events that form knot at centre of bowtie, 110*t*
examples of unacceptable risks and unwanted events from industry, 109*t*
Risk optimization approach, 34–35
Risk perceivers, 50–52
risk perception inputs and outputs, 51*f*
Risk perception, 51
inputs and outputs, 51*f*
Risk ranking
example
of impact categories used in, 27*f*
of likelihood categories used in, 27*f*
risk ranking matrix, 26*f*
exercise, 306, 309
Risk register, 101
Risk treatment, 21, 25, 103–121, 172
bowtie analysis, 105–108
control analysis, 114–118
control, 116*f*
layered bowtie for vehicle interactions, 119*f*
layers and types of control by function and response time, 117*f*
management of controls, 118–121, 120*f*
risk control activities by organisational level, 115*f*
types of controls, 118*t*
determining the scope of analysis, 109–111
identifying the possible consequences that could result from unwanted event, 112–113
example of consequences for unwanted vehicle interaction, 113*f*
examples of consequences result unwanted event occurs, 113*t*
identifying the threats that can cause unwanted event, 111
examples of threats for unwanted vehicle interaction, 112*f*
examples of threats that can lead to unwanted events, 111*t*
options for addressing unwanted events, 103*f*
overview, 103–104
residual risk determined by effectiveness of controls, 104*f*
of tank filling, 311–333
unwanted event characterisation, 108
unwanted event identification, 108–109
descriptions of unwanted events that form knot at the centre of bowtie, 110*t*
examples of unacceptable risks and unwanted events from industry, 109*t*
Road tanker filling, designing improvements in response to possible operator strategies for, 66*t*
Robots, 47–48
Root Cause Analysis (RCA), 133

S

Safe design approaches, 248
Safe Work Australia, 291–292
Safeguards. *See* Prevention controls
SAfER. *See* Strategies Analysis for Enhancing Resilience (SAfER)
Safety, 162
availability of safety PPE, 36
critical situation assessment factor analysis for tank filling, 326*t*
culture, 178–180
safety culture model, 178*f*
culture, 32
with human-centred design, improving, 321–333
instrumented systems, 176, 292, 312
management system, 181
managing safety risks, 166–178
approaches to managing process safety, 172*f*
defence-in-depth, 174–176
inherently safer design, 173–174
recent process safety accidents, 167*t*
resilience engineering, 177–178
process safety
consideration for filling bulk storage tanks, 183–186
effectiveness, 180–181
risks, 163–166
classifications of, 163*f*
industry-related health risks, 164*t*
personal and process safety trends over time, 166*f*
relationship between process safety and personal safety, 165*f*
silicosis, 164*f*

safety management systems, 181–183
overview of ISO45001, 181–183
overview of process safety management systems, 183
safety-related consequences, 319–320
Sanitising products, 262
SCADA systems. *See* Supervisory Control and Data Acquisition systems (SCADA systems)
Scrubbing system, 15
Secondary approvals, 199
Security risk
cyber security in process industries, 248–253
physical security in process industries, 245–248
processing plant security risks, 244*f*
from tank farm risk register, 249*f*
Seminal industry, 231–232
Sensors, 248–249
SHERPA. *See* Systematic Human Error Reduction and Prediction Approach (SHERPA)
Ship's crew, 90
Shutdown, 292
SIMP. *See* Social impact management plan (SIMP)
Simplification, 62, 311
Site documentation, 296–297
Situational awareness, 64
Skeleton bowtie, 314
Small-scale projects, 214–215
SMART. *See* Specific, measurable, assignable, realistic/relevant and time-bounded (SMART)
Social impact management plan (SIMP), 196, 208–210, 209*t*, 341
template, 344
Social licence to operate, 202–203, 202*f*
layers, 203*f*
Social risks, 195–198
impact assessment, 200–202
project lifecycle, 198–200
project stages and lifecycle processes, 198*f*
social licence to operate, 202–203
tools to identify, assess, treat and monitor, 204–210
Social values, 197
Space Shuttle Columbia and Fukushima Nuclear Power Plant accident, 46–47
SPEAR, 31–32
Specific, measurable, assignable, realistic/relevant and time-bounded (SMART), 154
SSOC. *See* State Shale Oil Company (SSOC)
Stakeholders, 52
engagement, 204
identification and analysis, 204–205
risk
and opportunity assessment template, 343*t*
assessment, 207–208
STAMP. *See* Systems-Theoretic Accident Modelling and Processes (STAMP)
Standardisation, 62
Standards Australia, 34
Start-up, 292
State Shale Oil Company (SSOC), 207
Steal gasoline, 245
Stealing chemicals, 245
Steam, 289
Storage tank, example of engineering drawing of, 297*f*
STPA. *See* Systems Theoretic Process Analysis (STPA); Systems-Theoretic Process Analysis (STPA)
Strategies Analysis for Enhancing Resilience (SAfER), 32–33, 49, 134, 150–154, 326
analysis, 324–326
method, 326
Street navigation systems, 276
Strengths, Weaknesses, Opportunities and Threats analysis (SWOT analysis), 92, 94*f*
Substitute hazard, 311
Substitution, 311
Sulphuric acid, 245
Sunshine Beach, 336
Supervisory Control and Data Acquisition systems (SCADA systems), 251
Supply chain, 258
overview of, 258*f*
resilience, 260–261
Supply chain risks
changing paradigm of supply chain risk management, 266
examples, 259*t*, 261–266
scope table for, 260*t*
management, 263
managing supply chain risks, 259–261
process industry supply chain risks, 258–259
on scope of risk and uncertainty management, 261*f*
Surat Gas Project, 210
Sustainability of processing industries, 3–4
Swiss Cheese Model, 32, 47
SWOT analysis. *See* Strengths, Weaknesses, Opportunities and Threats analysis (SWOT analysis)
Systematic Human Error Reduction and Prediction Approach (SHERPA), 31–32, 46
Systems Theoretic Process Analysis (STPA), 324
Systems-Theoretic Accident Modelling and Processes (STAMP), 47, 49, 134
based approaches, 32
Systems-Theoretic Process Analysis (STPA), 49

T

Tank filling, 303, 313
 case study hazard identification process, 296*f*
 identifying hazards, threats and consequences, 297–311
  example of FMECA, 308–310
  example of hazard and operability, 305–307
  example of hazard identification and analysis, 300–305
  risk assessment results using hazard identification, hazard and operability and FMECA, 311
 risk treatment, 311–333
  bowtie analysis, 313–318
  consequence and layers of protection analysis, 319–321
  defence in-depth analysis, 312
  fault tree and event tree analysis, 318–319
  improving safety with human-centred design, 321–333
 scope, 296–297
  example of engineering drawing of storage tank, 297*f*
  example scope for tank filling case study, 298*t*
  examples of bulk tank farm installations, 296*f*
  examples of tank overfilling incidents, 299*t*
Tank overfilling incidents, examples of, 299*t*
TapRoot, 133
Task-specific competence, 77
Tata soda ash plant, 265
Technique for Human Error Assessment (THEA), 31–32, 46
Technique for the Retrospective and Predictive Analysis of Cognitive Errors (TRACEr), 31–32, 46
Technological system controls, 120*t*
Technology adoption, 4*f*
Texas City Refinery explosion, 31–32, 58
Texas City refinery operators (2005), 58
THEA. *See* Technique for Human Error Assessment (THEA)
Thermal hazard, 300
Thermal radiation from pool fire, model results of, 321*f*
Three Mile Island, 47
 nuclear power plant, 46
Three-tiered system of knowledge, 10*f*
Time ranges required to increase capacity of microchip production, 264*f*
Timeline, 134–135
 analysis, 133
Tolerability, 99
Toxic gases, 248
TRACEr. *See* Technique for the Retrospective and Predictive Analysis of Cognitive Errors (TRACEr)
Trade waste specifications, 15
Trans-Adriatic Pipeline, 210
Trapped pressure, 289

U

U.K. Health and Safety Executive, 173
U.S. Chemical Safety Board and Safework Australia, 173
U.S. Department of Energy, 45–46
UK Buncefield oil storage depot fire, 58
UN Globally Harmonized System of Classification and Labelling of Chemicals (GHS), 287
Uncertainty, 3, 35, 90
Undergraduate engineering education, 72
Unique technical skills, 228
Unit price contracts, 234
Unstructured techniques, 90
Unwanted events, 29–31
 identification, 108–109
Unwanted vehicle interaction, example of consequences for, 113*f*
US Chemical Safety and Hazard Investigation Board, 290
US Chemical Safety Board, 291
User beware, 134

V

Vacuum, 289
Vapour cloud explosion, 6–7
Ventilation systems, 266
Verification activities, 118–119, 120*f*, 121*t*
Virtual reality (VR), 279–280
Volatile chemicals, 246
VR. *See* Virtual reality (VR)
VW emissions scandal, 82

W

Warren Centre for Advanced Engineering, 73, 80
Warren Report, 77–78
Wastewater treatment plant, 239
Water availability, 197
Water utility control system, 251
Wearable technologies, 47–48
Workplace hazardous chemicals
 classifying, 287–289
 and conditions, 287
 example of globally harmonized system label, 288*f*
 nine GHS classifications and labels, 288*t*
Workplace health, 235
World Economic Forum, 11–13

X

Xcel Energy hydroelectric plant, 236–237
Xcel's prequalification process, 236

Z

Zero risk, 35

CPI Antony Rowe
Eastbourne, UK
June 12, 2024